AF360834

Novel Natural-Based Biomolecules Discovery for Tackling Chronic Diseases

Novel Natural-Based Biomolecules Discovery for Tackling Chronic Diseases

Editor

Hang Fai (Henry) Kwok

MDPI • Basel • Beijing • Wuhan • Barcelona • Belgrade • Manchester • Tokyo • Cluj • Tianjin

Editor
Hang Fai (Henry) Kwok
Faculty of Health Sciences
University of Macau
Avenida de Universidade
Macau SAR
China

Editorial Office
MDPI
St. Alban-Anlage 66
4052 Basel, Switzerland

This is a reprint of articles from the Special Issue published online in the open access journal *Biomolecules* (ISSN 2218-273X) (available at: https://www.mdpi.com/journal/biomolecules/special_issues/natural-based_biomolecules).

For citation purposes, cite each article independently as indicated on the article page online and as indicated below:

LastName, A.A.; LastName, B.B.; LastName, C.C. Article Title. *Journal Name* **Year**, *Volume Number*, Page Range.

ISBN 978-3-0365-0386-8 (Hbk)
ISBN 978-3-0365-0387-5 (PDF)

Contents

About the Editor

Hang Fai (Henry) Kwok (Associate Professor, Biomedical Sciences/Consultant, Histopathology at the Faculty of Health Sciences University of Macau): Prof. Kwok obtained his first-class BSc (Hons) degree and PhD in Biomedical Sciences in the UK in 2000 and 2003, respectively. He then became a Knowledge Transfer Partnerships Fellow as a postdoctoral researcher at Queen's University Belfast, a Top UK Pharmacy School. After 4 years of postdoctoral training, he moved to the pharmaceutical industry as a Senior Scientist from 2007 to 2011. Then, Prof. Kwok returned to academia as a Senior Research Fellow in the Department of Oncology at the University of Cambridge, bringing together his interests in protease biochemistry research with biologics (antibody/peptide drug) development to pursue novel therapeutic and prognostic approaches in the treatment of cancer and other chronic diseases. Apart from protease and antibody/peptide drug research, Prof. Kwok is also interested in the isolation and characterization of novel bioactive molecules from sources in nature, including amphibian defensive skin secretions and reptile, scorpion and insect venoms, for their anticancer and/or antimicrobial therapeutic potential.

Editorial

Novel Natural-based Biomolecules Discovery for Tackling Chronic Diseases

Hang Fai Kwok [1,2]

[1]   Institute of Translational Medicine, Faculty of Health Sciences, University of Macau, Avenida de Universidade, Macau SAR, China; hfkwok@um.edu.mo
[2]   Cancer Center, Faculty of Health Sciences, University of Macau, Avenida de Universidade, Macau SAR, China

Received: 20 October 2020; Accepted: 19 November 2020; Published: 15 December 2020

In the last decade, natural-derived/-based biomolecules have continuously played an important role in novel drug discovery (as a prototype drug template) for potential chronic disease treatment. Many recent research studies have demonstrated that the development of natural peptide/protein-based, toxin-based, and antibody-based drugs can significantly improve the biomedical efficiency of disease-specific therapy.

The focus of this Special Issue of Biomolecules includes eleven papers: ten original research articles and one communication article from nine different countries/regions dealing with a broad range of the discovery and development of the natural biomolecules as potential medical therapy for tackling chronic diseases (e.g., cancer, diabetes, cardiovascular diseases, rheumatoid arthritis, and pain treatment)

Four cancer-related research articles by Wang Q. et al. [1], Miao Y. et al. [2], Martínez-García D. et al. [3], Thangaraj K. et al. [4], and Lai Y. et al. [5] demonstrate that several natural-based compounds/proteins could effectively influence the cancer formation and progression. These study findings unveil the relationship of the SUMO pathway and DAPK1 protein degradation, demonstrate the target modifications of novel protease could effectively and efficiently alter its anticancer bioactivity, study the survivin levels through potent STAT3 Inhibition in lung cancer, investigate the cell cycle arrest and mitochondria-mediated intrinsic apoptosis in colorectal carcinoma, and develop a Fucoidan-based drug delivery system by using hydrophilic anticancer polysaccharides to simultaneously deliver hydrophobic anticancer drugs, respectively.

For potential diabetes therapy, the fascinating study by Lee D. et al. [6] evaluates two isoflavonoids and a nucleoside which were isolated from the roots of *Astragalus membranaceus*. These bioactive compounds can improve insulin secretion in β-cells, representing the first step towards the development of potent antidiabetic drugs. Besides, the paper by Martínez-Navarro I. et al. [7] concludes that the anionic lipid environment and degree of solvation are critical conditions for the stability of segments with the propensity to form β-sheet structures. This situation will eventually affect the structural characteristics and stability of IAPP within insulin granules, thus modifying the insulin secretion.

Two articles deals with inflammatory-related diseases. The Lee J. S. et al. [8] study speculates on two characterized compounds, petasitesin A (1) and cimicifugic acid D (3), which are worthy of further pharmacological evaluation for their potential as anti-inflammatory drugs. In addition, Celiksoy V. et al. [9] aimed to examine punicalagin in combination with Zn (II), and demonstrate that this novel combination promotes anti-inflammatory and fibroblast responses to aid oral healing.

The article by Szűcs E. et al. [10] explored the biological effect of novel opioid peptide analogs incorporating L-kynurenine (L-kyn) and kynurenic acid (kyna) in place of native amino acids. This novel oligopeptide exhibits a strong antinociceptive effect after i.c.v. and s.c. administrations in in vivo tests, according to good stability in human plasma which has a potential for tackling pain syndromes.

Finally, the short communication paper by Maheshwari G. et al. [11] tested the hypothesis that monomethyl branched-chain fatty acids (BCFAs) and a lipid extract of Conidiobolus heterosporus (CHLE) can activate the nuclear transcription factor peroxisome proliferator-activated receptor alpha (PPARalpha). In conclusion, they showed that the monomethyl BCFA isopalmitic acid (IPA) IPA is a potent PPARalpha activator. CHLE activates PPARalpha-dependent gene expression in Fao cells, an effect that is possibly mediated by IPA.

Overall, this Special Issue describes important findings related to natural-derived/-based biomolecules for potential chronic diseases treatment by dysregulating several biological pathways/receptors. It also highlights the most recent progress on the knowledge and the clinical and pharmacological applications related to the most relevant areas of healthcare.

Acknowledgments: The editor is grateful to all the important authors who contributed to this Special Issue "Novel Natural-based Biomolecules Discovery for Tackling Chronic Diseases". They are also mindful that without the rigorous and selfless evaluation of the submitted manuscripts by external peer reviewers/expertise, this Special Issue could not have happened. Moreover, the editor (Kwok H. F.) thanks for the support from the Science and Technology Development Fund (FDCT) of Macau SAR [File no. 0055/2019/A1 and 019/2017/A1] and the Faculty of Health Sciences (FHS) University of Macau [File no. MYRG2015-00025-FHS]. Finally, the valuable contributions, organization, and editorial support of the MDPI management team and staff are greatly appreciated.

Conflicts of Interest: The author declares no conflict of interest.

References

1. Wang, Q.; Zhang, X.; Chen, L.; Weng, S.; Xia, Y.; Ye, Y.; Li, K.; Liao, Z.; Chen, P.; Alsamman, K.; et al. Regulation of the Expression of DAPK1 by SUMO Pathway. *Biomolecules* **2019**, *9*, 151. [CrossRef] [PubMed]
2. Miao, Y.; Chen, G.; Xi, X.; Ma, C.; Wang, L.; Burrows, J.F.; Duan, J.; Zhou, M.; Chen, T. Discovery and Rational Design of a Novel Bowman-Birk Related Protease Inhibitor. *Biomolecules* **2019**, *9*, 280. [CrossRef] [PubMed]
3. Martínez-García, D.; Pérez-Hernández, M.; Korrodi-Gregório, L.; Quesada, R.; Ramos, R.; Baixeras, N.; Pérez-Tomás, R.; Soto-Cerrato, V. The Natural-Based Antitumor Compound T21 Decreases Survivin Levels through Potent STAT3 Inhibition in Lung Cancer Models. *Biomolecules* **2019**, *9*, 361. [CrossRef] [PubMed]
4. Thangaraj, K.; Balasubramanian, B.; Park, S.; Natesan, K.; Liu, W.; Manju, V. Orientin Induces G0/G1 Cell Cycle Arrest and Mitochondria Mediated Intrinsic Apoptosis in Human Colorectal Carcinoma HT29 Cells. *Biomolecules* **2019**, *9*, 418. [CrossRef] [PubMed]
5. Lai, Y.-H.; Chiang, C.-S.; Hsu, C.-H.; Cheng, H.-W.; Chen, S.-Y. Development and Characterization of a Fucoidan-Based Drug Delivery System by Using Hydrophilic Anticancer Polysaccharides to Simultaneously Deliver Hydrophobic Anticancer Drugs. *Biomolecules* **2020**, *10*, 970. [CrossRef] [PubMed]
6. Lee, D.; Lee, D.H.; Choi, S.; Lee, J.S.; Jang, D.S.; Kang, K.S. Identification and Isolation of Active Compounds from Astragalus membranaceus that Improve Insulin Secretion by Regulating Pancreatic β-Cell Metabolism. *Biomolecules* **2019**, *9*, 618. [CrossRef] [PubMed]
7. Martínez-Navarro, I.; Díaz-Molina, R.; Pulido-Capiz, A.; Mas-Oliva, J.; Luna-Reyes, I.; Rodríguez-Velázquez, E.; Rivero, I.A.; Ramos-Ibarra, M.A.; Alatorre-Meda, M.; García-González, V. Lipid Modulation in the Formation of β-Sheet Structures. Implications for De Novo Design of Human Islet Amyloid Polypeptide and the Impact on β-Cell Homeostasis. *Biomolecules* **2020**, *10*, 1201. [CrossRef] [PubMed]
8. Lee, J.S.; Jeong, M.; Park, S.; Ryu, S.M.; Lee, J.; Song, Z.; Guo, Y.; Choi, J.-H.; Lee, D.; Jang, D.S. Chemical Constituents of the Leaves of Butterbur (Petasites japonicus) and Their Anti-Inflammatory Effects. *Biomolecules* **2019**, *9*, 806. [CrossRef] [PubMed]
9. Celiksoy, V.; Moses, R.L.; Sloan, A.J.; Moseley, R.; Heard, C.M. Evaluation of the In Vitro Oral Wound Healing Effects of Pomegranate (Punica granatum) Rind Extract and Punicalagin, in Combination with Zn (II). *Biomolecules* **2020**, *10*, 1234. [CrossRef] [PubMed]
10. Szűcs, E.; Stefanucci, A.; Dimmito, M.P.; Zádor, F.; Pieretti, S.; Zengin, G.; Vécsei, L.; Benyhe, S.; Nalli, M.; Mollica, A. Discovery of Kynurenines Containing Oligopeptides as Potent Opioid Receptor Agonists. *Biomolecules* **2020**, *10*, 284. [CrossRef] [PubMed]

Biomolecules **2020**, *10*, 1674

11. Maheshwari, G.; Ringseis, R.; Wen, G.; Gessner, D.K.; Rost, J.; Fraatz, M.A.; Zorn, H.; Eder, K. Branched-Chain Fatty Acids as Mediators of the Activation of Hepatic Peroxisome Proliferator-Activated Receptor Alpha by a Fungal Lipid Extract. *Biomolecules* **2020**, *10*, 1259. [CrossRef] [PubMed]

Publisher's Note: MDPI stays neutral with regard to jurisdictional claims in published maps and institutional affiliations.

Article

Regulation of the Expression of DAPK1 by SUMO Pathway

Qingshui Wang [1,†], Xiuli Zhang [1,†], Ling Chen [1,†], Shuyun Weng [1], Yun Xia [1], Yan Ye [1], Ke Li [1], Ziqiang Liao [1], Pengchen Chen [1], Khaldoon Alsamman [2], Chen Meng [1], Craig Stevens [3], Ted R. Hupp [4] and Yao Lin [1,*]

[1] Provincial University Key Laboratory of Cellular Stress Response and Metabolic Regulation, College of Life Sciences, Fujian Normal University, Fuzhou 350117, China; wangqingshui@fjnu.edu.cn (Q.W.); 15980239296@163.com (X.Z.); chenling654321@163.com (L.C.); wsy09080700@163.com (S.W.); xiayunnyyl@163.com (Y.X.); m18759141945@163.com (Y.Y.); 13107673087@163.com (K.L.); liaoziqiangcontact@163.com (Z.L.); yb77620@umac.mo (P.C.); mmenger@126.com (C.M.)
[2] Department of Clinical Laboratory Sciences, College of Applied Medical Sciences, Imam Abdulrahman bin Faisal University, Dammam 34212, Saudi Arabia; kmalsamman@iau.edu.sa
[3] School of Applied Sciences, Edinburgh Napier University, Edinburgh EH11 4BN, UK; C.Stevens@napier.ac.uk
[4] Institute of Genetics and Molecular Medicine, Cell Signaling Unit, CRUK p53 Transduction Group, University of Edinburgh, EH4 2XR EH4 2XR, UK; ted.hupp@ed.ac.uk
* Correspondence: yaolin@fjnu.edu.cn; Tel.: +86-(0)591-22860592
† These authors contributed equally to this work.

Received: 15 March 2019; Accepted: 15 April 2019; Published: 17 April 2019

Abstract: Death Associated Protein Kinase 1 (DAPK1) is an important signaling kinase mediating the biological effect of multiple natural biomolecules such as IFN-γ, TNF-α, curcumin, etc. DAPK1 is degraded through both ubiquitin-proteasomal and lysosomal degradation pathways. To investigate the crosstalk between these two DAPK1 degradation pathways, we carried out a screen using a set of ubiquitin E2 siRNAs at the presence of Tuberous Sclerous 2 (TSC2) and identified that the small ubiquitin-like molecule (SUMO) pathway is able to regulate the protein levels of DAPK1. Inhibition of the SUMO pathway enhanced DAPK1 protein levels and the minimum domain of DAPK1 protein required for this regulation is the kinase domain, suggesting that the SUMO pathway regulates DAPK1 protein levels independent of TSC2. Suppression of the SUMO pathway did not enhance DAPK1 protein stability. In addition, mutation of the potential SUMO conjugation sites on DAPK1 kinase domain did not alter its protein stability or response to SUMO pathway inhibition. These data suggested that the SUMO pathway does not regulate DAPK1 protein degradation. The exact molecular mechanism underlying this regulation is yet to be discovered.

Keywords: DAPK1; SUMO; SENP; protein degradation; post-translational modification

---

1. Introduction

Death-Associated Protein Kinase 1 (DAPK1) is an important serine/threonine kinase that is involved in multiple cellular processes such as apoptosis, autophagy, inflammation [1]. DAPK1 Plays a vital role in the anti-carcinogenic effects of many natural-based biomolecules, such as IFN-γ, TNF-α, curcumin, etc. [2,3]. Decreased expression of DAPK1 has been proven to be an unfavorable prognostic factor in bladder cancer, liver cancer and non-small cell lung cancer, etc. [4,5].

DAPK1 protein is composed of multiple functional domains including a catalytic kinase domain, a Ca2+/CaM auto-regulatory domain, eight ankyrin repeats, a Ras of Complex proteins (ROC)-C-terminus of ROC (COR) domain, a death domain and a serine-rich tail [6]. Dysregulation of DAPK1 expression or activity is often related to multiple diseases including cancer and stroke [7]. Recently, DAPK1 was also found to inhibit Hypoxia-inducible factor 1α (HIF-1α) in T cells [8], maintain epidermal tissue integrity through regulation of the microtubule cytoskeleton in C.elegans [9], and mediate pegylated Interferon-α (IFN-α)-induced suppression of hepatitis C virus (HCV) replication [10].

Expression of DAPK1 is often lost in cancers due to DNA methylation of the DAPK1 gene [11]. Due to DNA methylation, the expression of DAPK1 is lost in primary tumor samples of 26% of rectal cancer patients. Similarly, different degrees of DNA methylation of DAPK1 have also been found in lung cancer, leukemia, breast cancer, uterine cancer and prostate cancer [12–14]. Previous studies revealed that the expression of the DAPK1 protein does not match its expression of mRNA in some cancers, indicating that the regulation of DAPK1 expression is a complex process. The degradation of DAPK1 protein is controlled by both proteasomal and lysosomal degradation pathways [15]. Three ubiquitin E3s, Mind Bomb 1(Mib1) [16], C-terminus of Hsc70-interacting protein (CHIP) [17] and KLHL20-Cullin3-RBX1 complex [18], target DAPK1 for ubiquitin-proteasome system (UPS)-mediated degradation. The lysosomal degradation pathway of DAPK was found to be late compared to proteasome degradation. Proteins known to be involved in the DAPK1 protein of lysosomal degradation include s-DAPK, Tuberin (TSC2) and cathepsin B. The TSC complex, formed by two proteins (TSC1 (hamartin) and TSC2), is a major regulator of the mTORC1 signaling pathway [19]. In our previous work, we discovered that TSC2 and a splice variant of DAPK1 (s-DAPK1) induced the lysosomal degradation of DAPK1 [20,21]. Moreover, a lysosomal protease, cathepsin B, is able to cleave DAPK1 in response to Tumor Necrosis Factor Receptor 1 (TNFR-1) over-expression [22].

Growing evidence demonstrates that there is intense crosstalk between UPS and lysosome [23]. Ubiquitination on proteins such as p62 can lead to their degradation by both UPS and lysosome [23]. Although DAPK1 has been shown to be degraded by both degradation pathways, it is not clear whether the ubiquitin related signaling pathways contribute to its lysosomal degradation. To investigate the crosstalk between these two DAPK1 degradation pathways, we carried out a screen using a set of ubiquitin E2 siRNAs and identified that the small ubiquitin like molecule (SUMO) pathway regulates DAPK1 protein levels.

SUMO is a family of ubiquitin-related modifiers that can be post-translationally conjugated to various substrates [24]. Intracellular proteins can be modified by SUMO, which affects substrate protein localization, stability, protein modification, and protein-protein interactions [25]. Five different SUMO paralogues have been reported in vertebrates, named SUMO-1 to SUMO-5 [24]. SUMO-1 shares 45% homology with SUMO-2/3, and there are only two amino acids difference between SUMO-2 and SUMO-3 [26]. SUMO-4 encodes a 95-amino acid protein having an 86% amino acid homology with SUMO-2 [27]. SUMO5 is a novel SUMO variant and contains a protein-coding sequence of 306 nucleotides [28]. The covalent modification reaction of SUMO is catalyzed by a series of enzymes including E1 activating enzyme (SAE1/SAE2), E2 binding enzyme (UBC9) and E3 ligase enzyme [29]. The process of SUMOylation is dynamic and reversible. A family of SUMO specific proteases (SENPs) are capable of removing SUMO from attached substrates and responsible for SUMO maturation [24]. Family members of SENPs include SENP-1, SENP-2, SENP-3, SENP-5, SENP-6 and SENP-7. The SENP family can be divided into three subfamilies based on the degree of amino acid sequence homology, cell localization, and substrate preference. The first subfamily comprises SENP-1 and SENP-2, which are located on the nuclear membrane and have the widest selection of substrates, which participate in the deubiquitination of proteins modified by SUMO-1 and SUMO-2/3 [30,31]. The second subfamily is SENP-3 and SENP-5. They are mainly found in nucleoli and involved in the synthesis of ribosomes and the regulation of cell mitosis [32]. The third subfamily is SENP-6 and SENP-7. They are located in the nucleus and are essential for the formation of multi-cluster ubiquitin chains [33,34].

In summary, the degradation of DAPK protein is regulated by both proteasome and lysosome. The research of E2 on degradation of DAPK1 protein will help to better understand the regulation of DAPK protein degradation by ubiquitin and ubiquitin-like small molecule signaling pathways and to discover the link between ubiquitin-related small molecule and lysosomal degradation signaling pathways.

2. Materials and Methods

2.1. Cell Culture and Transfection

HEK293 (human embryonic kidney cell line) and HCT116 (Human colon carcinoma) cells were obtained from ATCC (Manassas, MD, USA). Cell lines were examined for mycoplasma contamination using Mycoplasma Detection Kit (Vazyme, Nanjing, Jiangsu, China). Both cells were cultured in DMEM medium (Invitrogen, Carlsbad, CA, USA) supplemented with 10% fetal bovine serum (FBS) and a penicillin and streptomycin mixture at 37 °C with 5% CO_2 in a humidified atmosphere. Before harvesting, cells were first washed twice with PBS and then scraped into 1 mL of PBS. PCDNA3-HA-DAPK1 was a gift from Ted R. Hupp (University of Edinburgh). Flag-SENP1 (Plasmid #17357, deposited by Edward Yeh), FLAG-SENP2 (Plasmid #18047, deposited by Edward Yeh), RGS-SENP3 (Plasmid #18048, deposited by Edward Yeh), RGS-SENP5 (Plasmid #18053, deposited by Edward Yeh), FLAG-SENP6 (Plasmid #18065, deposited by Edward Yeh), 3xFLAG-SENP7 (Plasmid #42886, deposited by Edward Yeh) and Flag TSC2 wt (Plasmid #12132, deposited by Cheryl Walker) were obtained from Addgene (Cambridge, UK). The DAPK1 mutant constructs were generated using the QuikChange Lightning Site-Directed Mutagenesis Kit from Vazyme (Nanjing, China). All plasmids were sequenced to verify the integrity of the constructs. UBE2A siRNA (E-009424-00-0005), UBE2B siRNA (E-009930-00-0005), UBE2C siRNA (E-004693-00-0005), UBE2D1 siRNA (E-009387-00-0005), UBE2D2 siRNA (E-010383-00-0005), UBE2D3 siRNA (E-008478-00-0005), UBE2E1 siRNA (E-008850-00-0005), UBE2E2 siRNA (E-031782-00-0005), UBE2E3 siRNA (E-008845-00-0005), UBE2G1 siRNA (E-010154-00-0005), UBE2G2 siRNA (E-009095-00-0005), UBE2H siRNA (E-009134-00-0005), UBE2I siRNA (E-004910-00-0005), UBE2J1 siRNA (E-007266-00-0005), UBE2J2 siRNA (E-008614-00-0005), UBE2L3 siRNA (E-010384-00-0005), UBE2M siRNA (E-004348-00-0005), UBE2N siRNA (E-003920-00-0005), UBE2NL siRNA (E-031625-00-0005), UBE2Q1 siRNA (E-008631-00-0005), UBE2R2 siRNA (E-009700-00-0005), UBE2S siRNA (E-009707-00-0005) and UBE2V2 siRNA (E-008823-00-0005) were purchased from Dharmacon (Lafayette, MA, USA). The transfection was performed using lipofectamine 2000 (Invitrogen, Carlsbad, CA, USA) according to the manufacturer's guidelines. To confirm that the difference in different lanes is due to the biological effect of the transfected plasmids, rather than technical differences, equal amounts of lacz in each plate were transfected as a co-transfection plasmid to balance the difference in transfection efficiency.

2.2. Western Blot

After harvesting, the cells were lysed in ice-cold lysis buffer (50 mM Tris (pH 8.0), 150 mM NaCl, 1% Nonidet P-40, 1% sodium deoxycholate, 1% SDS, 1× protease inhibitor mixture (Roche, Basel, Switzerland)) for 30 min and centrifuged at 4 °C, 13,000 rpm for 10 min to remove insoluble material. The soluble protein concentration was determined by Bradford assay. Protein samples (60 µg) were separated by SDS-PAGE and transferred to nitrocellulose blotting membranes (Bio-Rad, Hercules, CA, USA). The membranes were treated with block buffer (5% non-fat milk in 0.1% TBST (20 mM Tris-HCl pH 8.0, 150 mM NaCl, 0.1% Tween-20)) at room temperature for 1 h. The membranes were then incubated with primary antibodies overnight at 4 °C, then washing (3 × 12 min) in PBS/Tween 20, followed by incubating with secondary antibodies in blocking buffer at room temperature for 2 h. Finally, washing (3 × 12 min) in PBS/Tween 20 again. The signals were detected and measured using LICOR Odyssey system (LI-COR, Lincoln, NE, USA). All the western blots were repeated at least three times.

2.3. Statistical Analysis

Data were analyzed using Prism 5.0 software (Graphpad Software, Inc., La Jolla, CA, USA). Results are presented as the mean ± standard deviation of three independent experiments. The difference between the means were tested by the One-way ANOVA testing or Student's t-test, $p < 0.05$ was considered to indicate a statistically significant difference.

2.4. Prediction of SUMOylation Sites

In this study, prediction of the SUMOylation sites on the kinase domain on DAPK1 was performed using GPS-SUMO, which is a novel web server that can be used to predict potential SUMOylation sites (http://sumosp.biocuckoo.org/) [35].

2.5. Antibodies and Chemicals

Anti-GAPDH Antibody (2118) and anti-DAPK1 Antibody (3008) were purchased from Cell Signaling (Boston, MA, USA), anti-HA-Tag Antibody (902301) was purchased from Biolegend (San Diego, CA, USA), anti-flag Antibody (M20008) was purchased from Abmart (Shanghai, China), anti-SENP2 antibody (ab96865) was purchased from ABCAM (Cambridge, MA, USA), anti-Beta Galactosidase (β-gal) (*E. coli*) antibody (28449) was purchased from Rockland (Limerick, PA, USA). IRDye 800CW Goat-anti-Mouse (C60405-05), IRDye 680RD Goat-anti-Mouse (C60405-08), IRDye 800CW Goat-anti-Rabbit (C60607-15) and IRDye 680RD Goat-anti-Rabbit (C60329-15) were purchased from LI-COR (Lincoln, NE, USA).

MG132 was purchased from Calbiochem (LaJolla, NJ, USA) and used at 10 μM. Chloroquine and Cycloheximide (CHX) were purchased from Sigma (Louis, MO, USA) and used at 100 μM and 10 μg/mL, respectively. EST (E-64D) and Leupeptin were purchased from EMD Millipore Crop (Billerica, MA, USA) and used at 10 μg/mL and 200 μM respectively.

3. Results

To search the potential ubiquitin or ubiquitin-like regulatory pathways involved in TSC2-mediated DAPK1 protein degradation, a screen using an E2 siRNA library was carried out. Co-transfection of the siRNAs targeting three E2s (UBE2B, UBE2D1 and UBE2I) up-regulated HA-DAPK1 protein levels upon co-transfection of TSC2 (Figure 1A). Of these three E2s, UBE2I, also named UBC9, is the E2 for SUMO, which has been shown to participate in protein degradation [36]. Therefore, we co-transfected four different SENPs with TSC2 and HA-DAPK1. Co-transfection of SENPs enhanced the level of HA-DAPK1 protein (Figure 1B,C), but not to the level without TSC2 co-transfection, suggesting inhibition of SUMO pathway is not able to abrogate the effect of TSC2 towards DAPK1 protein levels.

Next, we co-transfected individual SUMO construct with HA-DAPK1. Co-transfection of neither SUMO construct resulted in significant down-regulation of HA-DAPK1 protein levels (Figure 2A). However, when SUMO-1 was co-transfected with SUMO-2 or SUMO-3, it significantly stimulated the reduction of the HA-DAPK1 protein levels, whereas co-transfection of SUMO-2 and SUMO-3 did not seem to pose additive effect (Figure 2B). Furthermore, all six known SENPs significantly enhanced the levels of HA-DAPK1 protein when co-transfected with HA-DAPK1 in HEK293T cells (Figure 2C) and HCT-116 cells (Figure 2D). Moreover, consistent with the exogenous expression data, when endogenous UBC9 was knocked down using siRNA, the endogenous DAPK1 protein also significantly increased (Figure 2E), suggesting SUMO pathway is able to regulate DAPK1 protein levels without simultaneously manipulation of TSC2-related pathway.

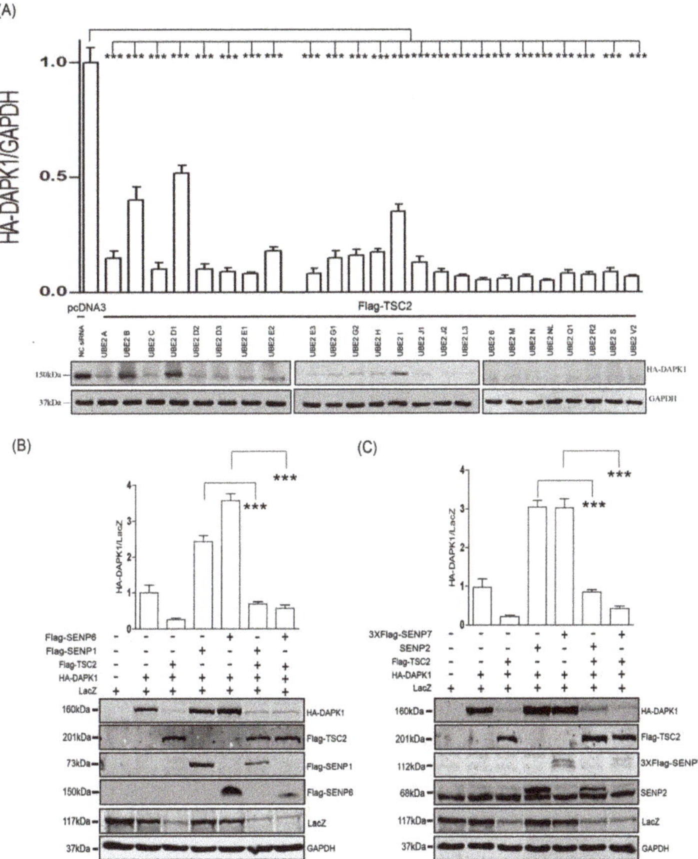

Figure 1. Inhibition of SUMO pathway partially restored DAPK1 protein levels at the presence of TSC2. (**A**) HEK293T was transfected with HA-DAPK1, Flag-TSC2, and different E2 siRNAs. (**B,C**) HEK293T cells were transfected with HA-DAPK1, LacZ, TSC2 and four different SENPs as indicated. Cell lysates were extracted 48 h post-transfection and immunoblotted with respective antibodies. The intensity of the bands was quantified using LICOR Odyssey software and represented by bar graphs. The experiments were repeated three times (n = 3), and representative images are presented. The representative western blot images are from different gels and each lane was loaded with the same amount of proteins. Data of triplicate assays are presented as mean ± S.D. * $p < 0.05$; ** $p < 0.01$; *** $p < 0.001$; NS, no significance. "+" indicated that the plasmid was transfected, "−" indicated that the plasmid was not transfected.

To further elucidate the underlying molecular mechanisms, a deletion series of DAPK1 constructs was created (Figure 3A) and co-transfected with three different SENPs. SENP1 (Figure 3B), SEP2 (Figure 3C) and SENP6 (Figure 3D) significantly enhanced the expression levels of all the deletion mutants, suggesting SUMO pathway regulates DAPK1 protein levels via the kinase domain. This is further confirmed when three SENPs displayed no effect towards the level of HA-DAPK1 (275–1430) lacking the kinase domain (Figure 3E).

Next, the HA-DAPK1 (1–364) was exposed to both proteasome and lysosome inhibitors. Only the proteasome inhibitor MG132 significantly enhanced the protein levels of HA-DAPK1 (1–364) in both HEK293T (Figure 4A) and HCT116 cells (Figure 4B), suggesting this kinase domain mutant HA-DAPK1 (1–364) is predominantly degraded via proteasome. In the protein stability assays using cycloheximide (CHX), MG132 significantly enhanced HA-DAPK1 (1–364) protein stability (Figure 4C,D,F), whereas co-transfection of SENP6 was unable to enhance the stability of HA-DAPK1 (1–364) protein (Figure 4C,E,F), indicating that the SUMO pathway does not regulate DAPK1 protein levels via protein degradation. This also suggested that the SUMO pathway was unlikely to control DAPK1 protein levels through direct conjugation.

Using the GPS-SUMO system, we identified two potential SUMO conjugation sites on HA-DAPK1 (1–364). Therefore, we mutated these two sites separately or simultaneously (Figure 5A). As expected, the mutation did not influence the effect of SENP6 on DAPK1 (1–364) (Figure 5B), supporting that the SUMO pathway does not regulate DAPK1 protein levels via direct conjugation. Next, we compared the protein stability of HA-DAPK1 (1–364) and the mutants with potential SUMO conjugation sties mutated. No mutants were able to enhance the protein stability of HA-DAPK1 (1–364) (Figure 5C–E).

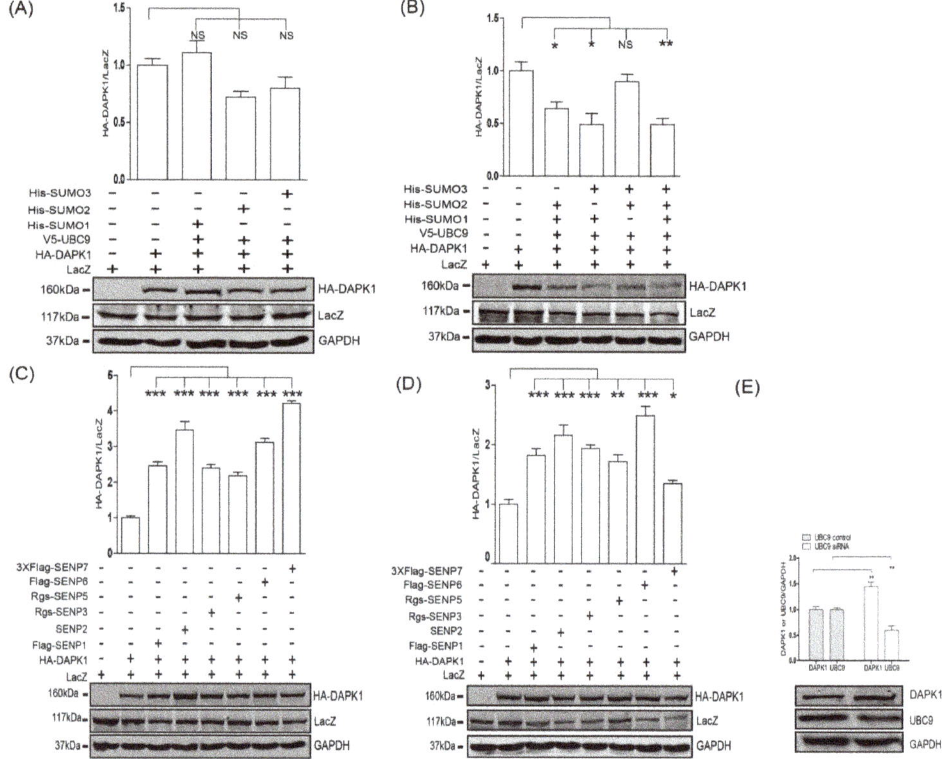

Figure 2. The SUMO pathway regulated the protein levels of DAPK1. HEK293T were transfected with (**A**) HA-DAPK1, LacZ, V5-UBC9, His-SUMO-1, His-SUMO-2 or His-SUMO-3, or (**B**) HA-DAPK1, LacZ, V5-UBC9, His-SUMO-1, His-SUMO-2 or His-SUMO-3, or (**C**) HA-DAPK1, LacZ and six different SENPs, or (**E**) control and UBC9 siRNA as indicated. HCT116 were transfected with (**D**) HA-DAPK1, LacZ and six different SENPs as indicated. Cell lysates were extracted 48 h post-transfection and immunoblotted with respective antibodies. The intensity of the bands was quantified using LICOR Odyssey software and represented by bar graphs. The experiments were repeated three times (n = 3) and representative images are presented. The representative western blot images are from different gels and each lane was loaded with the same amount of proteins. Data of triplicate assays are presented as mean ± S.D. * $p < 0.05$; ** $p < 0.01$; *** $p < 0.001$; NS, no significance. "+" indicated that the plasmid was transfected, "−" indicated that the plasmid was not transfected.

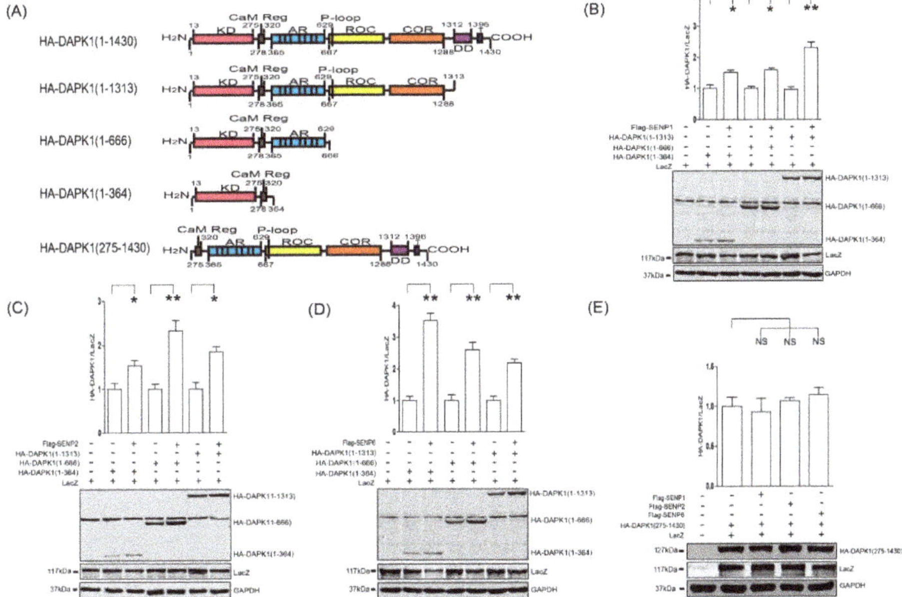

Figure 3. SENPs up-regulated DAPK1 protein levels via the 1-364 kinase domain. (**A**) A diagram illustrating the panel of DAPK1 deletion mutants. (**B–D**) The DAPK1 deletion mutants were co-transfected with LacZ and either Flag-SENP1 (**B**), Flag-SENP2 (**C**) or Flag-SENP6 (**D**), as indicated. (**E**) HA-DAPK (275–1430) mutant was co-transfected with LacZ and either Flag-SENP1, Flag-SENP2 or Flag-SENP6. Cell lysates were extracted 48 h post-transfection and immunoblotted with respective antibodies. The intensity of the bands was quantified using LICOR Odyssey software and represented by bar graphs. The experiments were repeated three times (n = 3) and representative images are presented. The representative western blot images are from different gels and each lane was loaded with the same amount of proteins. Data of triplicate assays are presented as mean ± S.D. * $p < 0.05$; ** $p < 0.01$; *** $p < 0.001$; NS, no significance. "+" indicated that the plasmid was transfected, "−" indicated that the plasmid was not transfected.

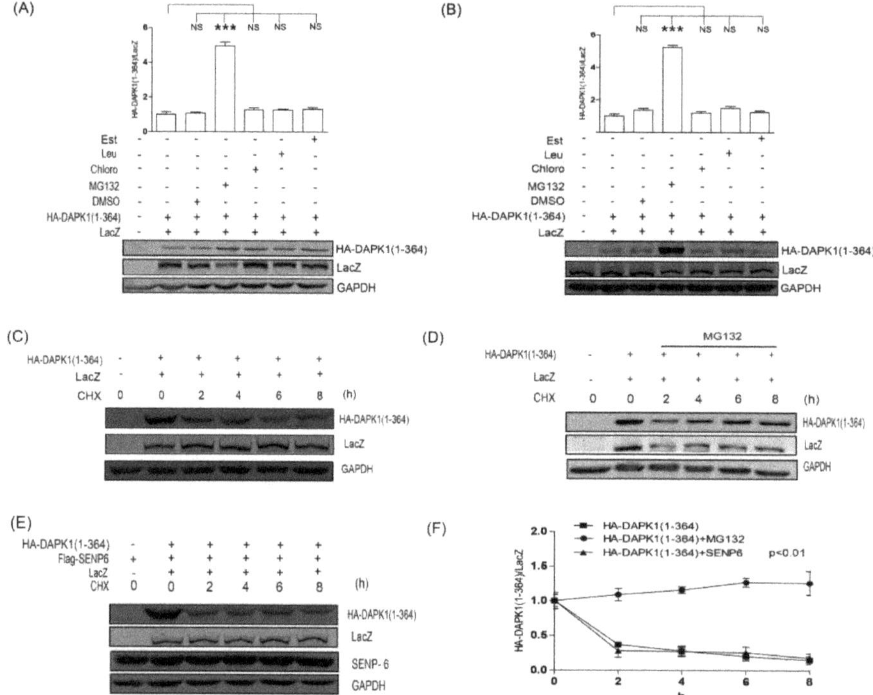

Figure 4. SUMO pathway did not regulate the protein degradation of HA-DAPK1 (1–364). (**A**) HEK293T cells or (**B**) HCT116 cells transfected with DAPK1 (1–364) and LacZ were exposed to MG132 (10 μM, 6 h) or leupeptin (200 μM), Est (10 μg/mL) and chloroquine (100 μM) for 24 h as indicated. (**C**) HEK293T transfected with LacZ, DAPK1(1–364) were exposed to 20 μg/mL CHX for 0–8 h as indicated. (**D**) HEK293T transfected with HA-DAPK1 (1–364) and LacZ were exposed to 10 μM MG132 and 20 μg/mL CHX for 0–8 h as indicated. (**E**) HEK293T transfected with LacZ, HA-DAPK1 (1–364) and Flag-SENP6 were exposed to 20 μg/mL CHX for 0–8 h as indicated. The statistical results of (**C–E**) are summarized in (**E**). Cell lysates were extracted and immunoblotted with respective antibodies. The intensity of the bands was quantified using LICOR Odyssey software and represented by trend lines. The experiments were repeated three times (n = 3) and representative images are presented. The representative western blot images are from different gels and each lane was loaded with the same amount of proteins. Data of triplicate assays are presented as mean ± S.D. * p <0.05; ** p < 0.01; *** p < 0.001; NS, no significance. "+" indicated that the plasmid was transfected, "−" indicated that the plasmid was not transfected.

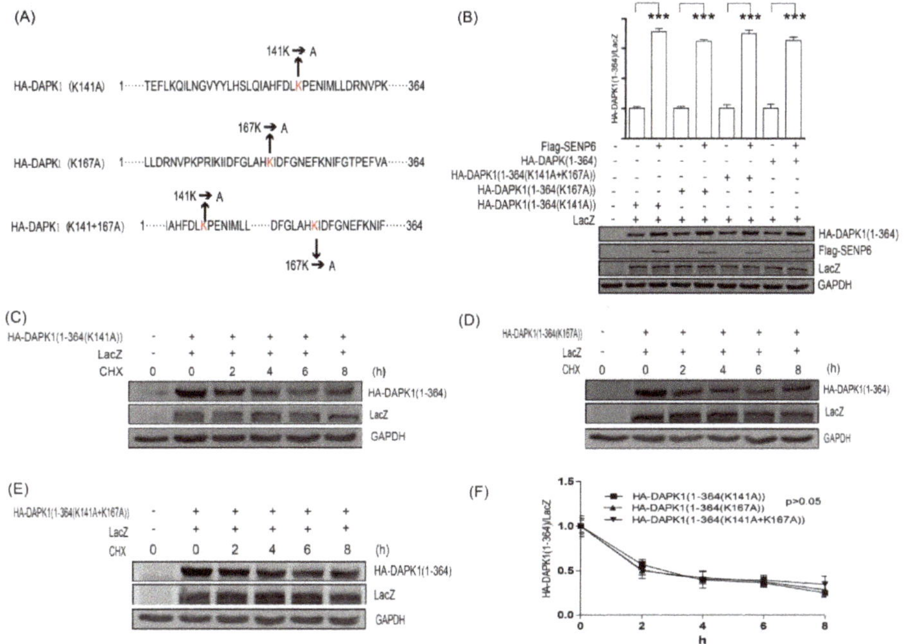

Figure 5. SUMO pathway did not regulate DAPK1 protein levels via direct conjugation. (**A**) A diagram illustrating the panel of DAPK1 (1–364) point mutants. (**B**) HEK293T cells were transfected with LacZ, SENP6 and the DAPK1 point mutants as indicated. (**C–E**) HEK293T transfected with LacZ, DAPK1 (1–364) point mutants were exposed to 20 μg/mL CHX for 0–8 h as indicated and the statistical results were summarized in (**F**). The intensity of the bands was quantified using LICOR Odyssey software and represented by bar graphs. The experiments were repeated three times (n = 3) and representative images are presented. The representative western blot images are from different gels and each lane was loaded with the same amount of proteins. Data of triplicate assays are presented as mean ± S.D. * $p < 0.05$; ** $p < 0.01$; *** $p < 0.001$; NS, no significance. "+" indicated that the plasmid was transfected, "−" indicated that the plasmid was not transfected.

4. Discussion

In our previous work, we discovered that TSC2 mediated the lysosomal degradation of DAPK1 via binding to the death domain of DAPK1 [20]. In this study, we discovered that inhibition of the SUMO pathway was able to enhance HA-DAPK1 protein levels at the presence of TSC2 co-transfection (Figure 1). However, the minimum domain of DAPK1 protein that the SUMO pathway is able to regulate is the kinase domain (Figure 3). The protein degradation of HA-DAPK1 (1–364) mutant is via proteasome (Figure 4). All these data suggest that the SUMO pathway regulates DAPK1 protein levels independent of TSC2. In this research, we co-transfected Lacz plasmid to balance the transfection variation and found that every single exogenous protein co-expressed with TSC2 shown decreased expression (Figure 1B,C). TSC2 is an important suppressor of mammalian target of rapamycin (mTORC1), which is a key regulator of translation and may be critical for exogenous protein levels. Therefore, when TSC2 was co-transfected, the expression of all other plasmids might be reduced due to inhibition of mTORC1 activity. The reason Glyceraldehyde-3-phosphate dehydrogenase (GAPDH) was stable could be because its translation is not mTORC1 dependent or its long half-life [37].

In two high-throughput mass spectrometry SUMOylation assays [38,39], DAPK1 protein was not found to be SUMOylated, suggesting DAPK1 may not be able to be SUMOylated, despite the presence of its potential SUMO conjugation sites. Our study is consistent with these two reports, but did not fully rule out the possibility that DAPK1 protein can be SUMOylated. It is only clear that even if DAPK1 protein can be SUMOylated, the SUMO conjugation is unlikely to affect the protein levels of DAPK1.

Although the SUMO pathway regulates DAPK1 protein levels, inhibition of the SUMO pathway does not change the protein stability of DAPK1 (Figure 4E). Therefore, the SUMO pathway probably regulates DAPK1 protein levels upstream of protein degradation. Considering that inhibition of the SUMO pathway enhanced the expression of exogenous HA-DAPK1, which only contains the coding region of DAPK1 mRNA, it is unlikely that SUMO pathway regulates DAPK1 protein levels through transcription. Recently, long noncoding RNAs (lncRNAs) have been shown to regulate gene expression at various levels, including chromatin modification, transcription and post-transcriptional processing [40]. Moreover, a new regulatory mechanism has been identified in which crosstalk between lncRNAs and mRNAs occurs by competing for shared microRNAs (miRNAs) response elements [41]. It is possible that the SUMO pathway regulates the expression of the miRNAs or lncRNAs targeting DAPK1, thus affect the levels of DAPK1 via these additional post-transcriptional regulatory pathways. It is also possible that the SUMO pathway regulates the expression levels or activity of proteins that are responsible for DAPK1 mRNA stability or translation.

In summary, our study discovered that the protein levels of DAPK1 can be regulated by the SUMO pathway. However, this regulation is not mediated via manipulation of DAPK1 protein degradation. The molecular mechanisms underlying this SUMO-mediated regulation of DAPK1 expression is still unclear. Further investigation is needed to elucidate this observation.

Author Contributions: Conceptualization, Q.W. and Y.L.; Data curation, X.Z. and L.C.; Methodology, X.Z., L.C., S.W., Y.X., Y.Y., K.L., Z.L., P.C. and C.M.; Resources, K.A., C.S. and T.R.H.; Writing–original draft, Q.W. and Y.L.; Writing–review & editing, Y.L.

Funding: This work was funded by the International S & T Cooperation Program of China (ISTCP, 2016YFE0121900), the Educational and Scientific Research Project for Young Scholars in Fujian Province (JAT170136), Natural Science Foundation of Fujian Province (2018J01723), scientific research innovation team construction program of Fujian Normal University (IRTL1702), United Fujian Provincial Health and Education Project for Tackling the Key Research (WKJ2016-2-27) and Fujian normal university (FZSKG2018008).

Conflicts of Interest: The authors have no conflicts of interest to declare.

References

1. Bialik, S.; Kimchi, A. The DAP-kinase interactome. *Apoptosis* **2014**, *19*, 316–328. [CrossRef]
2. Wu, B.; Yao, H.; Wang, S.; Xu, R. DAPK1 modulates a curcumin-induced G2/M arrest and apoptosis by regulating STAT3, NF-κB, and caspase-3 activation. *Biochem. Biophys. Res. Commun.* **2013**, *434*, 75–80. [CrossRef] [PubMed]
3. Yoo, H.J.; Byun, H.J.; Kim, B.R.; Lee, K.H.; Park, S.Y.; Rho, S.B. DAPk1 inhibits NF-κB activation through TNF-α and INF-γ-induced apoptosis. *Cell. Signal.* **2012**, *24*, 1471–1477. [CrossRef] [PubMed]
4. Xie, J.Y.; Chen, P.C.; Zhang, J.L.; Gao, Z.S.; Neves, H.; Zhang, S.D.; Wen, Q.; Chen, W.D.; Kwok, H.F.; Lin, Y. The prognostic significance of DAPK1 in bladder cancer. *PLoS ONE* **2017**, *12*, e0175290. [CrossRef]
5. Niklinska, W.; Naumnik, W.; Sulewska, A.; Kozłowski, M.; Pankiewicz, W.; Milewski, R. Prognostic significance of DAPK and RASSF1A promoter hypermethylation in non-small cell lung cancer (NSCLC). *Folia Histochem. Cytobiol.* **2009**, *47*, 275–280. [CrossRef]
6. Shiloh, R.; Bialik, S.; Kimchi, A. The DAPK family: A structure–function analysis. *Apoptosis* **2014**, *19*, 286–297. [CrossRef] [PubMed]
7. Huang, Y.; Chen, L.; Guo, L.; Hupp, T.R.; Lin, Y. Evaluating DAPK as a therapeutic target. *Apoptosis* **2014**, *19*, 371–386. [CrossRef] [PubMed]
8. Chou, T.F.; Chuang, Y.T.; Hsieh, W.C.; Chang, P.Y.; Liu, H.Y.; Mo, S.T.; Hsu, T.S.; Miaw, S.C.; Chen, R.H.; Kimchi, A. Tumour suppressor death-associated protein kinase targets cytoplasmic HIF-1α for Th17 suppression. *Nat. Commun.* **2016**, *7*, 11904. [CrossRef]
9. Chuang, M.; Hsiao, T.I.; Tong, A.; Xu, S.; Chisholm, A.D. DAPK interacts with Patronin and the microtubule cytoskeleton in epidermal development and wound repair. *Elife* **2016**, *5*, e15833. [CrossRef]

10. Liu, W.L.; Yang, H.C.; Hsu, C.S.; Wang, C.C.; Wang, T.S.; Kao, J.H.; Chen, D.S. Pegylated IFN-α suppresses hepatitis C virus by promoting the DAPK-mTOR pathway. *Proc. Natl. Acad. Sci. USA* **2016**, *113*, 14799–14804. [CrossRef]

11. Benderska, N.; Schneider-Stock, R. Transcription control of DAPK. *Apoptosis* **2014**, *19*, 298–305. [CrossRef] [PubMed]

12. Raval, A.; Tanner, S.M.; Byrd, J.C.; Angerman, E.B.; Perko, J.D.; Chen, S.S.; Hackanson, B.; Grever, M.R.; Lucas, D.M.; Matkovic, J.J. Downregulation of Death-Associated Protein Kinase 1 (DAPK1) in Chronic Lymphocytic Leukemia. *Cell* **2007**, *129*, 879–890. [CrossRef] [PubMed]

13. Ivanovska, J.; Zlobec, I.; Forster, S.; Karamitopoulou, E.; Dawson, H.; Koelzer, V.H.; Agaimy, A.; Garreis, F.; Söder, S.; Laqua, W.; et al. DAPK loss in colon cancer tumor buds: Implications for migration capacity of disseminating tumor cells. *Oncotarget* **2015**, *6*, 36774–36788. [CrossRef]

14. Chan, M.W.; Chan, L.W.; Tang, N.L.; Tong, J.H.; Lo, K.W.; Lee, T.L.; Cheung, H.Y.; Wong, W.S.; Chan, P.S.; Lai, F.M.; et al. Hypermethylation of multiple genes in tumor tissues and voided urine in urinary bladder cancer patients. *Clin. Cancer Res.* **2002**, *8*, 464–470.

15. Gallagher, P.J.; Blue, E.K. Post-translational regulation of the cellular levels of DAPK. *Apoptosis* **2014**, *19*, 306–315. [CrossRef]

16. Jin, Y.; Blue, E.K.; Gallagher, P.J. Control of death-associated protein kinase (DAPK) activity by phosphorylation and proteasomal degradation. *J. Biol. Chem.* **2006**, *281*, 39033–39040. [CrossRef]

17. Zhang, L.; Nephew, K.P.; Gallagher, P.J. Regulation of death-associated protein kinase. Stabilization by HSP90 heterocomplexes. *J. Biol. Chem.* **2007**, *282*, 11795–11804. [CrossRef] [PubMed]

18. Lee, Y.R.; Yuan, W.C.; Ho, H.C.; Chen, C.H.; Shih, H.M.; Chen, R.H. The Cullin 3 substrate adaptor KLHL20 mediates DAPK ubiquitination to control interferon responses. *EMBO J.* **2014**, *29*, 1748–1761. [CrossRef]

19. Laplante, M.; Sabatini, D. mTOR Signaling in Growth Control and Disease. *Cell* **2012**, *149*, 274–293. [CrossRef]

20. Yao, L.; Paul, H.; Susanne, P.; Jack, S.; Ted, H.; Craig, S. Tuberous sclerosis-2 (TSC2) regulates the stability of death-associated protein kinase-1 (DAPK) through a lysosome-dependent degradation pathway. *FEBS J.* **2011**, *278*, 354–370.

21. Lin, Y.; Stevens, C.; Harrison, B.; Pathuri, S.; Amin, E.; Hupp, T.R. The alternative splice variant of DAPK-1, s-DAPK-1, induces proteasome-independent DAPK-1 destabilization. *Mol. Cell. Biochem.* **2009**, *328*, 101–107. [CrossRef] [PubMed]

22. Lin, Y.; Stevens, C.; Hupp, T. Identification of a Dominant Negative Functional Domain on DAPK-1 That Degrades DAPK-1 Protein and Stimulates TNFR-1-mediated Apoptosis. *J. Biol. Chem.* **2007**, *282*, 16792–16802. [CrossRef]

23. Korolchuk, V.I.; Menzies, F.M.; Rubinsztein, D.C. Mechanisms of cross-talk between the ubiquitin-proteasome and autophagy-lysosome systems. *FEBS Lett.* **2010**, *584*, 1393–1398. [CrossRef]

24. Huang, C.J.; Wu, D.; Khan, F.A.; Huo, L.J. DeSUMOylation: An Important Therapeutic Target and Protein Regulatory Event. *DNA Cell Biol.* **2015**, *34*, 1–9. [CrossRef] [PubMed]

25. Kim, K.I.; Baek, S.H. Chapter 7 Small Ubiquitin-Like Modifiers in Cellular Malignancy and Metastasis. *Int. Rev. Cell Mol. Biol.* **2009**, *273*, 265–311.

26. Krumova, P.; Weishaupt, J.H. Sumoylation in neurodegenerative diseases. *Cell. Mol. Life Sci.* **2013**, *70*, 2123–2138. [CrossRef]

27. Bohren, K.M.; Varsha, N.; Song, J.H.; Gabbay, K.H.; David, O. A M55V polymorphism in a novel SUMO gene (SUMO-4) differentially activates heat shock transcription factors and is associated with susceptibility to type I diabetes mellitus. *J. Biol. Chem.* **2004**, *279*, 27233–27238. [CrossRef]

28. Liang, Y.C.; Lee, C.C.; Yao, Y.L.; Lai, C.C.; Schmitz, M.L.; Yang, W.M. SUMO5, a Novel Poly-SUMO Isoform, Regulates PML Nuclear Bodies. *Sci. Rep.* **2016**, *6*, 26509. [CrossRef]

29. Nan, S.L.; Marie-Claude, G.; Jaffray, E.G.; Hay, R.T. Characterization of SENP7, a SUMO-2/3-specific isopeptidase. *Biochem. J.* **2009**, *421*, 223–230.

30. Vertegaal, A.; Andersen, J.; Ogg, S.; Hay, R.; Mann, M.; Lamond, A. Distinct and overlapping sets of SUMO-1 and SUMO-2 target proteins revealed by quantitative proteomics. *Mol. Cell. Proteom.* **2006**, *5*, 2298–2310. [CrossRef]

31. Bacco, A.D.; Ouyang, J.; Lee, H.; Catic, A.; Ploegh, H.; Gill, G. The SUMO-specific protease SENP5 is required for cell division. *Mol. Cell. Biol.* **2006**, *26*, 4489–4498. [CrossRef] [PubMed]

32. Garone, L.; Fitton, J.E.; Steiner, R.F. Characterization of a family of nucleolar SUMO-specific proteases with preference for SUMO-2 or SUMO-3. *J. Biol. Chem.* **2006**, *281*, 15869–15877.

33. Chawon, Y.; Yonggang, W.; Debaditya, M.; Peter, B.; Nagamalleswari, K.; Alfred, Y.; Wilkinson, K.D.; Mary, D. Nucleolar protein B23/nucleophosmin regulates the vertebrate SUMO pathway through SENP3 and SENP5 proteases. *J. Cell Biol.* **2008**, *183*, 589–595.

34. Lima, C.D.; David, R. Structure of the human SENP7 catalytic domain and poly-SUMO deconjugation activities for SENP6 and SENP7. *J. Biol. Chem.* **2008**, *283*, 32045–32055. [CrossRef] [PubMed]

35. Zhao, Q.; Xie, Y.; Zheng, Y.; Jiang, S.; Liu, W.; Mu, W.; Liu, Z.; Zhao, Y.; Xue, Y.; Ren, J. GPS-SUMO: A tool for the prediction of sumoylation sites and SUMO-interaction motifs. *Nucleic Acids Res.* **2014**, *42*, W325–W330. [CrossRef] [PubMed]

36. Tatham, M.H.; Geoffroy, M.C.; Shen, L.; Plechanovova, A.; Hattersley, N.; Jaffray, E.G.; Palvimo, J.J.; Hay, R.T. RNF4 is a poly-SUMO-specific E3 ubiquitin ligase required for arsenic-induced PML degradation. *Nat. Cell Biol.* **2008**, *10*, 538–546. [CrossRef]

37. Franch, H.A.; Sooparb, S.; Du, J.; Brown, N.S. A mechanism regulating proteolysis of specific proteins during renal tubular cell growth. *J. Biol. Chem.* **2001**, *276*, 19126–19131. [CrossRef]

38. Hendriks, I.A.; Vertegaal, A.C. A comprehensive compilation of SUMO proteomics. *Nat. Rev. Mol. Cell Biol.* **2016**, *17*, 581. [CrossRef]

39. Hendriks, I.A.; Lyon, D.; Young, C.; Jensen, L.J.; Vertegaal, A.C.; Nielsen, M.L. Site-specific mapping of the human SUMO proteome reveals co-modification with phosphorylation. *Nat. Struct. Mol. Biol.* **2017**, *24*, 325–336. [CrossRef] [PubMed]

40. Wilusz, J.E.; Hongjae, S.; Spector, D.L. Long noncoding RNAs: Functional surprises from the RNA world. *Genes Dev.* **2009**, *23*, 1494–1504. [CrossRef]

41. Liu, X.H.; Sun, M.; Nie, F.Q.; Ge, Y.B.; Zhang, E.B.; Yin, D.D.; Kong, R.; Xia, R.; Lu, K.H.; Li, J.H.; et al. Lnc RNA HOTAIR functions as a competing endogenous RNA to regulate HER2 expression by sponging miR-331-3p in gastric cancer. *Mol. Cancer* **2014**, *13*, 92. [CrossRef] [PubMed]

Article

Discovery and Rational Design of a Novel Bowman-Birk Related Protease Inhibitor

Yuxi Miao [1,†], Guanzhu Chen [1,†], Xinping Xi [1], Chengbang Ma [1,*], Lei Wang [1], James F. Burrows [1], Jinao Duan [2], Mei Zhou [1,*] and Tianbao Chen [1]

[1]   Natural Drug Discovery Group, School of Pharmacy, Queen's University Belfast, Belfast, Northern Ireland BT7 1NN, UK
[2]   Jiangsu Key Laboratory for Traditional Chinese Medicine (TCM) Formulae Research, Nanjing University of Chinese Medicine, Nanjing 210046, China
*   Correspondence: c.ma@qub.ac.uk (C.M.); m.zhou@qub.ac.uk (M.Z)
†   The authors contributed equally to this work.

Received: 31 May 2019; Accepted: 12 July 2019; Published: 14 July 2019

Abstract: Anuran amphibian skin secretions are a rich source of peptides, many of which represent novel protease inhibitors and can potentially act as a source for protease inhibitor drug discovery. In this study, a novel bioactive Bowman-Birk type inhibitory hexadecapeptide of the Ranacyclin family from the defensive skin secretion of the Fukien gold-striped pond frog, *Pelophlax plancyi fukienesis*, was successfully isolated and identified, named PPF-BBI. The primary structure of the biosynthetic precursor was deduced from a cDNA sequence cloned from a skin-derived cDNA library, which contains a consensus motif representative of the Bowman-Birk type inhibitor. The peptide was chemically synthesized and displayed a potent inhibitory activity against trypsin (Ki of 0.17 µM), as well as an inhibitory activity against tryptase (Ki of 30.73 µM). A number of analogues of this peptide were produced by rational design. An analogue, which substituted the lysine (K) at the predicted P_1 position with phenylalanine (F), exhibited a potent chymotrypsin inhibitory activity (Ki of 0.851 µM). Alternatively, a more potent protease inhibitory activity, as well as antimicrobial activity, was observed when P^{16} was replaced by lysine, forming K^{16}-PPF-BBI. The addition of the cell-penetrating peptide Tat with a trypsin inhibitory loop resulted in a peptide with a selective inhibitory activity toward trypsin, as well as a strong antifungal activity. This peptide also inhibited the growth of two lung cancer cells, H460 and H157, demonstrating that the targeted modifications of this peptide could effectively and efficiently alter its bioactivity.

Keywords: amphibian Bowman-Birk inhibitor; Tat peptide; molecular cloning; antifungal; drug design; protease inhibitor

---

1. Introduction

Serine proteases are a widely studied group of proteins as they play various roles in healthy and diseased tissues. Serine protease inhibitors can also modulate a series of important biological processes, such as coagulation and inflammation, making them a focus for biomedical studies [1,2].

Plants are remarkable sources of the serine protease inhibitor, which can be grouped into at least 10 families. The Bowman-Birk family inhibitors (BBIs) are the best studied and most widely known among them. Identified in and isolated from soybean, they were the first to often be referred to as "classical BBI". However, subsequently multiple BBIs have been isolated from plants such as legumes and Gramineae [3–5].

The skin of frogs is the main organ involved in their defense system, which manufactures diverse bioactive peptides that possess cytolytic pharmacological activities [6], and as a result it is also an excellent source of protease inhibitors. To date, many BBIs isolated from amphibians have been reported, such as peptide leucine arginine (pLR) [7], peptide tyrosine arginine (pYR) [8], the Bowman-Birk-like trypsin inhibitor from *Huia versabilis* (HV-BBI) [9], Hejiang trypsin inhibitor (HJTI) [10], the Bowman-Birk-type inhibitor from *Odorrana schmackeri* (OSTI) [11], *Hylarana erythraea* chymotrypsin inhibitor (HECI) [12] and *Pelophyla esculentus Bowman-Birk proteinase inhibitor* (PE-BBI) [13]. Generally, the BBI peptides from amphibians possess a highly-conserved 11-residue canonical disulfide loop, which is different from plant BBIs. The structure of this peptide follows the consensus sequence, $CWTP_1SXPPXPC$, with P_1 representing the inhibitory active site and X indicating that various amino acids are found in these positions. This disulfide-bridged loop is considered a trypsin inhibitory loop (TIL), which has a significant trypsin inhibitory activity [14].

Based on previous studies, these amphibian BBIs not only have potent protease inhibitory activities, but also exhibit an antimicrobial activity. Antimicrobial peptides (AMPs) are attractive alternatives to producing novel antibiotics. However, their susceptibility to proteases appreciably limits the potential applications of most AMPs. Therefore, a bifunctional peptide possessing both antimicrobial and protease inhibitory activities, with a low cytotoxicity, could represent an ideal template for future clinical use [15].

In this report, a novel peptide from the defensive skin secretion of the Fukien gold-striped pond frog, *Pelophlax plancyi fukienesis*, was successfully isolated and identified, named PPF-BBI. It was shown to possess a potent trypsin and tryptase inhibitory activity with a high specificity. Several analogues were created by rational design, and a P_1 site F substituted analogue displayed a considerable and specific chymotrypsin inhibitory activity. A better antimicrobial activity was observed when P^{16} was replaced by Lys, and the addition of the cell-penetrating peptide Tat_{48-56} resulted in a peptide with a strong antifungal activity. Moreover, anti-proliferative effects on H157 and H460 were also observed.

2. Materials and Methods

2.1. Specimen Biodata and Secretion Harvesting

Specimens of the Fukien gold-striped pond frog, *Pelophelax plancyi fukienensis* (n = 3, snout-to-vent length 7 cm) were captured in Fuzhou City, Fujian Province, China. All frogs were adults, and skin secretion was obtained by a mild electrical stimulation on the dorsal skin surface of the frogs [16]. The secretion was collected by washing the skin using deionized water and was lyophilized after the liquid nitrogen freezing. The obtained secretion was stored at $-20\,°C$. This study is approved by the Nanjing University of Chinese Medicine Ethical Review Board-Approval Code: SYXK (SU) 2018-0048.

2.2. Identification of Precursor-Encoding cDNA from the Skin Secretion

The precursor encoding cDNA from the skin secretion was obtained as described previously [17]. Briefly, the 3′-RACE reactions employed a nested universal (NUP) primer (supplied with the kit) and a sense primer (S: 5′-GCIYTIMGIGGITGYTGGACIAA-3′) that was complementary to the amino acid sequence, ALRGCWTK-, of PPF-BBI. The RACE reactions were purified and cloned using a pGEM-T vector system (Promega Corporation, Madison, WI, USA) and sequenced using an ABI 3100 automated sequencer. The nucleotide sequence of PPF-BBI has been deposited in the GenBank database under the accession number MK965542.

2.3. Isolation and Identification of PPF-BBI from Skin Secretion

The isolation and identification of the mature peptide from the skin secretion using the RP-HPLC and LC-MS analyses were performed as outlined previously [18]. A molecular mass analysis of the contents contained in the HPLC fraction was achieved by use of a matrix-assisted laser desorption ionization time-of-flight (MALDI-TOF) mass spectrometer (Voyager DE, Perseptive Biosystems, Framingham, MA, USA). The major peptide within this fraction was subjected to MS/MS fragmentation sequencing using an LCQ-Fleet mass spectrometer (Thermo Fisher Scientific, San Jose, CA, USA).

2.4. Peptide Design and Solid Phase Peptide Synthesis of PPF-BBI, F^8-PPF-BBI, K^{16}-PPF-BBI, Tat-loop, Tat and Trypsin Inhibitory Loop

There are three design strategies based on the obtained parent peptide, one to alter the inhibitory specificity, one to enhance the antimicrobial activity, and one to enhance the drug delivery and cell targeting. In the first case, since synthetic work on the BBI-like peptides has focused mostly on the P_1 site, the lysine at the P_1 site was replaced by a phenylalanine (F^8-PPF-BBI; ALRGCWTFSIPPKPCP-NH$_2$) which confers chymotrypsin inhibitory specificity. In the second case, the last amino acid residue, proline, was substituted with a lysine to give a positive charge and achieve a more structural similarity with the members of the Ranacyclin family, which have an antimicrobial activity (K^{16}-PPF-BBI; ALRGCWTKSIPPKPCK-NH$_2$). The trypsin inhibitory loop (TIL, CWTKSIPPKPC), derived from the amphibian Bowman-Birk-type protease inhibitor, is found to have a potent trypsin inhibitory activity. Thus, in the last case, a short cell-penetrating peptide Tat$_{48-56}$ (RKKRRQRRR), which has been considered one of the most promising tools to improve the cellular delivery of therapeutic molecules [19–21], was added to the N-terminal of the TIL (Tat-loop; RKKRRQRRRCWTKSIPPKPC) by solid phase peptide synthesis. The Tat peptide and TIL were used in the antimicrobial assays for a comparison with the activity of Tat-loop. The TIL peptide was also involved in protease inhibitory assays to determine the influence on the inhibitory activity of the extended amino acid residues at both termini.

The novel-cloned cDNA-encoded peptide, wild-type PPF-BBI, and its analogues were synthesized by chemical synthesis using a Tribute peptide solid-phase synthesizer (Protein Technologies, Inc, Tucson, AZ, USA), as outlined previously [11]. The synthetic peptides were analyzed both by reverse phase HPLC and MALDI-TOF mass spectrometry to establish the degree of purity and the identity of the structure.

2.5. Trypsin, Chymotrypsin and Tryptase Inhibition Assay

The trypsin, chymotrypsin inhibition tests were performed as described previously [22]. 10 µL tryptase (1 mg/mL, Calbiochem, UK) was added to the wells of a micro-titer plate containing 180 µL substrate (Boc-Phe-Ser-Arg-NHMec, obtained from Bachem, UK) (50 µM) and 20 µL synthetic replicates (0.1–100 µM) in a tryptase buffer, pH7.6, containing 0.05 M Tris, 0.15 M NaCl, and 0.2% (*w/v*) polyethylene glycol 6000 (final volume 210 µL).

The rate of hydrolysis of the substrate was monitored by measuring the rate of increase of fluorescence due to the release of 7–amino–4–methylcoumarin (AMC) at 460 nm (excitation 360 nm) in a FLUOstar OPTIMA multi-well plate reader. The inhibition curves of the trypsin/chymotrypsin inhibition assay and tryptase inhibition assay were formed as outlined before [11,12].

2.6. Minimal Inhibitory Concentration (MIC) Assay and Minimal Batericidal Concentration (MBC) Assay

The MIC and MBC of the synthetic peptides were determined as previously described [17], using *S. aureus* (NCTC 10788), *E. coli* (NCTC 10418) and *C. albicans* (NCYC 1467), together with two species of resistant micro-organisms, methicillin-resistant *S. aureus* (MRSA) (ATCC 12493) and *P. aeruginosa* (ATCC 27853).

2.7. Membrane Permeability Assay

The membrane permeability assay was performed as described previously [20]. The peptides at concentrations of 1-fold MIC, 2-fold MIC and 4-fold MIC were mixed with a bacterial cell suspension. The membrane permeability rate was measured via the monitor of the fluorescent intensity of SYTOX Green Nucleic Acid Stain (Life Technologies, Glasgow, UK) by a Synergy HT plate reader with excitation at 485 nm and emission at 528 nm.

2.8. Secondary Structure Analysis through Circular Dichroism (CD)

The sample peptide solutions (50 μM) were prepared in a 1 mm high precision quartz cell (Hellma Analytics, Essex, UK) with 10 mM ammonium acetate and 50% TFE in 10 mM ammonium acetate buffer respectively. CD measurements were performed at 20 °C by a JASCO J-815 CD spectrometer (Jasco, Essex, UK) across the wavelength range of 190–250 nm. The scanning speed was 100 nm/min, the bandwidth was one nm, and the data pitch was 0.5 nm. The CD spectra were further analysed using the online software, BeStSel [23], and the proportion of different secondary structures were predicted.

2.9. MTT Assay

The MTT assay was carried out as described in a previous study [20], using a series of lung cancer cell lines, NCI-H157 (RRID: CVCL_0463), NCI-H460 (ATCC® HTB-177™), H838 (ATCC® CRL-5844™), and H23 (ATCC® CRL-5800™), as well as other cancer cell lines: HT-29 (ATCC® HTB-38™), PC-3 (ATCC® CRL-1435™), U251MG (ECACC-09063001), and the normal human dermal microvascular endothelium cell line HMEC-1 (ATCC® CRL-3243™). The anti-metabolite 5-fluorouracil (5-FU) was utilized as the positive control group.

2.10. Haemolysis Test

The haemolytic activity of each peptide was measured by incubating a range of final peptide concentrations from 512 to 1 μM in a two-fold dilution in a 2% suspension of horse erythrocytes, as described in a previous study [20].

2.11. Statitical Analysis

The data of all the bioactive assays were statistical analyzed using Prism 6 (GraphPad Prism Software, GraphPad, San Diego, CA. USA). The data points represent the average of three independent experiments with the error bars representing the standard error of the mean (SEM).

3. Results

3.1. Identification and Structural Determination of PPF-BBI

From a skin-derived cDNA library, a cDNA encoding a biosynthetic precursor of PPF-BBI was consistently and repeatedly cloned (Figure S1). The crude skin secretion of *Pelophlax plancyi fukienesis* was analysed by LC-MS. The retention time of the fragmentation indicated in Figure S2 showed a corresponding molecular weight (Figure S3A) to the prediction for PPF-BBI according to the putative peptide from the cloned cDNA. The elution fraction was analysed by an electrospray mass spectrometer, and the primary structure sequence was confirmed (Figure S3B,C). The open-reading frame consisted of 65 amino acid residues. The alignment of this peptide with other members of the Ranacyclin family of Bowman-Birk-type protease inhibitors indicates that it is structurally related as its sequence shows a high degree of conservation, as well as including a typical inhibition loop (Figure 1), which started with a 22-residue putative signal peptide at the N-terminus. After the acidic spacer of 24-amino acids, the deduced mature peptide of 16 residues at the C-terminus is present in a single copy (Figure 1). The sequence was preceded by two consecutive basic amino acids, Lys-Arg (KR), representing a typical processing site for endoproteolytic cleavage, and was immediately followed by a glycine residue amide donor.

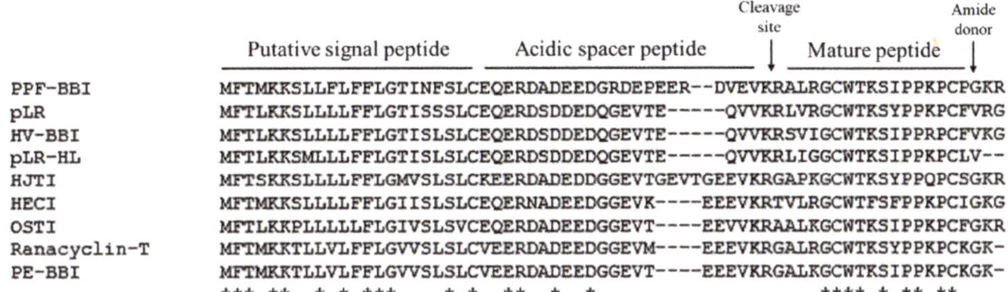

Figure 1. Multiple sequence alignment test results from Clustal Omega. Fully conserved residue indicated by asterisks.

3.2. Peptide Design

Three analogues were designed based on the natural peptide PPF-BBI (Table 1). Briefly, a substitution of phenylalanine at the P_1 site, F^8-PPF-BBI, was aimed to produce a chymotrypsin inhibitory peptide. The substitution of a lysine at the position 16 of the native peptide (K^{16}-PPF-BBI) enhanced the net positive charge, which might improve the ability to interact with the cell membrane. The Tat sequence was added at the N-terminus of the typical 11-mer trypsin inhibitory loop structure to increase the membrane penetration effect. All the analogues were chemically synthesized, purified by RP-HPLC and analysed by MALDI-TOF.

Table 1. The sequence and positive charge of *Pelophlax plancyi fukienesis* Bowman-Birk-type inhibitor (PPF-BBI) and its rational design analogues.

Peptide Name	Sequence	Positive Charge
PPF-BBI	ALRG CWTKSIPPKPC P-amide	+4
F^8-PPF-BBI	ALRG CWTKSIPPKPC P-amide	+3
K^{16}-PPF-BBI	ALRG CWTKSIPPKPC **K**-amide	+5
Tat-loop	RKKRRQRRR CWTKSIPPKPC	+10

The highly conserved loop is shaded, and the substituted sites are in bold.

3.3. Synthesis and Secondary Structure Analysis of PPF-BBI and its Analogues

PPF-BBI and the analogue peptides were successfully synthesized, impurities were removed by HPLC, and their identity was confirmed by MALDI-TOF. The secondary structures of all peptides were determined by circular dichroism spectroscopy (Figure 2). PPF-BBI, F^8-PPF-BBI and K^{16}-PPF-BBI revealed a broad negative band with the minimum around 200 nm, typical of an unfolded peptide in equilibrium with a β-sheet structure [24], except that Tat-loop exhibited the negative band close to 197 nm, which is considered to be a random coil. With the presence of 50% TFE, which is a secondary structure promoting the reagent, the negative minimum of PPF-BBI shifted from 200–203 nm to 206–210 nm and displayed a broad negative band. Additionally, F^8-PPF-BBI and K^{16}-PPF-BBI displayed the same red shift trend of the negative minimum. Since the spectra displayed a broad minimum spanning the region 200–210 nm and did not show positive bands above 210 nm, this suggests that the conformation of peptides is likely to consist of a mixture of secondary structures of β-sheet structure and random coil, which is consistent with previous studies [11,24–26]. Furthermore, the Tat peptide possess a random coil structure, the presence of which would not increase the helicity of the peptide [26].

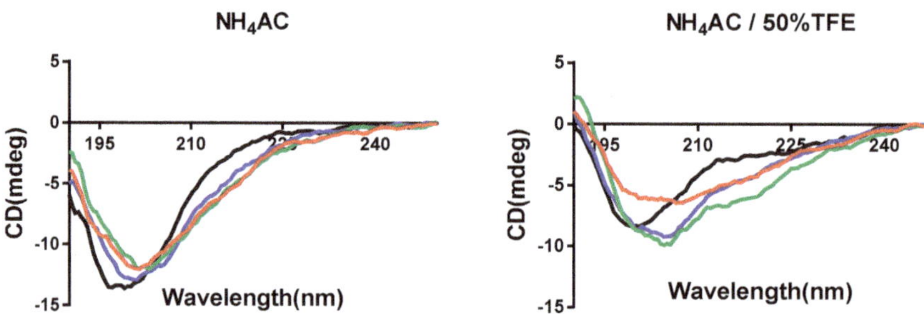

Figure 2. Secondary structures of PPF-BBI and the analogues. The CD spectra of peptides were measured in their free form (aqueous 10 mM NH_4AC buffer) and membrane-mimic 10 mM NH_4AC/50% TFE buffer, respectively (PPF-BBI, red; F^8-PPF-BBI, green; K^{16}-PPF-BBI, blue; Tat-loop, black).

3.4. Trypsin, Chymotrypsin and Tryptase Inhibitory Activity of PPF-BBI and its Analogues

PPF-BBI and its analogues were tested for inhibitory activity against trypsin, chymotrypsin and tryptase, respectively. The progress curves for the hydrolysis of the fluorogenic substrate were used to estimate an initial rate (Vi) to generate the Morrison plots. All of the progress curves and corresponding Morrison plots in the presence of each peptide of different concentrations are shown in Figure S4. Among these, both wild type PPF-BBI and K^{16}-PPF-BBI exhibited a potent trypsin inhibitory activity. However, K^{16}-PPF-BBI had a more potent inhibitory effect against tryptase than the parent peptide did. In addition, F^8-PPF-BBI only displayed a strong chymotrypsin inhibitory activity and lost the trypsin inhibitory activity. Interestingly, the trypsin inhibitory loop (TIL) and Tat-loop kept the trypsin inhibition activity but did not exhibit a tryptase inhibitory activity (Table 2).

Table 2. PPF-BBI and its analogues against trypsin, chymotrypsin and tryptase.

Peptide	Name	Ki (μM) of Trypsin	Ki (μM) of Tryptase	Ki (μM) of Chymotrypsin
ALRG CWTKSIPPKPC P-amide	PPF-BBI	0.17	30.73	N.I.*
ALRG CWTKSIPPKPC P-amide	F^8-PPF-BBI	N.I.*	N.I.*	0.85
ALRG CWTKSIPPKPC K-amide	K^{16}-PPF-BBI	0.112	9.67	N.I.*
RKKRRQRRR CWTKSIPPKPC	Tat-loop	0.607	N.I.*	N.I.*
CWTKSIPPKPC	TIL	0.741	N.I.*	N.I.*

The highly conserved loop is shaded, and the substituted sites are in bold. N.I.* means that no inhibition was observed.

3.5. Antimicrobial Activity

The antimicrobial activity of PPF and its analogues was tested against a representative set of microorganisms (Table 3). Both PPF-BBI and K^{16}-PPF-BBI displayed a mild activity against the tested microorganisms, although K^{16}-PPF-BBI exhibited a better bioactivity than the parent peptide against *S. aureus* and *C. albicans*. However, although its activity against the other microorganisms was similar to the parental peptide, Tat-loop showed a much more potent activity against *C. albicans*. The component parts of Tat-loop (separate Tat peptide and the TIL) were also tested, but had little activity on their own. To sum up, Tat-loop showed a potent activity against *C. albicans*, as well as exhibiting slightly more activity against MRSA and *P. aeruginosa*. K^{16}-PPF-BBI exhibited the best activity against *S. aureus* and was slightly better than PPF-BBI against all of the others except *E. coli*.

Table 3. The minimal inhibitory concentrations (μM) and minimal bactericide concentrations (μM) of PPF-BBI and the synthetic analogue peptides against microorganisms.

Microorganisms	MIC/MBC (μM)					
	PPF-BBI	**K^{16}-PPF-BBI**	**F^8-PPF-BBI**	**Tat**	**Tat-loop**	**TIL**
S. aureus	128/128	64/64	>512	512/512	128/128	>512
E. coli	128/128	128/128	>512	256/256	128/128	>512
C. albicans	512/512	128/128	>512	>512	4/8	>512
MRSA	>512	512/512	>512	>512	256/512	>512
P. aeruginosa	>512	512/512	>512	>512	256/256	>512

3.6. Membrane Permeability

The MIC and MBC results indicated that only Tat-loop had a potent activity on any of the tested microorganisms. Therefore, it was tested to determine if its impact against *C. albicans* was due to it impacting its membrane permeability. However, Tat-loop did not cause any membrane permeabilization, even at high concentrations (4-fold of its MIC versus *C. albicans*, Figure 3).

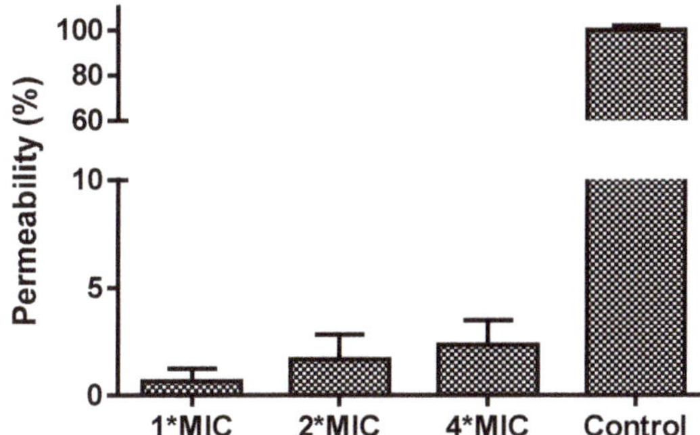

Figure 3. The cell permeability of *C. albicans* treated for 2 h by Tat-loop at 1-fold, 2-fold and 4-fold of MIC. The membrane permeabilized cells by 70% isopropanol were used as the positive control (100% permeability).

3.7. Anti-Cancer and Haemolytic Activity

PPF-BBI and its analogues were subjected to an MTT assay using a series of lung cancer cell lines (H460, H157, H23 and H838) and other cancer cell lines (HT29, PC-3, U251MG), and were also tested on a human normal cell (HMEC-1) (Figure 4a). Among these, only Tat-loop inhibited the growth of H460 and H157 at a concentration of 100 μM. All of the peptides showed a slight inhibition on HMEC-1. Furthermore, they also exhibited a low degree of haemolytic activity on horse erythrocytes (Figure 4b).

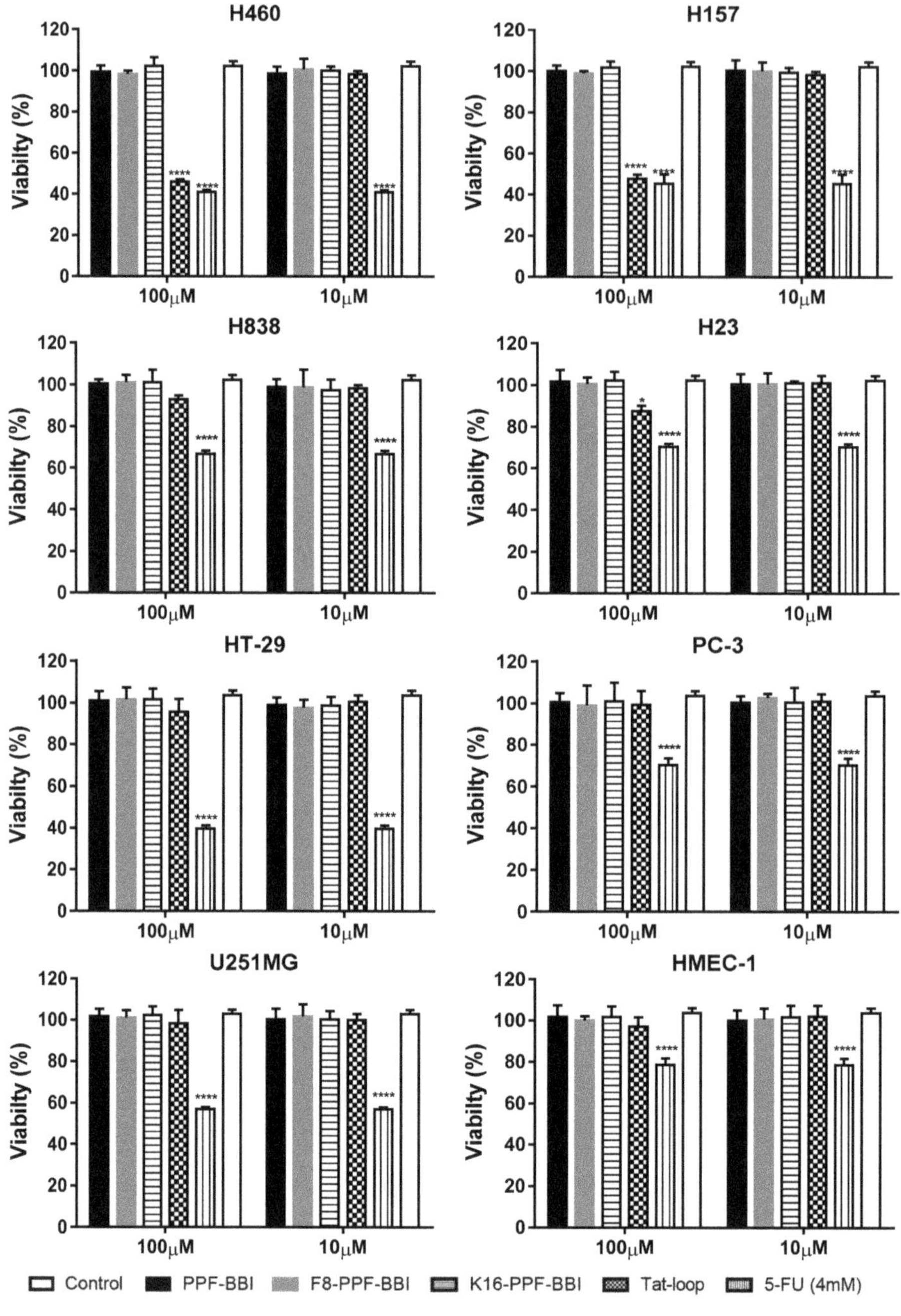

(a)

Figure 4. *Cont.*

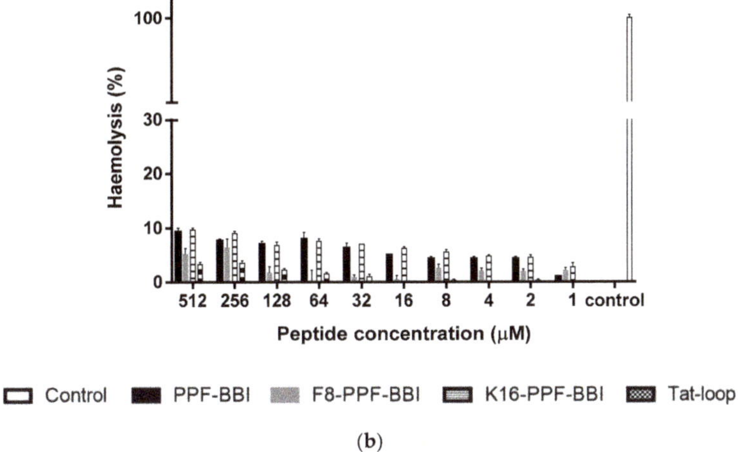

(**b**)

Figure 4. (**a**) The cell viability of the cancer cell lines H460, H157, H23, H838, HT-29, PC-3, U251MG and HMEC-1 at 4 mM 5-FU (stripe bar), 100 μM and 10 μM PPF-BBI, F^8-PPF-BBI, K^{16}-PPF-BBI and Tat-loop. The control represents the cell viability without any treatments. The statistical significance of difference was analyzed by a one-way ANOVA (* $p < 0.05$, **** $p < 0.0001$). (**b**) The haemolysis rates of PPF-BBI, F^8-PPF-BBI, K^{16}-PPF-BBI and Tat-loop on erythrocytes after being incubated for 4 h. The incubation of erythrocytes with 2% (*v/v*) Triton X-100 was designated as a positive control (100% haemolysis).

4. Discussion

In this study, we report the identification and bioactivity evaluations of PPF-BBI, a novel Bowman-Birk type protease inhibitor from the skin secretion of the Fukien gold-striped pond frog, *Pelophylax plancyi fukienensis*. In addition, we also examine the bioactivity of three rationally designed analogues of PPF-BBI, F^9-PPF-BBI, K^{16}-PPF-BBI and Tat-loop.

Like other BBI-type peptides, PPF-BBI has potent protease inhibitory activities. The specificity of inhibition is determined by whether the P_1 position residue can fit into the S1 pocket of protease. Based on previous reports, Lys as the P_1 position is optimal for trypsin inhibition, and Phe is optimal for chymotrypsin [11,27]. Similarly, PPF-BBI, which has a Lys at the P1 site, showed a strong trypsin inhibitory activity and substitution of the Lys at the P_1 position, with Phe leading to the elimination of the trypsin inhibition and giving rise to a chymotrypsin inhibitory activity. In the tryptase activity assay, PPF-BBI displayed a mild potency toward tryptase with a Ki value of 30.52 μM. K^{16}-PPF-BBI shows a three-fold better inhibition with a Ki value of 9.67 μM, but intriguingly, Tat-loop and TIL lost their inhibitory activity against tryptase, even though they retained their trypsin inhibitory activity. Indeed, this contradicts a previous study [28], which demonstrated that short BBI-derived cyclic peptides had an inhibitory activity against tryptase, even though tryptase is seen as unique due to its resistance to all known endogenous proteinase inhibitors [29]. As the data of K^{16}-PPF-BBI showed here, the substitution of Lys at the C-terminus improved the tryptase inhibitory activity, indicating that the C-terminal extensive residue might contribute to the binding between the BBI peptide and tryptase. Therefore, the lack of such an extension of TIL and Tat-loop could eliminate the affinity to the reactive pocket of tryptase, so that both cannot produce any inhibitory activity.

The mode of action of Ranacyclins is different from most known positively charged antimicrobial peptides. They bind and insert into both zwitterionic and negatively charged membranes, and they presumably form transmembrane pores without bacteria wall damage. Indeed, it has been reported that Ranacyclins E and T have a great potential as antimicrobials [30]. PPF-BBI and K^{16}-PPF-BBI shared a high sequence similarity with Ranacyclin members, and they were also found to have moderate effects on the tested microorganisms, as was expected. However, K^{16}-PPF-BBI was shown to have a better effect, which is possibly because one more lysine increases its positive charges, and it is easier to get it close to the negative groups of the cell membrane.

Tat, a cationic-rich cell-penetrating peptide derived from the HIV protein, has been used to conjugate with other compounds to enhance the cell penetrating activity. In the previous study, the design of Tat-fusion biopeptides demonstrated a remarkable improvement on their biological activities [20,21]. In the meantime, the BBI trypsin inhibitory loop was also considered as a drug template that was applied in some studies [14,31]. Therefore, a combination of the Tat peptide with TIL was conducted here, one that could attempt to introduce the cell-penetrating effect and that possessed an inhibition against trypsin-like activity intracellularly.

Interestingly, this is reflected in the significant increase of potency against the tested strains (especially *C. albicans* with MIC of 4 µM) of Tat-loop. However, this does not appear to be due to its impact upon the membrane, as treatment with this peptide does not induce changes in membrane permeability even at high concentrations (4-fold of its MIC versus *C. albicans*). Using the principle of plant protease inhibitors, it would be pertinent to evaluate the antifungal activity of Tat-loop, possibly because it might interfere with the trypsin-mediated activation of the chitin synthase zymogen and further affect the process of cell wall development [32].

In the report by Zhang et al. [12], a Bowman-Birk type chymotrypsin inhibitory peptide, HECI, and its analogue, K^9-HECI, exhibited great anti-proliferative potency against H157, PC-3 and MCF-7. Also, PE-BBI was reported to have an inhibition effect on several colon cancer cell lines [13]. In our study, Tat-loop suppresses the growth of the lung cancer cells H460 and H157. It was not clear whether Tat-loop could enter the nucleus and bind the target receptor in vitro; compared to the native peptide, Tat-loop exhibited a slight improvement of the anti-proliferative activity. Moreover, a low haemolytic activity was observed when Tat-loop was assessed, which further supports the data which indicates that it does not exhibit cytotoxicity. Similarly, no significant haemolytic activity was observed with the parental peptide and the other synthetic analogues. This indicates that the Tat-loop will not have toxicity issues and lays a solid foundation for further in vivo research.

In addition, PPF-BBI, F^9-PPF-BBI and K^{16}-PPF-BBI initiated a slight conversion from a random coil into a β-structure in the TFE buffer due to assisting the folding of the secondary structure [33]. However, Tat-loop did not show the band shift to 210 nm, but around 197 nm, which indicated that it mainly formed a random coil in both aqueous and TFE buffer [26]. We assume that BBI-related peptides might exhibit a certain degree of β-sheet structure involved in binding with enzyme; therefore, Tat-loop demonstrated a lower inhibitory activity against trypsin compared with the others. Although the spectra data showed slightly different results, the major part of the secondary structure of the BBI-related peptides is random coil and β-sheet, which is consistent with previous studies [11,25,34].

5. Conclusions

In summary, PPF-BBI is a naturally occurring peptide with a remarkable trypsin and tryptase inhibitory activity, as well as a moderate antimicrobial activity. This provided the basis for the rational design of further multifunctional protease inhibitors. Moreover, this is the first report investigating the addition of a cell-penetrating peptide to an amphibian skin-derived protease inhibitor. Tat-loop has high potency against *C. albicans*, which also results from its inhibition of trypsin, which might have potential towards fungal diseases. However, the mechanism by which Tat-loop impacts upon *C. albicans* is unclear, and further research is required to determine how this peptide exerts its impact upon this fungus.

Supplementary Materials: The following are available online at http://www.mdpi.com/2218-273X/9/7/280/s1, Figure S1: The nucleotide and translated open-reading frame amino acid sequence of cloned cDNA encoding the biosynthetic precursor of PPF-BBI from a skin secretion of *Pelophylax plancyi fukienensis*. Figure S2. Region of reverse-phase HPLC chromatogram of the skin secretions of *Pelopholax plancyi fukienesis*. Figure S3. The identification of PPF-BBI from the skin secretion of *Pelopholax plancyi fukienesis*. Figure S4. The identification of other peptides used in this study. Figure S5. The inhibitory activity of PPF-BBI and the analogues on trypsin, chymotrypsin and tryptase.

Author Contributions: Conceptualization, L.W., J.F.B., M.Z. and T.C.; Formal analysis, C.M.; Investigation, Y.M. and G.C.; Methodology, X.X., C.M. and J.D.; Project administration, T.C.; Resources, J.D.; Supervision, J.F.B. and M.Z.; Validation, X.X., C.M. and M.Z.; Visualization, Y.M. and G.C.; Writing–original draft, Y.M.; Writing–review & editing, X.X. and J.F.B.

Funding: This research received no external funding.

Conflicts of Interest: The authors declare no conflict of interest.

References

1. Stein, P.E.; Carrell, R.W. What do dysfunctional serpins tell us about molecular mobility and disease? *Nat. Struct. Mol. Boil.* **1995**, *2*, 96–113. [CrossRef]

2. Soualmia, F.; El Amri, C. Serine protease inhibitors to treat inflammation: A patent review (2011–2016). *Expert Opin. Ther. Patents* **2017**, *28*, 93–110. [CrossRef]

3. Norioka, S.; Ikenaka, T. Amino Acid Sequences of Trypsin-Chymotrypsin Inhibitors (A-I, A-II, B-I, and B-II) from Peanut (Arachis hypogaea)1: A Discussion on the Molecular Evolution of Legume Bowman-Birk Type Inhibitors. *J. Biochem.* **1983**, *94*, 589–598. [CrossRef] [PubMed]

4. Srikanth, S.; Chen, Z. Plant Protease Inhibitors in Therapeutics-Focus on Cancer Therapy. *Front. Pharmacol.* **2016**, *7*, 470. [CrossRef]

5. Dantzger, M.; Vasconcelos, I.M.; Scorsato, V.; Aparicio, R.; Marangoni, S.; Macedo, M.L.R. Bowman–Birk proteinase inhibitor from Clitoria fairchildiana seeds: Isolation, biochemical properties and insecticidal potential. *Phytochem.* **2015**, *118*, 224–235. [CrossRef]

6. Ma, J.; Luo, Y.; Ge, L.; Wang, L.; Zhou, M.; Zhang, Y.; Duan, J.; Chen, T.; Shaw, C. Ranakinestatin-PPF from the skin secretion of the Fukien gold-striped pond frog, Pelophylax plancyi fukienensis: A prototype of a novel class of bradykinin B2 receptor antagonist peptide from ranid frogs. *Sci. World J.* **2014**, *2014*. [CrossRef]

7. Salmon, A.L.; Cross, L.J.; Irvine, A.E.; Lappin, T.R.; Dathe, M.; Krause, G.; Canning, P.; Thim, L.; Beyermann, M.; Rothemund, S.; et al. Peptide leucine arginine, a potent immunomodulatory peptide isolated and structurally characterized from the skin of the Northern Leopard frog, Rana pipiens. *J. Biol. Chem.* **2001**, *276*, 10145–10152. [CrossRef] [PubMed]

8. Graham, C.; Irvine, A.E.; McClean, S.; Richter, S.C.; Flatt, P.R.; Shaw, C. Peptide Tyrosine Arginine, a potent immunomodulatory peptide isolated and structurally characterized from the skin secretions of the dusky gopher frog, Rana sevosa. *Peptide* **2005**, *26*, 737–743. [CrossRef]

9. Song, G.; Zhou, M.; Chen, W.; Chen, T.; Walker, B.; Shaw, C. HV-BBI—A novel amphibian skin Bowman–Birk-like trypsin inhibitor. *Biochem. Biophys. Res. Commun.* **2008**, *372*, 191–196. [CrossRef]

10. Wang, M.; Wang, L.; Chen, T.; Walker, B.; Zhou, M.; Sui, D.; Conlon, J.M.; Shaw, C. Identification and molecular cloning of a novel amphibian Bowman Birk-type trypsin inhibitor from the skin of the Hejiang Odorous Frog; Odorrana hejiangensis. *Peptide* **2012**, *33*, 245–250. [CrossRef]

11. Wu, Y.; Long, Q.; Xu, Y.; Guo, S.; Chen, T.; Wang, L.; Zhou, M.; Zhang, Y.; Shaw, C.; Walker, B. A structural and functional analogue of a Bowman–Birk-type protease inhibitor from Odorrana schmackeri. *Biosci. Rep.* **2017**, *37*. [CrossRef] [PubMed]

12. Zhang, L.; Chen, X.; Wu, Y.; Zhou, M.; Ma, C.; Xi, X.; Chen, T.; Walker, B.; Shaw, C.; Wang, L. A Bowman-Birk type chymotrypsin inhibitor peptide from the amphibian, Hylarana erythraea. *Sci. Rep.* **2018**, *8*, 5851. [CrossRef] [PubMed]

13. Lyu, P.; Ge, L.; Ma, R.; Wei, R.; McCrudden, C.M.; Chen, T.; Shaw, C.; Kwok, H.F. Identification and pharmaceutical evaluation of novel frog skin-derived serine proteinase inhibitor peptide-PE-BBI (Pelophylax esculentus Bowman-Birk inhibitor) for the potential treatment of cancer. *Sci. Rep.* **2018**, *8*, 14502. [CrossRef] [PubMed]

14. Li, J.; Zhang, C.; Xu, X.; Wang, J.; Yu, H.; Lai, R.; Gong, W. Trypsin inhibitory loop is an excellent lead structure to design serine protease inhibitors and antimicrobial peptides. *FASEB J.* **2007**, *21*, 2466–2473. [CrossRef] [PubMed]

15. Yan, X.; Liu, H.; Yang, X.; Che, Q.; Liu, R.; Yang, H.; Liu, X.; You, D.; Wang, A.; Li, J.; et al. Bi-functional peptides with both trypsin-inhibitory and antimicrobial activities are frequent defensive molecules in Ranidae amphibian skins. *Amino Acids* **2012**, *43*, 309–316. [CrossRef]

16. Tyler, M.J.; Stone, D.J.; Bowie, J.H. A novel method for the release and collection of dermal, glandular secretions from the skin of frogs. *J. Pharmacol. Toxicol. Methods* **1992**, *28*, 199–200. [CrossRef]

17. Gao, Y.; Wu, D.; Wang, L.; Lin, C.; Ma, C.; Xi, X.; Zhou, M.; Duan, J.; Bininda-Emonds, O.R.P.; Chen, T.; et al. Targeted Modification of a Novel Amphibian Antimicrobial Peptide from Phyllomedusa tarsius to Enhance Its Activity against MRSA and Microbial Biofilm. *Front. Microbiol.* **2017**, *8*, 951. [CrossRef]

18. Wu, D.; Gao, Y.; Wang, L.; Xi, X.; Wu, Y.; Zhou, M.; Zhang, Y.; Ma, C.; Chen, T.; Shaw, C. A Combined Molecular Cloning and Mass Spectrometric Method to Identify, Characterize, and Design Frenatin Peptides from the Skin Secretion of Litoria infrafrenata. *Mol.* **2016**, *21*, 1429. [CrossRef]

19. Brooks, H.; LeBleu, B.; Vivès, E. Tat peptide-mediated cellular delivery: Back to basics. *Adv. Drug Deliv. Rev.* **2005**, *57*, 559–577. [CrossRef]

20. Chen, X.; Zhang, L.; Wu, Y.; Wang, L.; Ma, C.; Xi, X.; Bininda-Emonds, O.R.; Shaw, C.; Chen, T.; Zhou, M. Evaluation of the bioactivity of a mastoparan peptide from wasp venom and of its analogues designed through targeted engineering. *Int. J. Boil. Sci.* **2018**, *14*, 599–607. [CrossRef]

21. Zhu, H.; Ding, X.; Li, W.; Lu, T.; Ma, C.; Xi, X.; Wang, L.; Zhou, M.; Burden, R.; Chen, T.; et al. Discovery of two skin-derived dermaseptins and design of a TAT-fusion analogue with broad-spectrum antimicrobial activity and low cytotoxicity on healthy cells. *PeerJ* **2018**, *6*, e5635. [CrossRef] [PubMed]

22. Lin, Y.; Hang, H.; Chen, T.; Zhou, M.; Wang, L.; Shaw, C. pLR-HL: A Novel Amphibian Bowman-Birk-type Trypsin Inhibitor from the Skin Secretion of the Broad-folded Frog, Hylarana latouchii. *Chem. Biol. Drug Des.* **2016**, *87*, 91–100. [CrossRef] [PubMed]

23. Micsonai, A.; Wien, F.; Kernya, L.; Lee, Y.-H.; Goto, Y.; Réfrégiers, M.; Kardos, J. Accurate secondary structure prediction and fold recognition for circular dichroism spectroscopy. *Proc. Natl. Acad. Sci. USA* **2015**, *112*, E3095–E3103. [CrossRef] [PubMed]

24. Anandalakshmi, V.; Murugan, E.; Leng, E.G.T.; Ting, L.W.; Chaurasia, S.S.; Yamazaki, T.; Nagashima, T.; George, B.L.; Peh, G.S.L.; Pervushin, K.; et al. Effect of position-specific single-point mutations and biophysical characterization of amyloidogenic peptide fragments identified from lattice corneal dystrophy patients. *Biochem. J.* **2017**, *474*, 1705–1725. [CrossRef] [PubMed]

25. Mulvenna, J.P.; Foley, F.M.; Craik, D.J. Discovery, Structural Determination, and Putative Processing of the Precursor Protein That Produces the Cyclic Trypsin Inhibitor Sunflower Trypsin Inhibitor 1. *J. Boil. Chem.* **2005**, *280*, 32245–32253. [CrossRef]

26. Eiríksdóttir, E.; Konate, K.; Langel, Ü.; Divita, G.; Deshayes, S. Secondary structure of cell-penetrating peptides controls membrane interaction and insertion. *Biochim. Biophys. Acta (BBA) Biomembr.* **2010**, *1798*, 1119–1128.

27. McBride, J.D.; Leatherbarrow, R.J. Synthetic peptide mimics of the Bowman-Birk inhibitor protein. *Curr. Med. Chem.* **2001**, *8*, 909–917. [CrossRef]

28. Scarpi, D.; McBride, J.D.; Leatherbarrow, R.J. Inhibition of human beta-tryptase by Bowman-Birk inhibitor derived peptides: Creation of a new tri-functional inhibitor. *Bioorg. Med. Chem.* **2004**, *12*, 6045–6052. [CrossRef]

29. Pereira, P.J.; Bergner, A.; Macedo-Ribeiro, S.; Huber, R.; Matschiner, G.; Fritz, H.; Sommerhoff, C.P.; Bode, W. Human beta-tryptase is a ring-like tetramer with active sites facing a central pore. *Nature* **1998**, *392*, 306–311. [CrossRef]

30. Mangoni, M.L.; Papo, N.; Mignogna, G.; Andreu, D.; Shai, Y.; Barra, D.; Simmaco, M. Ranacyclins, a New Family of Short Cyclic Antimicrobial Peptides: Biological Function, Mode of Action, and Parameters Involved in Target Specificity. *Biochemistry* **2003**, *42*, 14023–14035. [CrossRef]

31. Yu, H.; Wang, C.; Feng, L.; Cai, S.; Liu, X.; Qiao, X.; Shi, N.; Wang, H.; Wang, Y. Cathelicidin-trypsin inhibitor loop conjugate represents a promising antibiotic candidate with protease stability. *Sci. Rep.* **2017**, *7*, 2600. [CrossRef] [PubMed]

32. Chilosi, G.; Caruso, C.; Caporale, C.; Leonardi, L.; Bertini, L.; Buzi, A.; Nobile, M.; Buonocore, V.; Magro, P. Antifungal Activity of a Bowman-Birk-type Trypsin Inhibitor from Wheat Kernel. *J. Phytopathol.* **2000**, *148*, 477–481. [CrossRef]

33. Roccatano, D.; Colombo, G.; Fioroni, M.; Mark, A.E. Mechanism by which 2,2,2-trifluoroethanol/water mixtures stabilize secondary-structure formation in peptides: A molecular dynamics study. *Proc. Natl. Acad. Sci. USA* **2002**, *99*, 12179–12184. [CrossRef] [PubMed]
34. Matsumura, S.; Uemura, S.; Mihara, H. Fabrication of Nanofibers with Uniform Morphology by Self-Assembly of Designed Peptides. *Chem. A Eur. J.* **2004**, *10*, 2789–2794. [CrossRef] [PubMed]

 *biomolecules*

Article

The Natural-Based Antitumor Compound T21 Decreases Survivin Levels through Potent STAT3 Inhibition in Lung Cancer Models

David Martínez-García [1,2], Marta Pérez-Hernández [1,2], Luís Korrodi-Gregório [1], Roberto Quesada [3], Ricard Ramos [4], Núria Baixeras [5], Ricardo Pérez-Tomás [1,2] and Vanessa Soto-Cerrato [1,2,*]

[1] Department of Pathology and Experimental Therapeutics, Faculty of Medicine and Health Sciences, Universitat de Barcelona, 08905 Barcelona, Spain
[2] Oncobell Program, Institut d'Investigació Biomèdica de Bellvitge (IDIBELL), L'Hospitalet de Llobregat, 08908 Barcelona, Spain
[3] Department of Chemistry, Universidad de Burgos, 09001 Burgos, Spain
[4] Department of Thoracic Surgery and University of Barcelona, Hospital Universitari de Bellvitge, L'Hospitalet de Llobregat, 08907 Barcelona, Spain
[5] Department of Pathology, Hospital Universitari de Bellvitge-IDIBELL, L'Hospitalet de Llobregat, 08907 Barcelona, Spain
* Correspondence: vsoto@ub.edu; Tel.: +34-934-031-140

Received: 11 July 2019; Accepted: 10 August 2019; Published: 13 August 2019

Abstract: Lung cancer is the leading cause of cancer-related deaths worldwide; hence novel treatments for this malignancy are eagerly needed. Since natural-based compounds represent a rich source of novel chemical entities in drug discovery, we have focused our attention on tambjamines, natural compounds isolated from marine invertebrates that have shown diverse pharmacological activities. Based on these structures, we have recently identified the novel indole-based tambjamine analog 21 (T21) as a promising antitumor agent, which modulates the expression of apoptotic proteins such as survivin. This antiapoptotic protein plays an important role in carcinogenesis and chemoresistance. In this work, we have elucidated the molecular mechanism by which the anticancer compound T21 exerts survivin inhibition and have validated this protein as a therapeutic target in different lung cancer models. T21 was able to reduce survivin protein levels in vitro by repressing its gene expression through the blockade of Janus kinase/Signal Transducer and Activator of Transcription-3 (JAK/STAT3)/survivin signaling pathway. Interestingly, this occurred even when the pathway was overstimulated with its ligand interleukin 6 (IL-6), which is frequently overexpressed in lung cancer patients who show poor clinical outcomes. Altogether, these results show T21 as a potent anticancer compound that effectively decreases survivin levels through STAT3 inhibition in lung cancer, appearing as a promising therapeutic drug for cancer treatment.

Keywords: natural-based compound; anticancer therapy; lung cancer; survivin; apoptosis; STAT3

---

1. Introduction

Lung cancer is the leading cause of cancer-related deaths in both men and women worldwide, accounting for more than 2.1 million new cases and more than 1.8 million deaths estimated in 2018 [1]. This malignancy is broadly categorized into small cell lung cancer (SCLC) and non-small cell lung cancer (NSCLC), which represents the 85% of all lung cancers diagnosed. NSCLC is further divided in three major subtypes based on their histology: adenocarcinoma, squamous cell lung carcinoma (SQCLC) and large cell carcinoma, corresponding to 45–50%, 25–30% and 5–10% of all diagnosed NSCLC, respectively [2]. The current standard of care for lung cancer differs according to the tumor histological type, the stage of cancer, possible side effects, and overall patient health. Considering this, surgical resection, radiotherapy, conventional platinum-based doublet chemotherapy (cisplatin generally in combination with pemetrexed/Alimpta® or gemcitabine), immunotherapy (anti PD-1, mostly nivolumab/Opdivo®)), and targeted therapy are the main options to treat lung cancer [3]. Nevertheless, despite all the available therapeutic options, the five-year survival rate of lung cancer is low, 18.6%, according to the National Cancer Institute (NCI). Therefore, novel therapeutic strategies should be developed to increase the therapeutic options available for the treatment of this malignancy.

Natural-based compounds constitute an important research area for cancer drug discovery, with numerous compounds showing therapeutic potential in most cancer types. Interestingly, over 70% of anticancer compounds in clinical use derive from natural products, such as the marine organism-derived compounds cytarabine (Cytosar), travectedin (Yondelis), eribulin mesylate (Halaven) and the conjugated antibody brentuximab vedotin (Acentris) [4]. In particular, marine organisms are gaining interest for providing a huge array of biologically active metabolites for the development of new anticancer agents. The natural alkaloids tambjamines, originally isolated from marine invertebrates, have shown a wide spectrum of pharmacological properties [5]. In this regard, we have demonstrated that the indole-based tambjamine analog 21 (T21) exerts a potent anticancer effect in vitro and a significant therapeutic effect, with a favorable safety profile, in vivo in lung cancer mice models [6]. Moreover, T21 was able to modulate apoptotic protein levels, including survivin. As a member of the inhibitor of apoptosis (IAP) family, survivin plays an important role in tumorigenesis, metastasis and therapy resistance by promoting cell division and inhibiting apoptosis [7]. Furthermore, survivin is overexpressed in cancer cells, while in most normal finally differentiated tissues is almost undetectable [8]. Altogether, these features advocate survivin as an ideal therapeutic target to treat cancer and hence, T21 may be a promising future chemotherapeutic agent. In fact, several molecular approaches that block survivin expression and/or function are emerging as promising therapeutic strategies in cancer by sensitizing tumor cells to apoptosis, minimally affecting non-tumor cells [9].

Therefore, in this work we have deeply analyzed, for the first time, the molecular mechanism of action by which a recently described natural-derived compound called T21 inhibits survivin, inducing its anticancer effects in vitro as well as in in vivo mice models, validating survivin as a promising therapeutic target for lung cancer treatment.

2. Materials and Methods

2.1. Human Samples

Fresh squamous cell lung carcinoma tissue and adjacent non-tumor lung tissue samples were obtained from patients during resection surgery at Bellvitge University Hospital in Barcelona, Spain. The study was conducted in accordance with the Declaration of Helsinki ethical guidelines and informed consent was obtained from all patients included in the study. All study protocols were approved by the Clinical Research Ethics Board of Bellvitge University Hospital and by the local Ethics Committee (PR003/13). The histological typing was confirmed by the Pathology Department at the aforementioned Hospital. All the human tissue samples were preserved in RNAlater™ (Cat#76104, Qiagen) and stored at liquid nitrogen before being processed.

2.2. Reagents

Tambjamine-21 analogue (T21) was synthesized as previously reported [6], dissolved at 10 mmol/L in dimethyl sulfoxide (DMSO) and stored at −20 °C. Cycloheximide (CHX; Cat#C7698-IG) from Sigma-Aldrich was dissolved in ethanol at a stock solution of 100 mg/mL and stored at −20 °C. Interleukin 6 (IL-6; Cat#IL006) was purchased from EMD Millipore and dissolved in 1× phosphate-buffered saline (PBS; Cat#02-020, Biological Industries, Beit Haemek, Israel) with calcium and magnesium supplemented with 0.1% bovine serum albumin (BSA; Cat#A7906, Sigma-Aldrich,

St Louis MO, USA) at a stock solution of 100 µg/mL and stored at −80 °C. Hoechst 33,342 (Cat# B2261) was purchased from Sigma-Aldrich.

2.3. Antibodies

The antibodies used in this study were obtained from the following sources: anti-survivin (71G4B7, Cat#2808), anti-XIAP (3B6, Cat#2045), anti-phospho-JAK1 (Y1034/1035; Cat#3331), anti-phospho-STAT3 (Y705; D3A7, Cat#9145), anti-cleaved PARP (Cat#5625T), anti-cleaved caspase 3 (Cat#9664) and anti-phospho-JAK2 (Y1007/1008; Cat#3771) from Cell Signaling Technology Inc. (Beverly, MA, USA); anti-actin (I-19, Cat#sc-1616), anti-GAPDH (0411, Cat#sc-47724), anti-JAK1 (B-3, Cat#376996), anti-phospho-STAT3 (Y705; B-7, Cat#sc-8059), anti-STAT3 (F-2, Cat#sc-8019), from Santa Cruz Biotechnology Inc. (Santa Cruz, CA, USA); anti-vinculin (Cat#V-4505) from Sigma-Aldrich. Antibody binding was detected with donkey anti-mouse IgG-HRP (Cat#A16017), donkey anti-rabbit IgG-HRP (Cat#A16029) and donkey anti-goat IgG-HRP (Cat#A15999) from Thermo Fisher Scientific Inc. (Waltham, MA, USA); Alexa Fluor™ 488-conjugated donkey anti-mouse (Cat#A31572, Molecular Probes, Eugene, OR, USA) was used for antibody binding detection in immunofluorescence assays.

2.4. Cell Lines and Culture Conditions

Human cell lines SW900 and H520 (squamous lung carcinoma), A549 (lung adenocarcinoma), DMS53 (small cell lung carcinoma) and HFL-1 (lung fibroblasts), were obtained from the American Type Culture Collection (ATCC). SQCLC, adenocarcinoma and lung fibroblasts cells were cultured (passage number 10–25) in Roswell Park Memorial Institute medium (RPMI, Cat# 01-104), Dulbecco's modified Eagle's medium (DMEM, Cat#01-055) and Ham's F-12 (Cat#01-095) (Biological Industries), respectively. All of them were supplemented with 10% heat-inactivated fetal bovine serum (FBS Gibco™; Cat#10270106, Life Technologies), 100 units/mL penicillin, 100 µg/mL streptomycin, and 2 mM L-glutamine (all from Biological Industries). Non-essential amino acids (NEAA; Cat#X0557, Biowest; 1:100) were also used for HFL-1 culture (Biological Industries). 15 mM of HEPES buffer solution (Cat#03-025, Biological Industries) was also used for H520 culture. Cells were grown at 37 °C in a humidified incubator (Thermo Fisher Scientific Inc.) with 5% CO_2 atmosphere. The cells were mycoplasma tested using a standard PCR technique after thawing.

2.5. Gene Expression Analysis

Gene expression levels of BIRC5 were evaluated by Reverse Transcription quantitative-PCR (RT-qPCR) analysis. Total RNA was isolated and purified from 30 mg of frozen tissue samples using the column-based RNeasy Mini Kit (Qiagen) and following the manufacturer's standard protocol. Total RNA concentration and purity were checked in a nano spectrophotometer (Implen GmbH, Munchen, Germany) and integrity was analyzed using an Agilent 2100 Bioanalyzer (Agilent Technologies, Santa Clara, CA, USA). Samples with higher RNA integrity number (RIN) were selected (7 lung cancer samples and their paired normal tissue samples from the same patient). RNA amounts from selected tissue samples were equally pooled to create a sample (Σnon-tumoral and Σtumoral) and their concentration, purity and integrity were re-checked. For the reverse transcription, 1 µg of total RNA was used for cDNA synthesis using a mixture of random hexamers and oligo-dT primers and following the RT2 First Strand Kit protocol (Qiagen). Then, reverse transcription was confirmed through actin beta (ACTB) gene amplification by standard PCR procedure (BIOTAQ DNA Polymerase; BIOLINE). For survivin expression analysis in A549 and SW900 cells, 1.25×10^5 cells/mL were seeded and after 24 h they were treated in absence or presence of T21 for 6 or 16 h (IC_{50} concentrations). RNA was purified and cDNA obtained as described above. Specific oligonucleotide primers and probes for BIRC5 (Hs00153353_m1), and ACTB (Hs99999903_m1), were purchased as Assay-on-Demand Gene Expression Products (Applied Biosystems). TaqMan PCR reactions were performed on cDNA samples using TaqMan Universal PCR Master Mix (Applied Biosystems, Fosters city, CA, USA) and ABI PRISM 7900 HT Fast

Real-Time PCR system (Applied Biosystems). Gene expression levels were quantified and normalized using ACTB as a house keeping gene and relative mRNA expression was calculated in relation to the healthy samples. Ct values were determined using ExpressionSuite software (version 1.0.3, Applied Biosystems) and are presented as mean ± SD of three independent experiments.

2.6. Cell Viability Assays

Cell viability was evaluated using the methylthiazoletetrazolium (MTT, Sigma-Aldrich, Merck KGaA) colorimetric assay. Cells were harvested (10^5 cells/mL) in 96-well plates and allowed to grow overnight. At the following day, T21 was added to the cells at different ranging concentrations (0.8–100 μmol/L) or vehicle solution (DMSO, Sigma-Aldrich, Merck KGaA) to control cells. Cells were incubated for 24 h and after the treatment period, 10 μL of MTT (5 mg/mL) were added and the plates were incubated for 2 h at 37 °C. Crystals were dissolved in 100 mL of DMSO and reading was done in a spectrophotometer at 570 nm using a multiwell plate reader (Multiskan FC, Thermo Fisher Scientific Inc.). Cell viability and inhibitory concentration (IC) values were obtained using GraphPad Prism V5.0 for Windows (GraphPad Software). All data are shown as the mean value ± SD of three independent experiments.

2.7. Clonogenic Assay

A549 cells at 10^5 cells/mL were seeded (1 mL) in 24-well plate and incubated overnight to allow attachment. Cells were treated for 24 h with T21 at 2.5–10 μM and the same percentage of DMSO was added to the control cells. Then, cells were counted and 200 viable cells were seeded in a final volume of 3 mL in a 6-well plate and allowed to growth for 1 week. Medium was removed and cell colonies were fixed coloured with a mixture of glutaraldehyde (6% *v/v*) and crystal violet (0.5% *w/v*) for 20 min at room temperature and were counted.

2.8. Western Blot Analysis

For the evaluation of the molecular effects after 48 h of survivin silencing, A549 and H520 cells were seeded in 6-wells plates at a density of 1.25×10^5 cells/mL in a volume of 2 mL of medium without antibiotics. The day after, cells at 70–90% of confluence were transfected with 250 pmol of small interfering RNA (siRNA) against survivin (Cat#4390824; Thermo Fisher Scientific Inc.) or scrambled siRNA (Cat#4390843; Thermo Fisher Scientific Inc.) using Lipofectamine® 2000 reagent (Thermo Fisher Scientific Inc.) and following manufacturer's standard protocol.

For the study of T21 cellular effects, A549, SW900, H520 and DMS53 were seeded in 100-mm cell culture plates (1.25×10^5 cells/mL) and allowed to grow for 24 h. Then they were treated with different inhibitory concentrations (IC) of T21 compound (IC_{25}, IC_{50} and IC_{75} values in μM) for 24 h or the IC_{50} value during different time periods (4, 8, 16, 24 h).

For protein synthesis inhibition, A549 and SW900 were seeded in a 6-well plate (1.25×10^5 cells/mL) and allowed to grow for 24 h. Then they were treated with CHX at 100 μg/mL for 30 min followed by T21 treatment at IC_{50} during 24 h.

For the stimulation of STAT3 pathway with IL-6, A549 cells were seeded in 60-mm culture plates (1.25×10^5 cells/mL) for 24 h, and then were starved in serum-free medium for at least another 12 h. Next, cells were stimulated with IL-6 at different concentrations (1, 5 and 10 ng/mL) for 30 min. Similarly, after starvation, A549 cells were pretreated with IC_{50} T21 for 4 h followed by 5 ng/mL of IL-6 stimulation for 30 min.

In all experiments, whole cell lysates, from the selected tissue samples or from cultured cells, were prepared with ice cold lysis buffer containing 0.1% SDS, 1% NP-40, 0.5% sodium deoxycholate, 50 mmol/L sodium fluoride, 40 mmol/L β-glycerophosphate, 200 μmol/L sodium orthovanadate, 1 mmol/L phenylmethylsulfonyl fluoride (all from Sigma-Aldrich), and protease inhibitor cocktail (Cat#11836170001, Roche Diagnostics) in 1× PBS followed by its homogenization, using a tissue grinder (Cat#431-0100, VWR International) in case of the tissue samples. Protein concentration was determined by BCA protein assay (Cat#23225, Pierce™, Thermo Fisher Scientific Inc.) using BSA protein (Sigma-Aldrich) as a standard. For western blot analysis, 40–50 µg of protein extract were first separated by SDS-PAGE and transferred to Immobilon-P polyvinylidene difluoride (PVDF) membranes (EMD Millipore, Merck KGaG). Membranes were blocked in either 5% non-fat dry milk or BSA, both diluted in Tris-buffered saline (TBS)-Tween (50 mmol/L Tris-HCl pH 7.5, 150 mmol/L NaCl, 0.1% Tween-20) for 1 h and then incubated overnight with primary antibodies, according to the manufacturer's instructions. Actin, vinculin or GAPDH (Glyceraldehyde 3-phosphate dehydrogenase) were used as gel loading controls. The results shown are representative of Western blot data analysis obtained from at least three independent experiments. Images were captured on an Image Quant LAS 500 (GE Healthcare) using ECL™ Western blotting detection reagent (Cat#RPN2106, Amersham, GE Healthcare) and band densitometries were retrieved using the Image Studio Lite software (v5.2, LI-COR Biosciences).

2.9. Immunofluorescence Staining

HFL-1, A549 and SW900 cells (1.25×10^5 cells/mL) were seeded in a 12-well plate containing FBS-coated glass coverslips for 24 h. For the IL-6 experiment, A549 cells (1.25×10^5 cells/mL) were seeded in a 12-well plate containing FBS-coated glass coverslips for 24 h and then were starved in serum-free medium for at least another 12 h. Then, cells were pretreated with IC_{50} T21 for 4 h followed by 5 ng/mL of IL-6 stimulation for 30 min. Next, all cells were washed twice with 1× PBS and fixed with 4% paraformaldehyde for 20 min. Fixed cells were permeabilized by 0.2% Triton X-100 and then blocked with 1% BSA in 1× PBS for 1 h. Cells were incubated overnight at 4 °C with anti-survivin antibody at a dilution of 1:500 or anti-phospho-STAT3 (Cell Signaling) at a dilution of 1:100. Cells were then washed with 1× PBS and incubated with Alexa Fluor™ 488-conjugated goat anti-rabbit (Molecular Probes) at 1:400 dilution for 1 h at room temperature. At the same time, the cell nuclei were stained with 2 µg/mL hoechst 33,342 (Cat# B2261, Sigma-Aldrich). Afterwards, coverslips were washed with 1× PBS and were placed on the slides using Mowiol™ (Sigma-Aldrich). The immunofluorescence images were acquired using a Carl Zeiss LSM 880 spectral confocal laser scanning microscope (Carl Zeiss Microscopy GmbH, Jena, Germany) equipped with a multiline argon laser (458 nm, 488 nm and 514 nm), 405nm and 561nm diode lasers and 633 nm He/Ne laser (Centres Científics i Tecnològics, Universitat de Barcelona, Bellvitge Campus, Barcelona, Spain) using a 63× oil immersion objective (1.4 numerical aperture) an image resolution of 1024×1024 pixels. Representative images from three independent experiments are shown.

2.10. Immunohistochemistry Analysis

For in vivo studies, five-week-old female Crl:NU-Foxn1nu mice strain (Envigo) were used to generate a subcutaneous xenograft model. All animal studies were approved by the Autonomic Ethic Committee (Generalitat de Catalunya) under the protocol 9111. DMS53 cells (4.5×10^6 cells) suspended in a 1:1 solution of RPMI1460:Matrigel (BD Bioscience) were implanted subcutaneously in the flank of mice. Mice bearing homogenous subcutaneous tumors (approximately 150–200 mm^3) were randomly allocated to two treatment groups ($n = 7$/treatment) and intraperitoneally administrated with T21 (diluted in 7.5% DMSO/0.8% Tween-80) at a dose of 6 mg/kg in alternated days during 20 days. After the final dose of the treatment, animals were sacrificed, tumors dissected out and embedded in paraffin for immunohistochemistry staining as follows: 4 µm sections were cut, deparaffinized and after antigen retrieval in 10 mmol/L sodium citrate buffer with 0.05% Tween-20 in the microwave at sub-boiling temperature (95–98 °C) for 20 min, slides were washed 2 times with distilled H_2O (dH_2O) of 5 min each. Endogen peroxidase was blocked by incubation in 3% H_2O_2 for 5 min at room temperature following by washing steps for 5 min twice, with dH_2O and PBS. Slides were blocked with normal goat serum in a 1:30 dilution for 1 h at room temperature and incubated with anti-survivin antibody diluted 1:400 in PBS overnight at 4 °C in a wet chamber. Afterwards, slides were washed 3 times in PBS for 5 min each and incubated with secondary antibody coupled with HRP at 1:100 dilution in PBS for 1 h, at room temperature. Then, slides were washed 3 times with PBS 0.1% Tween-20 for 5 min each and signal was developed by incubation with DAB (3,3′-diaminobenzidine) (Cat#D8001, Sigma) for 10 min at room temperature. Finally, slides were washed for 5 min with dH$_2$O, counterstain with Hematoxylin (Cat#A3865, PanReac AppliChem, Barcelona, Spain), dehydrated, and mounted with DPX (Cat#100579, Merck, Madrid, Spain). Samples were observed in a Nikon Eclipse E800 microscope and images were taken with the camera ProgRes CFscan.

2.11. Statistical and Data Mining Analyses

For statistical analysis of single point qPCR results and western blot data, t-Student test and one-way ANOVA with post hoc Tukey analysis, were carried out using the Statgraphics plus 5.1 statistical Software, respectively. Statistically significant differences, $p < 0.05$, $p < 0.01$ and $p < 0.001$, are represented by *, ** and ***, respectively.

3. Results

3.1. Survivin Validation as A Promising Therapeutic Target in All Lung Cancer Subtypes

Since the anticancer T21 compound potently decreases the protein levels of the antiapoptotic protein called survivin, expression levels of this potential therapeutic target were evaluated in tumor samples from SQCLC patients through both specific qPCR analysis and survivin protein immunodetection (Figure 1A,B). In accordance with other tumors, BIRC5 was confirmed to be up-regulated more than 19 times in SQCLC samples using qPCR and consequently, survivin expression in tumor samples was much higher than in non-tumor samples. Furthermore, we wanted to analyze whether this protein may also be a therapeutic target for all the most prevalent lung cancer histological subtypes. Hence, we analyzed survivin overexpression in the adenocarcinoma cell line A549, the small cell lung cancer cell line DMS53 as well as in the SQCLC cell lines, SW900 and H520. All four tumor cell lines overexpress survivin compared to non-tumor human lung fibroblasts HFL-1 (Figure 1C; Figure S1). Interestingly, survivin has a dual role in cancer, both promoting cell cycle progression and inhibiting apoptosis [8]. Therefore, in order to validate survivin as an effective therapeutic target in lung cancer, we proposed to perform a loss-of-function assay by inhibiting survivin expression. In this context, A549 and H520 cells were transfected with a siRNA against survivin mRNA or a control siRNA designed to have no specific target in the cell. After 48 h of transfection, there was no activation of apoptosis in cells exposed to control siRNA; however, transfection with siRNA against survivin resulted in a marked activation of caspase 3 in A549 as well as a cleavage of their substrate poly (ADP-ribose) polymerase (PARP) (Figure 1D). In the case of transfected H520, where the silencing of survivin was less evident, the increase of the activated form of caspase 3 was more subtle, but also significant (Figure 1D). Cleaved PARP was also increased in H520 with silenced survivin. Finally, we wanted to elucidate whether the X-linked inhibitor of apoptosis protein (XIAP), a survivin partner in apoptosis inhibition, may also be affected after survivin silencing, since its stability against ubiquitin-dependent degradation is linked to survivin binding [10]. In our cellular models, XIAP also showed a significant reduction of its protein levels after 48 h of survivin downregulation, although less pronounced than survivin decrease, both in A549 and H520 cells (Figure 1D,E). Therefore, survivin appears as a potential therapeutic target in all lung cancer subtypes, suggesting the promising use of compounds targeting survivin for lung cancer treatment.

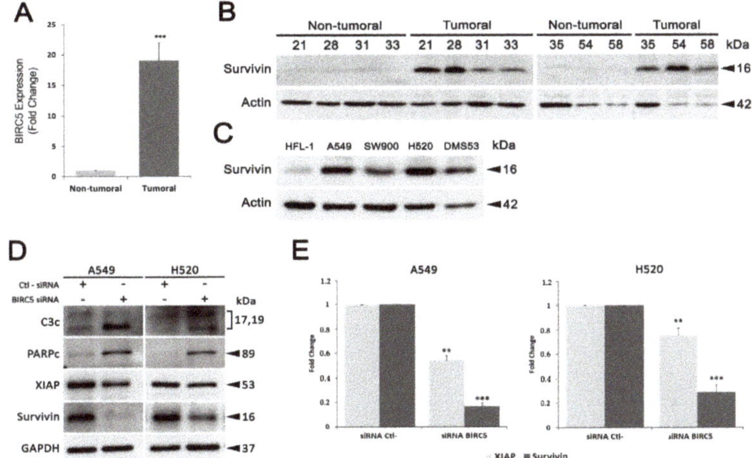

Figure 1. Survivin expression levels in lung cancer tumor samples and survivin functional validation in cellular models. (**A**) qPCR validation of BIRC5 expression levels. Fold changes of gene expression were calculated using β-actin as the housekeeping gene. (**B,C**), survivin levels were analyzed using whole cell lysates from patient tissue samples and lung cell lines by Western blot analysis. (**D,E**), After 48 h of transfection with siRNA against survivin mRNA, the expression of survivin, XIAP and pro-apoptotic proteins was analyzed by Western blot analysis in A549 and H520 cell lines. Protein levels were normalized with their respective loading controls in each blot. Results were obtained from at least three independent experiments. Bars represent the mean ± SD. Statistically significant results are indicated as *, *p*-value < 0.05; **, *p*-value < 0.01 and ***, *p*-value < 0.001. C3c, cleaved caspase 3; PARPc, cleaved PARP.

3.2. Indole-Based Tambjamine Analog 21 (T21) Downregulates Survivin in Lung Cancer In Vitro and In Vivo Models

To extend the reported cytotoxic properties of T21 (Figure 2A), the effect on cell viability of H520 squamous cell lung cancer cells was also evaluated after treatment with different concentrations of T21 for 24 h, showing similar results than in the other studied cancer cells (Figure 2B). Moreover, the ability to survive and form colonies after T21 treatment was evaluated in a clonogenic assay in A549 cells (Figure S2). These cells were treated with T21 during 24 h, counted, and 200 viable cells were seeded in a new plate with fresh medium for one week. The survival capacity significantly decreased in cells after T21 treatment, especially in those treated with 10 μM, corroborating the potent cytotoxic effect of this compound.

In order to characterize in detail the molecular mechanism of action of this anticancer compound, we analyzed whether T21 was able to inhibit survivin in different lung cancer histological subtypes. For that purpose, the inhibitory effect of T21 on survivin levels was evaluated in non-small cell lung cancer SW900, H520 and A549 cells as well as in the small cell lung cancer DMS53 cells. All cells were treated with T21 at different cell viability inhibitory concentrations for 24 h (IC_{25}, IC_{50} and IC_{75} concentrations) (Figure 2B). The results showed a sharp and significant dose-dependent decrease of survivin in all four cell lines (Figure 2C,D). Furthermore, T21 was also able to decrease, in a dose-dependent manner, the protein levels of XIAP in A549, H520 and to a lesser extent in SW900 and DMS53. Hence, these results suggest that the anti-apoptotic proteins survivin and XIAP decrease in a dose-dependent manner after T21 treatment allowing the induction of apoptosis.

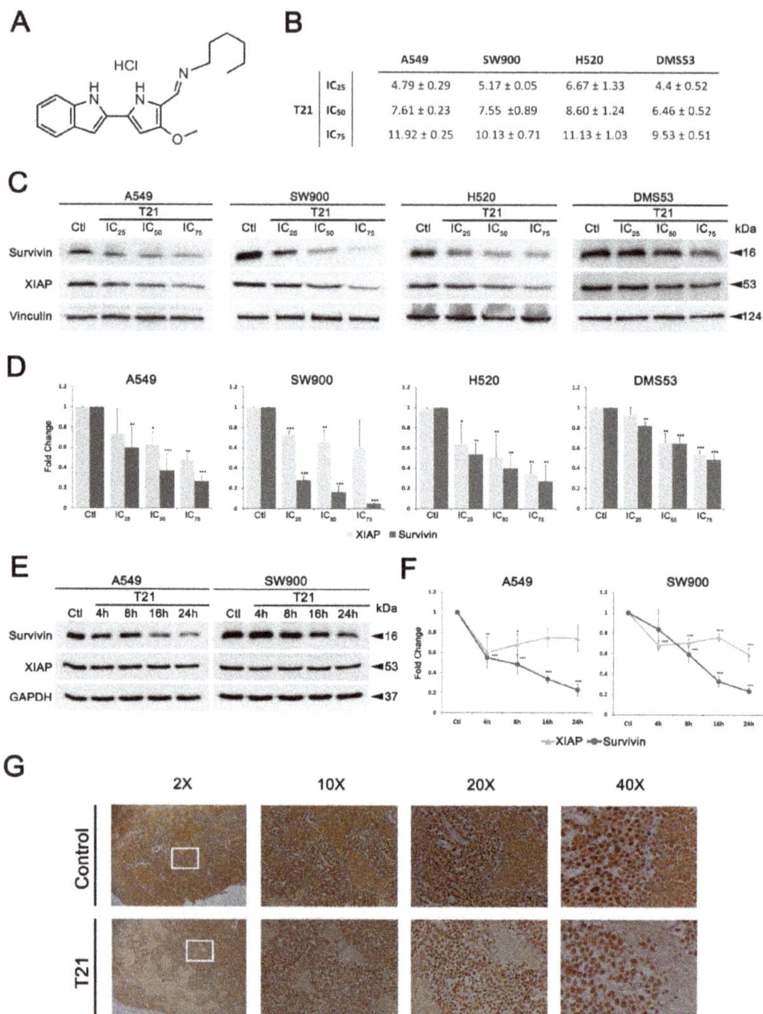

Figure 2. T21 decreases survivin levels in vitro and in vivo. (**A**) T21 chemical structure. (**B**) T21 inhibitory concentration (IC) values calculated in cell viability assays performed in several lung cancer cell lines. Some IC_{50} values were previously published in Manuel-Manresa et al. [6]. Values represent mean ± SD in μM. (**C,D**) After 24 h of treatment with IC_{25}, IC_{50}, IC_{75} values of T21 (Table S1), the expression of survivin and XIAP was analyzed by Western blot analysis in A549, SW900, H520 and DMS53 cell lines. (**E,F**) A549 and SW900 cells were treated with T21 at IC_{50} during different time periods (4, 8, 16, 24 h) followed by survivin and XIAP expression analysis through Western blot analysis. Results were obtained from at least three independent experiments. Bars represent the mean ± SD. Statistically significant results are indicated as *, *p*-value < 0.05; **, *p*-value < 0.01 and ***, *p*-value < 0.001. (**G**) Subcutaneous tumors from implanted DMS53 cells in Crl:NU-Foxn1nu mice were treated with 6 mg/kg of T21 every other day for 20 days. Tumor samples were stained against survivin by immunohistochemistry. Representative illustrations are shown at 2×, 10×, 20× and 40× magnifications, the scale bars correspond to 500, 500, 100 and 100 μm, respectively.

To further investigate whether the effect of T21 over survivin and XIAP expression was simultaneous or not, SW900 and A549 were treated at their IC_{50} values for different time periods (4, 8, 16 and 24 h). The results showed a progressive and sharp decrease of survivin in a time-dependent manner in A549 and SW900, reaching a significant decrease at 24 h. Conversely, XIAP moderately decreased its protein levels at 4 h and were stabilized over time in both cell lines (Figure 2E,F). Hence, these results suggest that T21 potently inhibits survivin protein levels in a dose and time-dependent manner, whilst a moderate XIAP inhibition is observed.

On the other hand, we proposed to analyze the decrease of survivin levels after T21 treatment in subcutaneous tumors implanted in the flank of mice that were treated intraperitoneally with 6 mg/kg of T21 every other day for 20 days. As observed in Figure 2G, T21 was able to considerably reduce the protein levels of survivin in tumor samples from treated mice, which showed a decrease in tumor growth. This result corroborates the molecular changes induced by T21 in vitro showing that T21 decreases the expression of survivin also in tumors. Additionally, these results confirm that T21 has reached the tumors and suggest that survivin may be a good biomarker for T21 treatment efficacy.

3.3. T21 Reduces Survivin Levels via Gene Transcription Repression in Lung Cancer Cells

In order to identify the cellular mechanism that T21 was triggering to decrease survivin levels, A549 and SW900 cell lines were treated with T21 IC_{50} for 6 and 16 h and BIRC5 gene expression was analyzed by RT-qPCR. T21 was able to downregulate BIRC5 at both incubation periods albeit being more evident at 16 h, indicating that T21 represses survivin gene expression (Figure 3A).

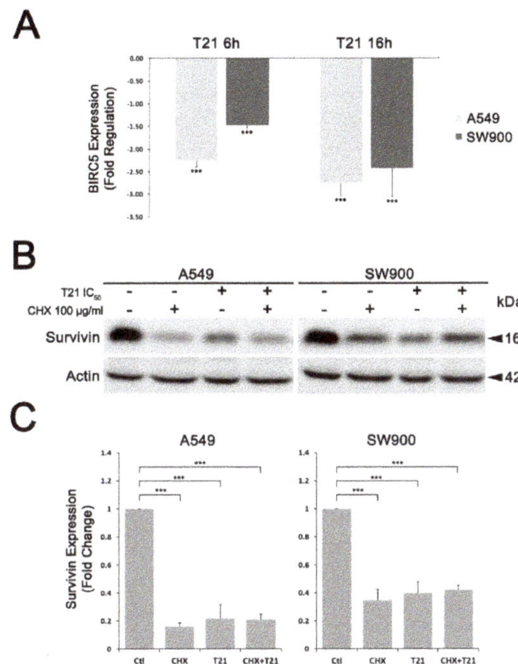

Figure 3. T21 reduces survivin levels through transcriptional repression. (**A**) A549 and SW900 cells were treated with T21 at IC_{50} for 6 and 16 h and BIRC5 gene expression was analyzed by RT-qPCR. (**B,C**) survivin levels were assessed by Western blot analysis in A549 and SW900 cells exposed to CHX at 100 µg/mL for 30 min followed by T21 treatment at IC_{50} during 24 h. Results were obtained from at least three independent experiments. Bars represent the mean ± SD. Statistically significant results are indicated as *, *p*-value < 0.05; **, *p*-value < 0.01 and ***, *p*-value < 0.001.

To further study whether the reduction of survivin levels observed after T21 exposure was only due to a transcriptional repression, SW900 and A549 were treated with CHX, an inhibitor of protein synthesis. After 30 min of CHX treatment, cells were treated with T21 at IC_{50} for 24 h. As observed in Figure 3B,C, survivin levels in both cell lines treated with T21 plus CHX were similar to CHX treated cells. Hence, T21 was inducing the same effect that only inhibiting the de novo protein synthesis, since no additional degradation was observed compared to CHX treated cells. Therefore, these results suggest that the reduction of survivin levels after T21 exposure is mainly due to a transcriptional repression and T21 is not triggering direct protein degradation.

3.4. T21 Suppresses STAT3 Phosphorylation via JAK/STAT3 Pathway Inhibition in Lung Cancer Cells

Among other signaling pathways, the JAK/STAT3 has been described to be involved in increasing survivin expression levels and promoting tumorigenesis after aberrant activation in cancer cells [11]. In this regard, we first evaluated STAT3 activity in A549 and SW900 lung cancer cells, compared to normal cells (HFL-1) (Figure 4A), observing a significant increase in basal STAT3 signaling in cancer cells, accompanied by higher levels of survivin expression.

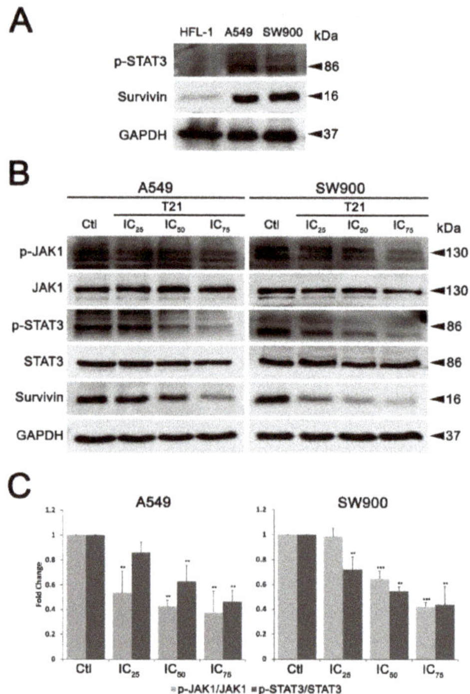

Figure 4. STAT3 pathway analysis. (**A**) STAT3 signaling in lung normal and cancer cells. Levels of phospho-STAT3 and survivin were analyzed by western blot in HFL-1, A549 and SW900 cells. (**B,C**) JAK/STAT3 pathway inactivation after T21 treatment. After 24 h of treatment with the IC_{25}, IC_{50}, IC_{75} values of T21 (Table S1), the expression and phosphorylation levels of several members of JAK1/STAT3 pathway were analyzed by Western blot in A549 and SW900 cells. Results were obtained from at least three independent experiments. Bars represent the mean ± SD. Statistically significant results are indicated as *, *p*-value < 0.05; **, *p*-value < 0.01 and ***, *p*-value < 0.001.

Then, in order to assess whether this pathway could be involved in survivin regulation after T21 treatment, STAT3 phosphorylation at Tyr705 (p-STAT Y705) was evaluated in SW900 and A549 cells after T21 exposure at their IC_{25}, IC_{50} and IC_{75} for 24 h. This phosphorylation allows STAT3 dimerization, nucleus translocation from cytoplasm, and DNA binding [12]. Both cell lines showed a significant reduction in net phosphorylation of STAT3 at Tyr705 after T21 treatment in a dose-dependent manner (Figure 4B,C), similarly to that observed in survivin levels. Since STAT3 can be phosphorylated by JAK, upon IL-6 stimulation and gp-130 receptor activation, phosphorylation status of this upstream STAT3 activator was also evaluated. JAK1 showed a significant reduction on its phosphorylation state at Tyr1034/1035 (p-JAK1 Y1034/1035). Similarly, JAK2 phosphorilation decreased after T21 treatment (Figure S3). These results suggest that T21 inhibits STAT3 phosphorylation through JAK/STAT3 pathway, which may reduce the gene expression of survivin provoking the observed protein level decrease.

3.5. T21 Blocks IL-6-Induced STAT3 Phosphorylation in A549

To examine whether T21 was also able to block STAT3 signaling after stimulating with one of its upstream ligands, we first analyzed the IL-6 effects on STAT3 phosphorylation in lung cancer cells. A549 cells, after starvation for at least 12 h to decrease the JAK/STAT signaling pathway, were stimulated with IL-6 at different concentrations for 30 min. STAT3 was activated by phosphorylation in a dose-dependent manner after IL-6 stimulation from 1 to 10 ng/mL (Figure 5A). In turn, the total STAT3 expression level was not altered with IL-6 stimulation. Next, in the same culture conditions after starvation, A549 cells were pretreated with T21 at IC_{50} for 4 h followed by 5 ng/mL of IL-6 stimulation for 30 min. The results showed that IL-6 treatment induced STAT3 phosphorylation, but this induction was almost totally repressed by T21 in pretreated cells, showing that T21 pretreatment impedes activation of this signaling pathway by IL-6 (Figure 5B,C).

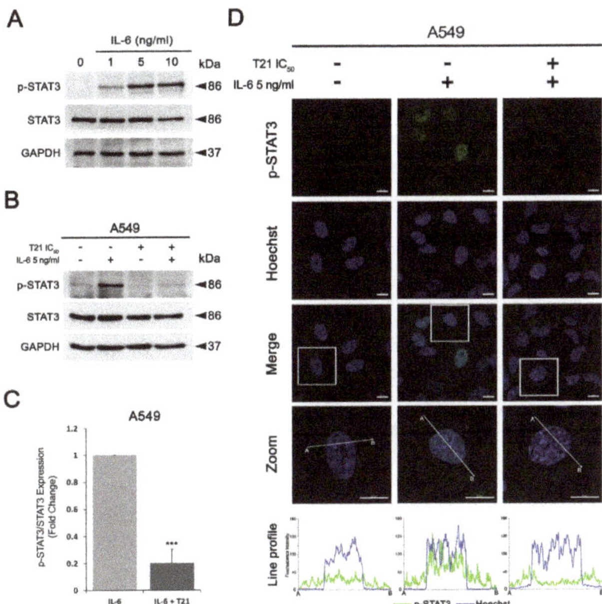

Figure 5. T21 blocks STAT3 activity after IL-6 stimulation. (**A**) A549 cells were starved, then treated with different concentrations of IL-6 and phosphorylation levels of STAT3 were assessed by Western blot analysis. (**B,C**) after starvation, A549 were pretreated with IC_{50} T21 for 4 h followed by 5 ng/mL of IL-6 stimulation for 30 min. Phosphorylation levels of STAT3 were then assessed by Western blot analysis. Bars represent the mean $\pm$ SD. Statistically significant results are indicated as *, p-value < 0.05; **, p-value < 0.01 and ***, p-value < 0.001. (**D**) A549 cells were pretreated with IC_{50} T21 for 4 h, after starvation, followed by 5 ng/mL of IL-6 stimulation for 30 min. Subcellular localization of phospho-STAT3 (Y705) was assessed by immunofluorescence staining. The framed regions are zoomed in the bottom images. The line profiles of phospho-STAT3 (Y705) and Hoechst signals were measured by ZEN 2.3 blue edition (Carl Zeiss) software. (Scale bars correspond to 10 μm).

Additionally, to investigate whether T21 can block the IL-6-induced translocation of STAT3 from cytoplasm to nucleus, A549 cells were pretreated with T21 at IC_{50} for 4 h followed by 5 ng/mL of IL-6 stimulation for 30 min. After cell fixation and immunofluorescence staining, phosphorylated STAT3 at Tyr705 was increased and translocated into nucleus due to IL-6 stimulation (Figure 5D). By contrary, cells stimulated with IL-6 after T21 treatment showed lower levels of phosphorylated STAT3 as well as no significant translocation into the nucleus. Altogether, the above results suggest that T21 is able to prevent IL-6-induced STAT3 phosphorylation and its nuclear translocation; hence T21 blocks the JAK/STAT3 signaling pathway in lung cancer cells.

4. Discussion

Although there have been significant advances in lung cancer management in recent years, lung cancer overall survival remains very low [13]. Hence, more efforts are needed to identify, design and develop new compounds aimed at treating lung cancer. The development of new technologies are aiding to find promising lead candidates from natural products, which have demonstrated to be a major source for drug discovery along history [14]. Natural products and their bioactive derivatives from animals, plants, fungi, and microorganisms, among others, have widely been studied for therapeutic use, being the morphine the first commercial plant-derived product in 1826. Interestingly, about a quarter of all Food and Drug Administration (FDA) and/or the European Medical Agency (EMA) approved drugs are directly or indirectly plant based [15]. Natural-based products have also been an important source of several clinically useful anti-cancer agents, as the well-known antineoplastic paclitaxel that derives from endophytic fungi isolated from plants and is used for the treatment of breast, ovarian and lung cancer. In this regard, we have recently described a marine organism-derived small molecule called indole-based tambjamine analog 21 (T21), which possesses a potent antitumor effect through the induction of apoptosis [6]. In accordance with our findings, several studies also demonstrated the potential of compounds derived from marine organism as anti-cancer agents, being some of these compounds in clinical trials and others already approved for clinical use [4]. Various examples of approved antineoplastic analogs derived from marine organisms include cytarabine (Cytosar), an antimetabolite drug used for the treatment of various types of leukemia; trabectedin (Yondelis), a DNA alkylator for soft sarcoma treatment; and the antimitotic compounds that inhibits microtubule dynamics eribulin mesylate (Halaven) and brentuximab vedotin (Acentris), which are used in breast cancer and Hodgkin's lymphoma treatment, respectively. Regarding the cell death induced by T21, this pro-apoptotic compound significantly decreases survivin levels, also in in vivo studies, inducing simultaneously a significant decrease in tumor volume without any obvious toxicity in mice [6]. As a member of the inhibitor of apoptosis protein (IAP) family, survivin (protein encoded by BIRC5) plays an important role in tumorigenesis, metastasis and therapy resistance [16,17]. Survivin is highly expressed during embryonic and fetal development, but is almost undetectable in most normal finally differentiated tissues [8]. In cancer cells, survivin is overexpressed, being associated with poor prognosis in many human neoplasms. Moreover, survivin also participates in complex molecular signaling cascades cancer-related, being therefore crucial for carcinogenesis [18]. Altogether, the fact that survivin is overexpressed in tumors as well as its key biological roles that promote carcinogenesis and chemoresistance, makes survivin a promising therapeutic target to treat cancer [8]. In these sense, putative survivin antagonists under study are showing promising antitumoral potential, such as YM155, a small-molecule inhibitor that targets and suppresses specifically the activity of the survivin promoter [19]. Nevertheless, despite showing good clinical results in combination regimens, YM155 and other survivin inhibitors under study present modest activity as a single agent, which may be attributed to incomplete or transient survivin inhibition.

In order to characterize in depth the molecular mechanism by which this compound inhibits survivin, we analyzed its effects on several lung cancer cells. Here, we show that T21 significantly reduced the levels of survivin in all survivin overexpressing cells. Interestingly, T21 also induced a decrease of survivin levels in vivo, suggesting an acceptable metabolism rate limiting biotransformation of this drug before exerting its effects. On the other hand, T21 was also able to reduce the levels of XIAP, another IAP protein which interacts and forms a complex with survivin to inhibit the effector caspases [20]. In this context, it has also been reported that this IAP-IAP complex enhances XIAP stability against ubiquitin-dependent degradation [10]. Hence, T21 effect over survivin expression and protein levels might be affecting the stability of XIAP, thus avoiding caspase inhibition and finally promoting apoptosis. Furthermore, we showed how the effect of T21 on survivin was not due to direct protein degradation but survivin gene repression. These results clearly categorize T21 as an indirect inhibitor of survivin, like herceptin, lapatinib or SD-1029, among others, which downregulates survivin expression targeting key cellular signaling pathways involved in the expression of this protein [19,21–23].

In this regard, survivin gene regulation can be triggered by several signaling pathways, such as PI3K/AKT, MAPK/ERK and JAK/STAT pathways [7]. Among them, STAT3 has been described as an important signaling mediator in malignant diseases by promoting the expression of genes involved in cancer proliferation, cell survival, immune suppression, inflammation, and metastasis [24,25]. Persistent activation of STAT3, which is found in 22–65% of NSCLC [26], induces survivin expression and thereby prevents apoptosis, which potentially contributes to resistance to chemotherapy [11]. In this study, we found that T21 treatment could inhibit the STAT3 Tyr705 phosphorylation in NSCLC cells. Interestingly, it has been proved that the acidification of the cytosol in A549 cells triggers a rapid dephosphorylation of STAT3 at Tyr705 [27]. In this regard, T21 significantly decreases the intracellular pH in A549 cells after 1 h, promoting the acidification of the cytosol [28]. Hence, the rapid dephosphorylation of STAT3 at Tyr705 observed in our A549 cells may be a consequence of this cytosolic acidification produced by T21. Moreover, since the transcriptional activity of STAT3 is closely associated with this phosphorylation, the inhibition of the JAK/STAT3 pathway may be the cause of the observed downregulation of survivin in our model. In fact, the repression of STAT3 transcriptional activity after T21 treatment is also supported by our previous results, where other STAT3 downstream genes, such as Bcl-2, Bcl-xL or Mcl-1 were also modified after treatment [6]. Furthermore, our results are consistent with previous studies that already described a reduction of survivin in glioblastoma, leukemia, NSCLC, breast and ovary cancers when JAK/STAT3 pathway was inhibited [23,29–32].

On the other hand, STAT3 activation is commonly triggered by the binding of growth factors and cytokines to their specific receptors. IL-6, a pleiotropic cytokine, has been associated with STAT3 activation after its binding to the gp130 receptor, which in turn, recruits and triggers the activation of JAK [33]. JAK can then phosphorylate STAT3 on Try705 residue, promoting its homodimerization and translocation to the nucleus, followed by gene expression regulation [34]. Data from previous studies have confirmed that elevated IL-6 levels are observed in patients with a variety of cancers [35]. In particular, IL-6 levels have been found increased in nearly 40% of lung cancer patients [36]. Furthermore, high levels of circulating IL-6 were associated with poor responses and worst survival outcomes in NSCLC patients treated with chemotherapy [37]. In this study, we confirmed that IL-6 was able to phosphorylate STAT3 at Tyr705 as well as promote its nuclear translocation. Conversely, both activation and nuclear translocation of STAT3 were repressed after T21 treatment, suggesting that T21 may help to overcome the chemoresistance of tumors with high IL-6 levels.

5. Conclusions

We have investigated in detail the molecular mechanisms of action by which the novel anticancer drug T21 inhibits the antiapoptotic protein survivin in vitro as well as in vivo and have validated this protein as a promising therapeutic target for lung cancer treatment. As a result, we have demonstrated that JAK/STAT3 signaling pathway is the molecular mechanism involved in the inhibition of survivin by T21. T21 regulates survivin by transcriptional gene repression, blocking the phosphorylation of STAT3 and its nuclear translocation. Moreover we have shown that STAT3 inhibition by T21 is effective even in the presence of IL-6 overstimulation. These findings reinforce the potential of T21 as a novel chemotherapeutic agent for lung cancer treatment, especially for lung cancer patients with high levels of IL-6, who show poor outcomes in the clinics. Altogether, T21 is a novel potent inhibitor of STAT3 transcription factor that significantly decreases survivin levels, which may sensitize cancer cells to apoptosis and may enhance the apoptotic effect induced by traditional chemotherapeutic drugs in lung cancer patients.

Supplementary Materials: The following are available online at http://www.mdpi.com/2218-273X/9/8/361/s1, Figure S1: Survivin overexpression validation by immunofluorescence in HFL-1, A549 and SW900 cells. Figure S2: Effects of T21 on cell viability by a clonogenic assay. Figure S3: JAK2 activity after T21 treatment.

Author Contributions: Conceptualization: D.M.-G. and V.S.-C.; Data curation: D.M.-G., M.P.-H. and L.K.-G.; Formal analysis: D.M.-G., M.P.-H, L.K.-G. and V.S.-C.; Funding acquisition: R.Q., R.P.-T. and V.S.-C.; Investigation: D.M.-G., M.P.-H., L.K.-G., R.Q., R.R., N.B. and V.S.-C.; Methodology: D.M.-G., M.P.-H., L.K.-G., R.Q. and V.S.-C.; Project administration: R.P.-T. and V.S.-C.; Resources: R.Q., R.R., N.B., R.P.-T. and V.S.-C.; Supervision: R.P.-T. and V.S.-C.; Validation: D.M.-G., R.P.-T. and V.S.-C.; Writing—original draft: D.M.-G., M.P.-H. and V.S.-C.; Writing—review & editing: D.M.-G., L.K.-G., R.Q., R.R., N.B., R.P.-T. and V.S.-C.

Funding: This research has been partially supported by Instituto de Salud Carlos III [Grant PI18/00441] (Co-funded by European Regional Development Fund. ERDF, a way to build Europe); and Consejería de Educación Junta de Castilla y León [Grant BU092U16]. This work was also supported by a postdoctoral program from the University of Barcelona in collaboration with Obra Social de la Fundació Bancaria "La Caixa" to L.K.-G. and M.P.-H.; and a predoctoral fellowship awarded from the Government of Catalonia through L'Agència de Gestió d'Ajuts Universitaris i de Recerca (AGAUR; FI-DRG 2016) to D.M.-G.

Acknowledgments: We thank CERCA Programme/Generalitat de Catalunya for institutional support. We also thank Benjamín Torrejón and Beatriz Barroso from CCiTUB (Centres Científics i Tecnològics Universitat de Barcelona, Barcelona, Spain) for technical assistance.

Conflicts of Interest: The authors declare no conflict of interest.

References

1. Bray, F.; Ferlay, J.; Soerjomataram, I.; Siegel, R.L.; Torre, L.A.; Jemal, A. Global cancer statistics 2018: GLOBOCAN estimates of incidence and mortality worldwide for 36 cancers in 185 countries. *CA Cancer J. Clin.* **2018**, *68*, 394–424. [CrossRef] [PubMed]

2. Zappa, C.; Mousa, S.A. Non-small cell lung cancer: Current treatment and future advances. *Transl. Lung Cancer Res.* **2016**, *5*, 288–300. [CrossRef] [PubMed]

3. Carrera, P.M.; Ormond, M. Current practice in and considerations for personalized medicine in lung cancer: From the patient's molecular biology to patient values and preferences. *Maturitas* **2015**, *82*, 94–99. [CrossRef] [PubMed]

4. Jimenez, P.C.; Wilke, D.V.; Costa-Lotufo, L.V. Marine drugs for cancer: Surfacing biotechnological innovations from the oceans. *Clinics* **2018**, *73*, e482s. [CrossRef] [PubMed]

5. Carbone, M.; Irace, C.; Costagliola, F.; Castelluccio, F.; Villani, G.; Calado, G.; Padula, V.; Cimino, G.; Cervera, J.L.; Santamaria, R.; et al. A new cytotoxic tambjamine alkaloid from the Azorean nudibranch Tambja ceutae. *Bioorg. Med. Chem. Lett.* **2010**, *20*, 2668–2670. [CrossRef] [PubMed]

6. Manuel-Manresa, P.; Korrodi-Gregório, L.; Hernando, E.; Villanueva, A.; Martínez-García, D.; Rodilla, A.M.; Ramos, R.; Fardilha, M.; Moya, J.; Quesada, R.; et al. Novel Indole-based Tambjamine-Analogues Induce Apoptotic Lung Cancer Cell Death through p38 Mitogen-Activated Protein Kinase Activation. *Mol. Cancer Ther.* **2017**, *16*, 1224–1235. [CrossRef] [PubMed]

7. Chen, X.; Duan, N.; Zhang, C.; Zhang, W. Survivin and Tumorigenesis: Molecular Mechanisms and Therapeutic Strategies. *J. Cancer* **2016**, *7*, 314–323. [CrossRef]

8. Kanwar, J.R.; Kamalapuram, S.K.; Kanwar, R.K. Survivin Signaling in Clinical Oncology: A Multifaceted Dragon. *Med. Res. Rev.* **2013**, *33*, 765–789. [CrossRef]

9. Garg, H.; Suri, P.; Gupta, J.C.; Talwar, G.P.; Dubey, S. Survivin: A unique target for tumor therapy. *Cancer Cell Int.* **2016**, *16*, 1243. [CrossRef]

10. Dohi, T.; Okada, K.; Xia, F.; Wilford, C.E.; Samuel, T.; Welsh, K.; Marusawa, H.; Zou, H.; Armstrong, R.; Salvesen, G.S.; et al. An IAP-IAP Complex Inhibits Apoptosis. *J. Boil. Chem.* **2004**, *279*, 34087–34090. [CrossRef]

11. Gritsko, T.; Williams, A.; Turkson, J.; Kaneko, S.; Bowman, T.; Huang, M.; Nam, S.; Eweis, I.; Diaz, N.; Sullivan, D.; et al. Persistent Activation of Stat3 Signaling Induces Survivin Gene Expression and Confers Resistance to Apoptosis in Human Breast Cancer Cells. *Clin. Cancer Res.* **2006**, *12*, 11–19. [CrossRef]
12. Bromberg, J.F.; Wrzeszczynska, M.H.; Devgan, G.; Zhao, Y.; Pestell, R.G.; Albanese, C.; Darnell, J.E. Stat3 as an oncogene. *Cell* **1999**, *98*, 295–303. [CrossRef]
13. Hirsch, F.R.; Scagliotti, G.V.; Mulshine, J.L.; Kwon, R.; Curran, W.J.; Wu, Y.-L.; Paz-Ares, L. Lung cancer: Current therapies and new targeted treatments. *Lancet* **2017**, *389*, 299–311. [CrossRef]
14. Dias, D.A.; Urban, S.; Roessner, U. A Historical Overview of Natural Products in Drug Discovery. *Metabolites* **2012**, *2*, 303–336. [CrossRef]
15. Patridge, E.; Gareiss, P.; Kinch, M.S.; Hoyer, D. An analysis of FDA-approved drugs: Natural products and their derivatives. *Drug Discov. Today* **2016**, *21*, 204–207. [CrossRef]
16. Andersen, M.H.; Svane, I.M.; Becker, J.C.; Straten, P.T. The Universal Character of the Tumor-Associated Antigen Survivin. *Clin. Cancer Res.* **2007**, *13*, 5991–5994. [CrossRef]
17. Khan, S.; Aspe, J.R.; Asumen, M.G.; Almaguel, F.; Odumosu, O.; Acevedo-Martinez, S.; De Leon, M.; Langridge, W.H.R.; Wall, N.R. Extracellular, cell-permeable survivin inhibits apoptosis while promoting proliferative and metastatic potential. *Br. J. Cancer* **2009**, *100*, 1073–1086. [CrossRef]
18. Rodel, F.; Sprenger, T.; Kaina, B.; Liersch, T.; Rodel, C.; Fulda, S.; Hehlgans, S. Survivin as a prognostic/predictive marker and molecular target in cancer therapy. *Curr. Med. Chem.* **2012**, *19*, 3679–3688. [CrossRef]
19. Martínez-García, D.; Manero-Rupérez, N.; Quesada, R.; Korrodi-Gregório, L.; Soto-Cerrato, V. Therapeutic strategies involving survivin inhibition in cancer. *Med. Res. Rev.* **2018**, *39*, 887–909. [CrossRef]
20. Eckelman, B.P.; Salvesen, G.S.; Scott, F.L. Human inhibitor of apoptosis proteins: Why XIAP is the black sheep of the family. *EMBO Rep.* **2006**, *7*, 988–994. [CrossRef]
21. Zhu, H.; Zhang, G.; Wang, Y.; Xu, N.; He, S.; Zhang, W.; Chen, M.; Liu, M.; Quan, L.; Bai, J.; et al. Inhibition of ErbB2 by Herceptin reduces survivin expression via the ErbB2-β-catenin/TCF4-survivin pathway in ErbB2-overexpressed breast cancer cells. *Cancer Sci.* **2010**, *101*, 1156–1162. [CrossRef]
22. Xia, W. Regulation of Survivin by ErbB2 Signaling: Therapeutic Implications for ErbB2-Overexpressing Breast Cancers. *Cancer Res.* **2006**, *66*, 1640–1647. [CrossRef]
23. Duan, Z.; Bradner, J.E.; Greenberg, E.; Levine, R.; Foster, R.; Mahoney, J.; Seiden, M.V. SD-1029 Inhibits Signal Transducer and Activator of Transcription 3 Nuclear Translocation. *Clin. Cancer Res.* **2006**, *12*, 6844–6852. [CrossRef]
24. Carpenter, R.L.; Lo, H.-W. STAT3 Target Genes Relevant to Human Cancers. *Cancers* **2014**, *6*, 897–925. [CrossRef]
25. Bowman, T.; Garcia, R.; Turkson, J.; Jove, R. STATs in oncogenesis. *Oncogene* **2000**, *19*, 2474–2488. [CrossRef]
26. Jiang, R.; Jin, Z.; Liu, Z.; Sun, L.; Wang, L.; Li, K. Correlation of Activated STAT3 Expression with Clinicopathologic Features in Lung Adenocarcinoma and Squamous Cell Carcinoma. *Mol. Diagn. Ther.* **2011**, *15*, 347–352. [CrossRef]
27. Liu, B.; Palmfeldt, J.; Lin, L.; Colaço, A.; Clemmensen, K.K.B.; Huang, J.; Xu, F.; Liu, X.; Maeda, K.; Luo, Y.; et al. STAT3 associates with vacuolar H$^+$-ATPase and regulates cytosolic and lysosomal pH. *Cell Res.* **2018**, *28*, 996–1012. [CrossRef]
28. Soto-Cerrato, V.; Manuel-Manresa, P.; Hernando, E.; Calabuig-Fariñas, S.; Martínez-Romero, A.; Dueñas, V.F.; Sahlholm, K.; Knopfel, T.; García-Valverde, M.; Rodilla, A.M.; et al. Facilitated Anion Transport Induces Hyperpolarization of the Cell Membrane that Triggers Differentiation and Cell Death in Cancer Stem Cells. *J. Am. Chem. Soc.* **2015**, *137*, 15892–15898. [CrossRef]
29. Premkumar, D.R.; Jane, E.P.; Pollack, I.F. Cucurbitacin-I inhibits Aurora kinase A, Aurora kinase B and survivin, induces defects in cell cycle progression and promotes ABT-737-induced cell death in a caspase-independent manner in malignant human glioma cells. *Cancer Biol. Ther.* **2015**, *16*, 233–243. [CrossRef]
30. Stella, S.; Tirro, E.; Conte, E.; Stagno, F.; Di Raimondo, F.; Manzella, L.; Vigneri, P. Suppression of Survivin Induced by a BCR-ABL/JAK2/STAT3 Pathway Sensitizes Imatinib-Resistant CML Cells to Different Cytotoxic Drugs. *Mol. Cancer Ther.* **2013**, *12*, 1085–1098. [CrossRef]
31. Aoki, Y.; Feldman, G.M.; Tosato, G. Inhibition of STAT3 signaling induces apoptosis and decreases survivin expression in primary effusion lymphoma. *Blood* **2003**, *101*, 1535–1542. [CrossRef]

32. Yan, X.; Li, P.; Zhan, Y.; Qi, M.; Liu, J.; An, Z.; Yang, W.; Xiao, H.; Wu, H.; Qi, Y.; et al. Dihydroartemisinin suppresses STAT3 signaling and Mcl-1 and Survivin expression to potentiate ABT-263-induced apoptosis in Non-small Cell Lung Cancer cells harboring EGFR or RAS mutation. *Biochem. Pharmacol.* **2018**, *150*, 72–85. [CrossRef]

33. Garbers, C.; Aparicio-Siegmund, S.; Rose-John, S. The IL-6/gp130/STAT3 signaling axis: Recent advances towards specific inhibition. *Curr. Opin. Immunol.* **2015**, *34*, 75–82. [CrossRef]

34. Yu, H.; Jove, R. The STATs of cancer—New molecular targets come of age. *Nat. Rev. Cancer* **2004**, *4*, 97–105. [CrossRef]

35. Kumari, N.; Dwarakanath, B.S.; Das, A.; Bhatt, A.N. Role of interleukin-6 in cancer progression and therapeutic resistance. *Tumor Biol.* **2016**, *37*, 11553–11572. [CrossRef]

36. Yanagawa, H.; Sone, S.; Takahashi, Y.; Haku, T.; Yano, S.; Shinohara, T.; Ogura, T. Serum levels of interleukin 6 in patients with lung cancer. *Br. J. Cancer* **1995**, *71*, 1095–1098. [CrossRef]

37. Chang, C.H.; Hsiao, C.F.; Yeh, Y.M.; Chang, G.C.; Tsai, Y.H.; Chen, Y.M.; Huang, M.S.; Chen, H.L.; Li, Y.J.; Yang, P.C.; et al. Circulating interleukin-6 level is a prognostic marker for survival in advanced nonsmall cell lung cancer patients treated with chemotherapy. *Int. J. Cancer* **2013**, *132*, 1977–1985. [CrossRef]

 *biomolecules*

Article

Orientin Induces G0/G1 Cell Cycle Arrest and Mitochondria Mediated Intrinsic Apoptosis in Human Colorectal Carcinoma HT29 Cells

Kalaiyarasu Thangaraj [1,2], Balamuralikrishnan Balasubramanian [3], Sungkwon Park [3], Karthi Natesan [2,4], Wenchao Liu [5] and Vaiyapuri Manju [2,*]

[1] Department of Microbiology and Biotechnology, Bharath Institute of Higher Education and Research, Tamilnadu 600045, India

[2] Molecular Oncology Lab, Department of Biochemistry, Periyar University, Tamilnadu 636011, India

[3] Department of Food Science and Biotechnology, College of Life Science, Sejong University, Seoul 05006, Korea

[4] Genomic Division, National Academy of Agricultural Science, RDA, Jeollabuk 560500, Korea

[5] Department of Animal Science, College of Agriculture, Guangdong Ocean University, Zhanjiang 524088, China

* Correspondence: manjucb11@gmail.com

Received: 1 August 2019; Accepted: 26 August 2019; Published: 27 August 2019

Abstract: Colorectal carcinoma is one of the utmost diagnosed cancer with a steep increase in mortality rate. The incidence has been increasing in developing countries like India due to a westernization life style. Flavonoids have been explored widely for its various pharmacological activity including antitumor activity. Orientin, an analogue of luteolin (citrus flavonoid) isolated from rooibos and tulsi leaves is also expected to deliver significant antitumor activity similar to that of luteolin. The present study anticipates exploring the antitumor activity of orientin against colorectal carcinoma cells (HT29). Orientin exhibited remarkable cytotoxicity and antiproliferative activity against HT29 cells, which is clearly evident from tetrazolium based cytotoxicity and lactate dehydrogenase release assays. Orientin induce G0/G1 cell cycle arrest and regulates cyclin and cyclin-dependent protein kinases in order to prevent the entry of the cell cycle to the S phase. Annexin V-FITC (V-Fluorescein Isothiocyanate) dual staining reveals the apoptotic induction ability of orientin. The Bcl-2 family proteins along with the inhibitor of apoptotic proteins were regulated and the tumor suppressor p-53 expression have been decreased. In conclusion, our results proposed that orientin could be a potent chemotherapeutic agent against colorectal cancer after ascertaining their molecular mechanisms.

Keywords: colorectal cancer; orientin; cell cycle arrest; Bcl-2 family proteins; apoptosis

1. Introduction

Colorectal cancer (CRC) is the third most recurrently diagnosed cancer and fourth most likely cause of cancer mortality worldwide with increasing incidence in recent years [1]. The incidence of CRC accounts for more than 8% of total cancer incidence among the cancers that occurs in both men and women [2]. Despite the rapidity in drug development, antagonistic screening and promoted public health awareness, the global burden of CRC is anticipated to rise almost 60% by 2030 [3]. Therefore, much more effective therapeutic strategies are requisite in the present era in the treatment of CRC. In accordance with this, numerous reports have highlighted the role of flavonoids in inducing apoptotic signaling pathways and thereby thwarting the progression of CRC. A meta-analysis reveals that, the increase in consumption of few dietary flavonoids highly correlated with a decreased risk of colon and rectal cancer [4]. Flavonoids such as quercetin, luteolin, kaempferol, apigenin, epigallocatechin, hesperetin, naringenin and pelargonidin have been reported earlier to possess a protective effect against the development of recurrent adenomas and colon cancer [5]. The heterogeneous CRC involves multiple cascades of events accelerated by an array of oncogenes and tumor suppressor genes in the transformation of colonic epithelium into metastatic carcinoma [6]. The cell cycle includes an array of events that drives the proliferation of cells through a sequence of checkpoints that repairs DNA damage, genetic derangements and incomplete DNA replication. The cell cycle checkpoints protect transformed cells under genotoxic and replicative stress, thereby shielding the integrity of the genome [7]. The cell cycle progression involves the heterodimeric interaction of cyclins and cyclin-dependent protein kinases (CDK), which regulate the activities of target proteins by phosphorylation.

Reactive oxygen species (ROS), a heterogeneous group associates in the regulation of many cellular physiological and pathological processes. Mitochondrial respiration is the prime endogenous source of ROS and almost 90% of ROS are generated by mitochondria [8]. The enhanced production of intracellular ROS can pose a serious menace to cells by causing lipid peroxidation, protein oxidation, damage to nucleic acids and modulation of cellular functions such as proliferation, differentiation, growth and activation of programmed cell death pathway and ultimately leading to cell death. One of the possible mechanisms could be the intervention in the intracellular signaling pathways regulating cell survival and apoptosis [9]. Apoptosis plays a pivotal role in cellular proliferation, differentiation, senescence and death with typical characteristics of membrane blebbing, shrinkage of nucleus, condensation of chromatin, fragmentation of DNA and formation of apoptosis [10]. The programmed cell death process involving a highly complex cascade of cellular events occurs in either way, intrinsic and extrinsic depending upon the trigger of the death inducing signals [11]. Meanwhile, both signaling pathways induce apoptotic effector molecules to induce apoptosis. The apoptotic mechanism is orchestrated by the evolutionarily conserved B-cell leukemia/lymphoma 2 (Bcl-2) family proteins such as pro-apoptotic (e.g., Bax and Bak) and anti-apoptotic (e.g., Bcl-2 and Bcl-XL). Indeed, the ratio between Bcl-2 and Bax helps decide in part, the cell susceptibility to death inducing signals [12]. The pro-apoptotic proteins act in such a way that mitochondria permeabilizes, releases cytochrome C and subsequently activates the caspase cascade to trigger apoptosis mediated by mitochondria [13]. The death inducing ligands bind specifically to the cell surface receptors and instigate extrinsic apoptosis through intracellular activation of caspase-8 resulting in cell death [9]. Several reports have been demonstrated that the Bcl-2 and Bcl-XL are widely over expressed in different forms of cancers [14]. Therefore, the development of anticancer drugs to inhibit antiapoptotic proteins of Bcl-2 family serves to be a promising strategy for treating cancer [15].

Orientin (luteolin-8-C-glucoside), a water soluble glycosyl flavonoid rich in rooibos tea, isolated from *Ocimum sanctum*, *Phyllostachys* species, *Passiflora* species and *Trollius* species [16]. It exerts various pharmacological activities such as antioxidant, anti-inflammatory, neuroprotective, cardioprotective and antitumor effects [17,18]. Earlier studies reported that orientin exerts cytotoxicity in esophageal cancer EC109 cells [19] and MCF-7 breast cancer cells [20]. However, there is limited information on the effect of orientin against CRC in vitro and the putative mechanisms of cytotoxicity induced by orientin also remain unknown. The present study investigates the influence of orientin on proliferation, cell cycle arrest and apoptosis in human CRC cells (HT29) and to explore the underlying mechanisms involved in the pharmacological actions of orientin.

2. Materials and Methods

2.1. Materials and Reagents

Orientin, RPMI-1640 medium, DMSO, antibiotic antimycotic solution, trypsin–EDTA solution and MTT dye were procured from Sigma Chemicals, MO. FBS was purchased from Gibco-BRL, MD. The primary antibodies anti-Bcl-2 (#15071S), Bax (#2772S), Bcl-xL (#2764S), Bid (#2003S), procaspase-3 (#9662S), cleaved caspase-3 (#9661S), procaspase 9(#9508S), cleaved caspase 9 (#9509S), cytochrome C (#11940S), Smac/DIABLO (#2954S), AIF(#4642S), p21 (#2946S), p53 (#9282S), p-Rb (#9307S) or p-H2AX (#2577S) were procured from Cell Signaling Technology, MA. Anti-cyclin B1 (sc-245), CDK1 (sc-53219), CDC2 (sc-54), MDM2 (sc-965), PARP (sc-56196), cleaved PARP (sc-56196), X-linked inhibitor of apoptotic proteins (XIAP; sc-55550), survivin (sc-17779) or β-actin (sc-47778) antibodies, and HRP conjugated secondary antibodies (sc-2359) were purchased from Santa Cruz Biotechnology, CA. All further chemicals used in this study were of reagent or analytical grade and obtained from commercial suppliers.

2.2. Cell Culture Maintenance and Treatment

The HT29 cell lines were procured from National Center for Cell Sciences, India. Cells were cultured in RPMI-1640 medium supplemented with 10% FBS and 2-mM L-glutamine, 100 U/mL antibiotic antimycotic solution and maintained at 37 °C in CO_2 (5%) incubator with 95% humidity. Orientin stock solution was prepared in DMSO (0.1%) and stored at −20 °C until use.

2.3. Tetrazolium Based Cell Viability Assay

The HT29 colon cells were treated with orientin and irinotecan (3.125 to 100 μM). Cell viability after 24 h was determined based on a MTT assay. Briefly, HT29 cells (3×10^3 cells/well) were seeded in a 96-well plate and left overnight to get adhere. After removal of the medium, 200 μL of fresh medium added per well, containing 10 mmol/L HEPES (pH 7.4). 50 μL MTT was added and the plate was incubated for 2–4 h at 37 °C in the dark. After removal of spent medium, DMSO (200 μL) and Sorensen's glycine buffer (25 μL) were added to the wells. The absorbance at 570 nm was determined using an ELISA plate reader (BioRad, Richmond, CA, USA). Meanwhile, the cytotoxicity of orientin on normal epithelial cells, Vero (normal kidney epithelial cell), was also evaluated.

2.4. Morphological Observation and Cell-Cycle Analysis

The HT29 cells (4×10^3 cells/coverslip) were grown and treated (24 h) with orientin (3.125 to 100 μM) and further dissolved in methanol/acetic acid solution (3:1, *v/v*). The effect of orientin HT29 cells morphology was observed by bright-field inverted light microscope (400×). The cell-cycle distribution was analyzed by flow cytometry as described earlier [21].

2.5. Annexin V-FITC/PI Apoptotic Assay

Annexin V-fluorescein isothiocyanate/propidium iodide dual staining was employed to distinguish early and late apoptotic cells (BD Pharmingen, San Jose, CA, USA). Fluorescein isothiocyanate -conjugated Annexin V was used to quantify the loss of asymmetry of phosphatidylserine on cell membranes involved in apoptosis while propidium iodide differentiates the early apoptotic, the late apoptotic and necrotic cells [22]. Briefly, HT29 cells (3×10^5 cells/well) were seeded in a 6-well plate and treated with the orientin for 48 h. Cells were gently washed twice with phosphate buffered saline accompanied by tryptic digestion (0.25% trypsin/EDTA) and washed yet again with PBS. All the cells including the floating and adherent ones were harvested, pooled and incubated with dual stain for 15 min on ice under dark. The cells were counted by a flow cytometer (Becton-Dickinson, San Jose, CA, USA).

2.6. Measurement of Intracellular ROS

The intracellular ROS was measured based on its conversion of 2′,7′-Dichlorodihydrofluorescein diacetate in to fluorescent DCFH. HT29 cells (1×10^5 cells/mL) were seeded in black colored bottom plate with 96 wells and allowed to adhere overnight. After being treated with orientin (6.25, 12.5 and 25 μM), the cells were washed twice with PBS. Then the cells were incubated with 20 μM of DCFH-DA solution at 37 °C for 30 min and suspended in 200 μL of PBS. The qualitative analysis of ROS generation was carried out by a fluorescence microscope (Nikon Eclipse, Tokyo, Japan) with an excitation filter around 510–590 nm (40×) and estimated with Image J software (Version 1.45, NIH, Bethesda, MD, USA).

2.7. Western Blotting

Briefly, HT29 cells (1×10^6 cells/mL) were treated for 24 h with Orientin (6.25, 12.5 and 25 μM). Cells were then washed with PBS and lysed using ice-cold RIPA (radio immuno precipitation assay) buffer. The separation of protein was carried out by subjecting samples on to 12% or 16% resolving polyacrylamide gels and electroblotting (25 mA) for 1 h at room temperature. After transferring the proteins to PVDF (polyvinylidene fluoride) membrane (Pierce, Rockford, IL, USA), the membranes were blocked for 1 h with a blocking buffer at room temperature. The primary antibodies namely anti-Bax (1:250), Bcl-2 (1:500), Bcl-xL (1:500), Bid (1:1000), procaspase-9 (1:1000), cleaved caspase-9 (1:1000), procaspase-3 (1:2000), cleaved caspase-3 (1:2000), mitochondrial cytochrome C, cytosolic cytochrome C, mitochondrial Smac/DIABLO (1:1000), cytosolic Smac/DIABLO, cyclin D1 (1:1000), cyclin E (1:1000), CDK2 (1:500), CDK4 (1:500), PARP (1:1000), cleaved PARP (1:1000), XIAP, Survivin, p-Rb (1:500), p53 (1:500), p-21$^{\text{WAF1/CIP1}}$ (1:1000), p-H2AX (1:1000), COX-IV (1:1000) or β-actin (1:5000), were incubated at 4 °C overnight. HRP-conjugated polyclonal secondary antibody (1:3000) and Chemiluminescence Plus detection components were used.

2.8. Statistical Analysis

All the experimental data were evaluated using SPSS 16.0 and expressed as mean ± SD. One way ANOVA was carried out and the probability values of <0.0001, <0.01 and <0.05 were considered as significant.

3. Results

3.1. Orientin Exhibits Cytotoxicity in HT29 Cells

Orientin and irinotecan (CPT-11) treated and untreated control HT29 cells were studied for determining the inhibition of cell viability using the tetrazolium based cytotoxicity assay. In addition, the cytotoxicity of orientin was also assessed in normal epithelial Vero cells. The percentage of viability was calculated at different concentrations (3.125 to 100 μM) for 24 h. Orientin exhibited potent dose dependent antiproliferative effect against HT29 cells as shown in Figure 1. Irinotecan (CPT-11) acting as the positive control demonstrated significant anti-proliferative potential against HT-29 cells. GI50 (50% growth inhibition) of orientin and irinotecan was found to be 12.55 and 5.19 μM, respectively, whereas in normal cells, no such cytotoxic effect was observed with the selected doses. The experimental findings revealed a potent inhibitory effect of orientin on carcinogenic HT29 cells without toxicity in normal epithelial Vero cells. The tetrazolium based antiproliferative study revealed that orientin significantly inhibits HT29 cell viability in a dose-dependent mode.

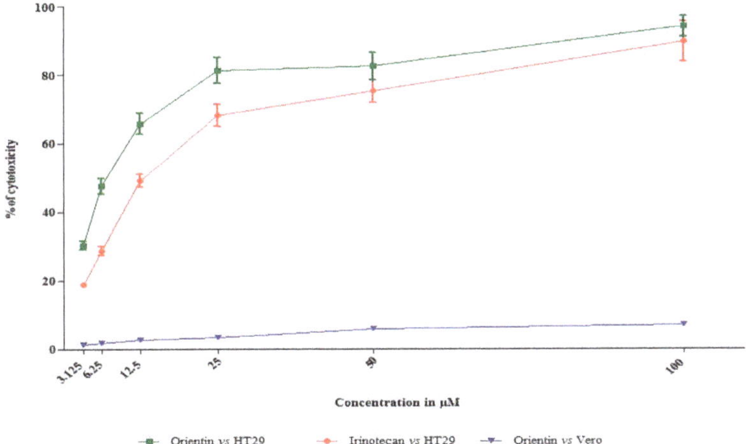

Figure 1. Effect of orientin on colorectal carcinoma and normal epithelial cell viability. Change in HT29 and normal epithelial cell viability with treatment of different concentrations of orientin and irinotecan were observed. The results were represented as mean ± SD of three independent parallel measurements.

In additional to quantify the extent of cellular toxicity in HT29 cells, a lactate dehydrogenase release assay was performed. Orientin significantly constrained the growth of HT29 cells and rendered cytotoxicity, which was clearly evident by the increasing amount of lactate dehydrogenase released in a dose dependent mode (Figure 2).

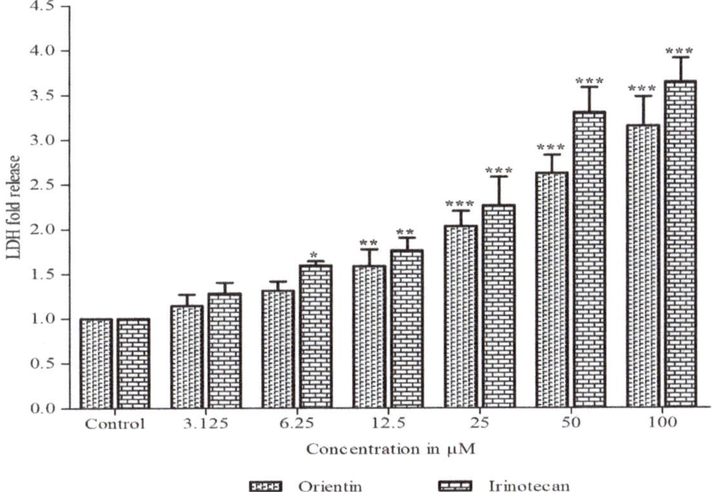

Figure 2. Effect of orientin on lactate dehydrogenase (LDH) release. The release of LDH in HT-29 cells treated with different concentration of orientin and irinotecan for 24 h. Results are expressed as the mean ± SD of three independent experiments. * $p < 0.05$, ** $p < 0.01$ and *** $p < 0.001$ vs. control.

3.2. Morphological Changes Induced in HT29 Cells by Orientin

The morphology of HT29 cells treated with orientin (3.125–100 µM) was observed by inverted light microscopy. The untreated cells were found to be healthy and polyhedral in shape with a distinct cytoskeleton. Orientin treated cell loss their normal architecture, found to be rounded and shrunken in nature. The increased number of detached cells with the increasing concentration reveals the apoptotic effect of orientin. The light microscopic observations showed typical variations in cell morphology after 24 h exposure (Figure 3). This includes the cell shrinkage from its polyhedral origin, rounded off, membrane blebbing and detachment of cells making substantiation for apoptosis [21]. The overall findings of cytotoxic assays and morphological observations obviously state that orientin exerted a significant antiproliferative effect against HT29 cells dose dependently. Thus, based on the cytotoxicity and morphological observations, further experiments were carried out at 12.5 µM (GI50) and 6.25 µM (sub GI50) concentrations of orientin and compared with 5 µM (GI50) of standard irinotecan.

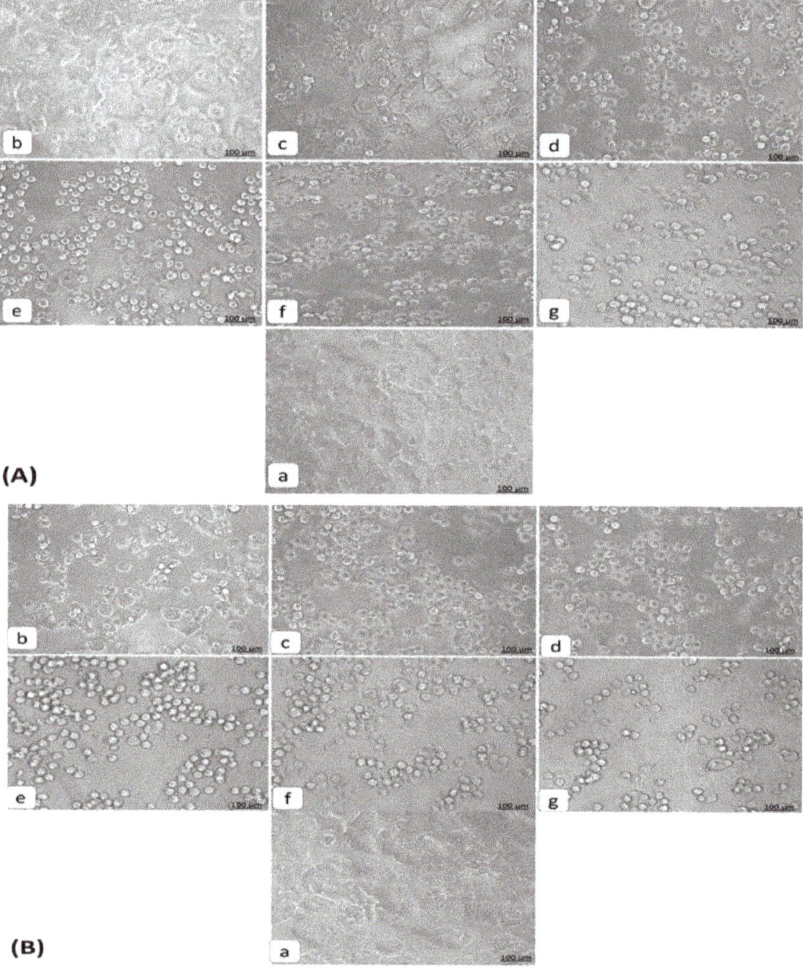

Figure 3. Morphological changes observed in HT29 cells on exposure to orientin and irinotecan. HT29 cells treated with different concentrations of orientin (**A**) and irinotecan (**B**) were observed after 24 h by inverted bright field microscopy (100 µm). (**a**) Untreated control, (**b**) 3.125 µM, (**c**) 6.25 µM, (**d**) 12.5 µM, (**e**) 25 µM, (**f**) 50 µM and (**g**) 100 µM. The treated cells were found to be shrunken, rounded off and detached from the layer.

3.3. Initiation of the G0/G1 Phase Cell Cycle Arrest by Orientin

The molecular mechanism of orientin induced anti-proliferation in HT29 cells was determined by their effects on cell cycle progression using PI. Flow cytometry analyses of cell cycle distribution in (6 μM and 12.5 μM) orientin and irinotecan treated cells were illustrated in Figure 4. Orientin was shown to block the cell progression at G0/G1 phase itself. After treatment with orientin, a substantial increase in G0/G1 phase population was observed when compared with control cells. Our data revealed that the percentage of untreated cells in the sub-G1, G0/G1, S and G2/M phases were determined to be 8.39%, 35.56%, 15.31% and 13.03%, respectively; whereas GI50 of orientin treated cells were 3.03%, 78.21%, 16.49% and 2.55%, respectively (Figure 4). The results suggest that orientin arrest at G0/G1 phase and the cell population was increased and this could be responsible for the anti-proliferation of HT29 cells.

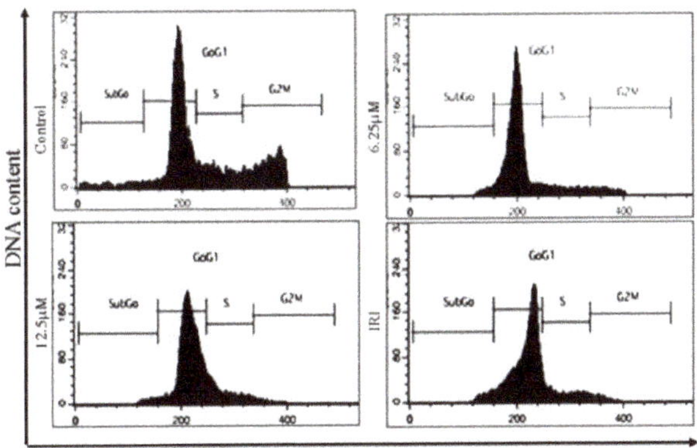

Figure 4. Effect of orientin on cell cycle phase distribution in HT29 cells. HT-29 cells were treated with different concentrations of orientin and analyzed after 24 h for DNA content by flow cytometry. Histogram representing the PI (propidium iodide) staining (FL2-A) of orientin treated cells. The data shown are representative of three independent experiments with similar findings. The significant differences from control were indicated by * $p < 0.05$, ** $p < 0.01$ and *** $p < 0.0001$ (considered as statistically significant).

3.4. Orientin Induced p21$^{WAF1/CIP1}$ Mediated G0/G1 Arrest in HT29 Cells

To determine the probable mechanism underlying the cell cycle arrest by orientin in HT29 cells, the change in the regulatory proteins of cell cycle such as cyclins (D1 and E), CDK (CDK2 and CDK4), p21$^{WAF1/CIP1}$ and p-Rb at the G0/G1 phase was observed using an immunoblotting analysis. As shown in Figure 5, orientin significantly (* $p < 0.05$, ** $p < 0.01$ and *** $p < 0.001$) decreased the expression of cyclins, CDK and concomitantly increased the level of cyclin D-CDK complexes inhibitor, p21[WAF1/CIP1] dose dependently. The decreased level of cyclin D-CDK2 and cyclin E-CDK-4 complexes made evident orientin induced cell cycle arrest at G0/G1 phase in HT-29 cells. p-Rb, the primary target of the G1 kinases was also found to be significantly decreased dose dependently upon treatment with orientin (** $p < 0.01$ and *** $p < 0.001$).

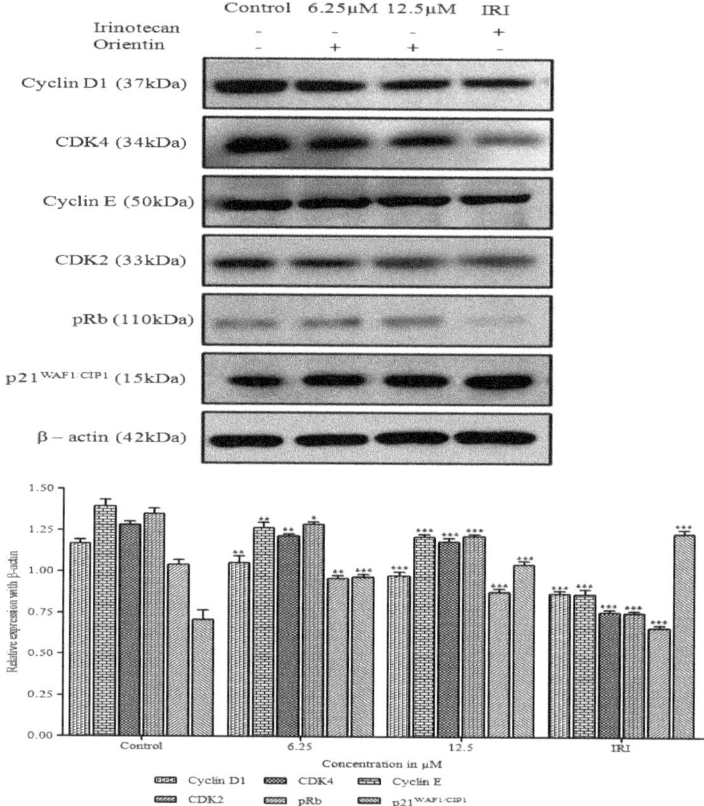

Figure 5. Effect of orientin on cell cycle regulatory proteins. HT29 cells were treated with orientin (6.25 and 12.5 µM) and irinotecan for 24 h. The immunoblot analysis of cyclins, CDKs and their inhibitors involved in G0/G1 phase were carried out. Orientin reduced the expression of cyclins, pRb and CDKs, in contrast, increased the level of p21[WAF1/CIP1]. β-actin was used as an internal control. Quantitative expression of proteins shown after normalization to β-actin. The data presented are the mean ± SD of results from three independent experiments (* $p < 0.05$, ** $p < 0.01$ and *** $p < 0.001$ vs. control).

To determine the probable mechanism underlying the cell cycle arrest by orientin in HT29 cells, the change in the regulatory proteins of cell cycle such as cyclins (D1 and E), CDK (CDK2 and CDK4), p21[WAF1/CIP1] and p-Rb at the G0/G1 phase was observed using an immunoblotting analysis. As shown in Figure 5, orientin significantly (* $p < 0.05$, ** $p < 0.01$ and *** $p < 0.001$) decreased the expression of cyclins, CDK and concomitantly increased the level of cyclin D-CDK complexes inhibitor, p21[WAF1/CIP1] dose dependently. The decreased level of cyclin D-CDK2 and cyclin E-CDK-4 complexes made evident orientin induced cell cycle arrest at the G0/G1 phase in HT-29 cells. p-Rb, the primary target of the G1 kinases was also found to be significantly decreased dose dependently upon treatment with orientin (** $p < 0.01$ and *** $p < 0.001$).

3.5. Orientin Induces Apoptosis in HT29 Cells

To further substantiate that their anti-proliferative activities occur due to apoptosis, the quantitative assessment of apoptosis was determined by Annexin V-FITC/PI staining by flow cytometry (Figure 6). HT29 cells treated with orientin for 24 h were stained with Annexin-V and PI consecutively. The live cell population and the cells undergoing early apoptosis (Annexin$^+$/PI$^-$), late apoptosis (Annexin$^+$/PI$^+$) and necrosis (Annexin$^-$/PI$^+$) were quantified. The findings revealed that orientin treated HT29 cells underwent apoptosis and showed a significant ($p < 0.05$) decrease in the proportion of live cells and enhanced the translocation of phosphatidyl serine (Annexin V positive cells) dose dependently. The percentage of cells undergoing early and late apoptosis on exposure to orientin at 6 and 12.5 µM were found to be 11%, and 36.9%, respectively, which is comparable to that of irinotecan (44.7%). However, the level of cells undergoing necrosis was limited making obvious that orientin exhibited an anti-proliferative effect with the increasing rate of apoptosis.

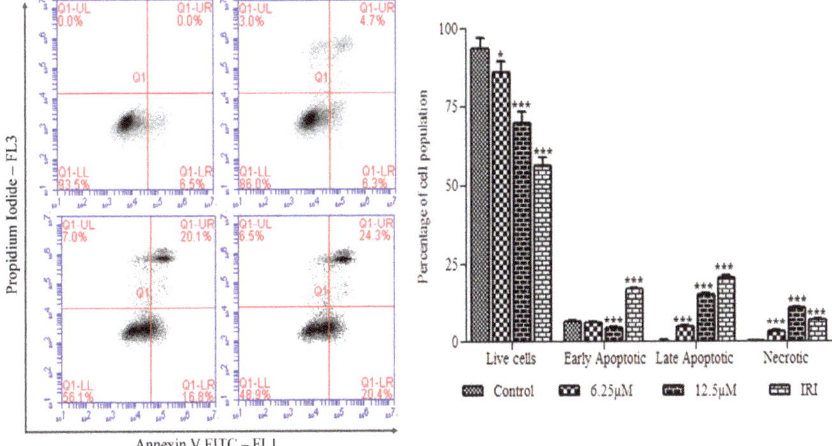

Figure 6. Effect of orientin on apoptosis in HT29 cells. Cells were treated with 6.25 and 12.5 µM orientin and irinotecan for 24 h, and the induction of cell apoptosis was examined by flow cytometry. LR represents early apoptotic cells and UR represents late apoptotic cells. Percentage of early and late apoptotic cells is shown in the bar graph. Data are shown as mean ± SD of three parallel independent experiments. * $p < 0.05$ and *** $p < 0.001$ vs. control.

3.6. Intracellular Accumulation of ROS by Orientin in HT29 Cells

The enhancement of oxidative stress by triggering intracellular ROS generation could be a potential therapeutic strategy for treating colon cancer. Although, ROS are involved in multiple signaling cascades of tumor development, its excessive generation could lead to causing severe harm to DNA and proteins, leading to apoptosis [23]. To verify whether orientin induces apoptosis by triggering ROS generation in HT29 cells, treated cells were stained with CM-H2DCFDA, a non-fluorescent derivative of 2,7-dichlorofluorescein. Orientin significantly increased (** $p < 0.01$ and *** $p < 0.001$) intracellular ROS levels in HT29 cells in a dose dependent manner as revealed by the increase in intensity (Figure 7). The increased intracellular accumulation of ROS in treated cells was observed at both 6.25 µM and 12.5 µM similar to that of irinotecan.

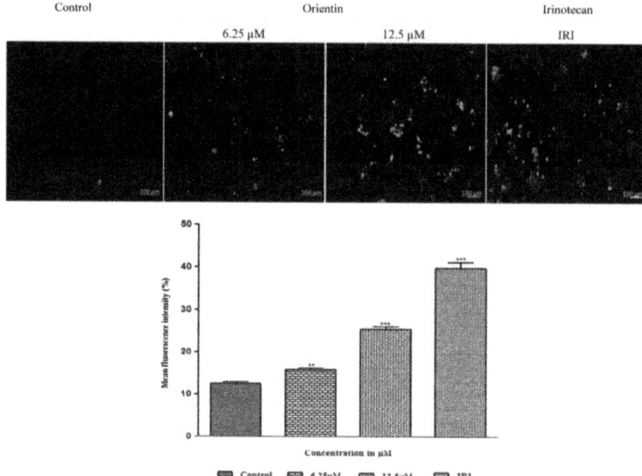

Figure 7. Intracellular reactive oxygen species (ROS) accumulation in HT29 cells by orientin. The fluorescent microscopic images of intracellular ROS generated in HT-29 cells treated with different concentrations of orientin and standard irinotecan for 24 h. The fluorescent images were taken with a fluorescence microscope and analyzed with Image J software. Data are shown as mean ± SD of three parallel independent experiments. ** $p < 0.01$ and *** $p < 0.001$ vs. control.

3.7. Orientin Modulates Bcl-2 Family Proteins

The mechanisms underlying orientin-induced apoptosis in HT29 cells was ascertained by the western blot analysis of apoptosis-related proteins. After being treated with 6.25 and 12.5 µM orientin, the expression of Bcl-2 and Bcl-XL proteins were significantly reduced in a concentration dependent manner (* $p < 0.05$, ** $p < 0.01$ and *** $p < 0.001$, respectively). The proapoptotic Bax and Bid was significantly increased in HT29 cells in a concentration dependent manner (** $p < 0.01$ and *** $p < 0.001$) when compared to control. As shown in Figure 8, the reduced Bcl-XL expression accompanied by increased expression of Bax and Bid illustrates the loss of mitochondrial membrane potential in orientin treated HT29 cells.

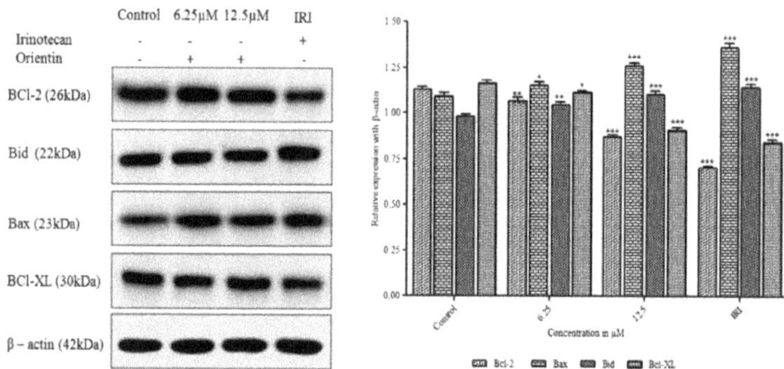

Figure 8. Effect of orientin on expression of Bcl-2 family proteins. HT-29 cells were treated with orientin (6.25 and 12.5 µM) and irinotecan for 24 h, respectively. Orientin increases proapoptotic Bax and Bid proteins and decreases the expression of anti-apoptotic Bcl-2 and Bcl-XL proteins. β-actin was used as an internal control. Quantitative expression of proteins shown after normalization to β-actin. Data are represented as the mean ± SD of three independent experiments (* $p < 0.05$, ** $p < 0.01$ and *** $p < 0.001$ vs. control).

3.8. Cytochrome C Release and Translocation of Smac/DIABLO by Orientin

As shown in Figure 9, orientin significantly induced the release of mitochondrial cytochrome C to cytosol in a concentration dependent manner (*** $p < 0.001$). The mitochondrial cytochrome C was reduced significantly when compared with untreated cells (*** $p < 0.001$), in contrast, cytosolic cytochrome C was gradually increased with an increase in concentration. The pro-apoptotic mitochondria derived Smac/DIABLO protein was also found to be translocated from mitochondria to cytosol significantly (*** $p < 0.001$). Our results, demonstrated that orientin induced disintegration of mitochondrial transmembrane potential in HT29 cells and pore formation might be due to the dimerization or activation of pro-apoptotic proteins thereby accompanying the release of caspases that were prominently responsible for apoptosis.

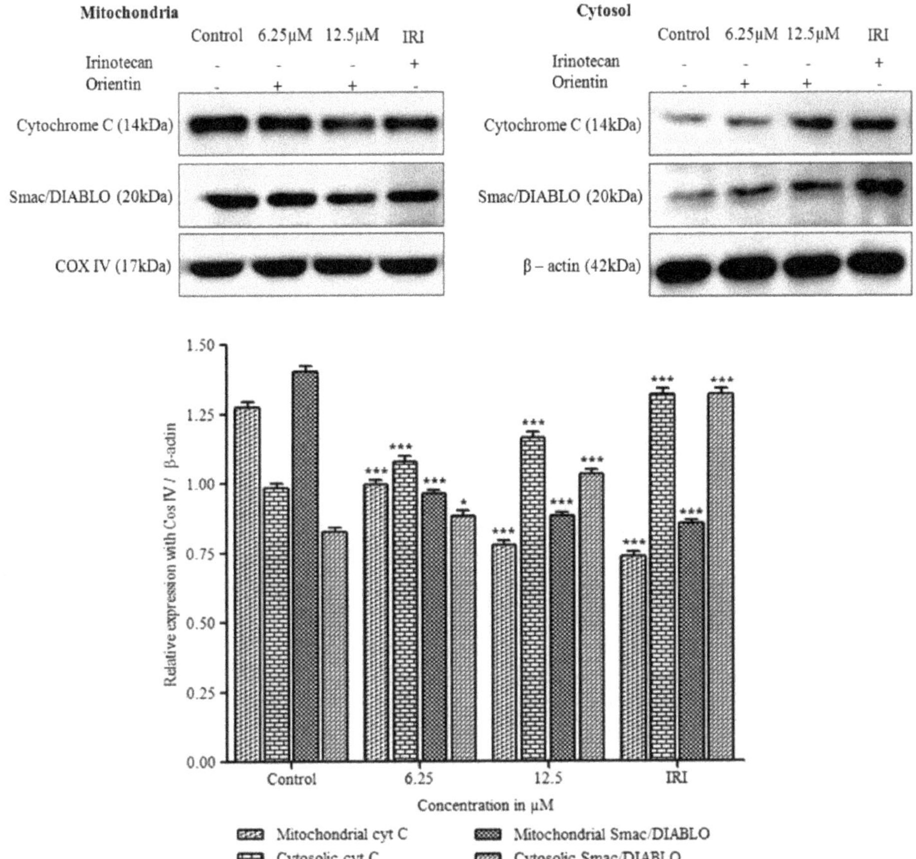

Figure 9. Effect of orientin on cytochrome C release and Smac/DIABLO protein. Orientin induces the translocation of cytochrome C and Smac/DIABLO from mitochondria to cytosol in HT29 cells. β-actin was used as an internal control. Quantitative expression of proteins has shown after normalization to β-actin. Values are represented as the mean ± SD of three independent experiments (* $p < 0.05$, ** $p < 0.01$ and *** $p < 0.001$ vs. control).

3.9. Orientin Activates Caspase Cascade and Induces PARP Cleavage

Apoptosis can be either intrinsic or extrinsic depending upon the death receptors activation or the release of mitochondrial membrane potential regulatory proteins. Cytochrome C release and increased cytosolic Smac/DIABLO indicates the caspase cascade activation and inhibition of apoptotic proteins. To determine the pathway of orientin triggered apoptosis, the immunoblot analysis of caspase-3 (initiator caspases), caspase-9 (effector caspases) and their cleaved forms were carried out.

As shown in Figure 10, caspase-3 and caspase-9 were found to be increased dose dependently in orientin treated cells (* $p < 0.05$ and ** $p < 0.001$), apparently, cleaved caspase-3 and caspase-9 were also found to be increased gradually with increasing concentration. Further, examined the effect of orientin on PARP cleavage, the hallmark of apoptosis. The expression of cleaved PARP increased significantly (* $p < 0.05$ and ** $p < 0.001$) in a concentration dependent manner after orientin treatment. These results suggested that orientin induced intrinsic apoptosis via caspase activation and cleavage of PARP.

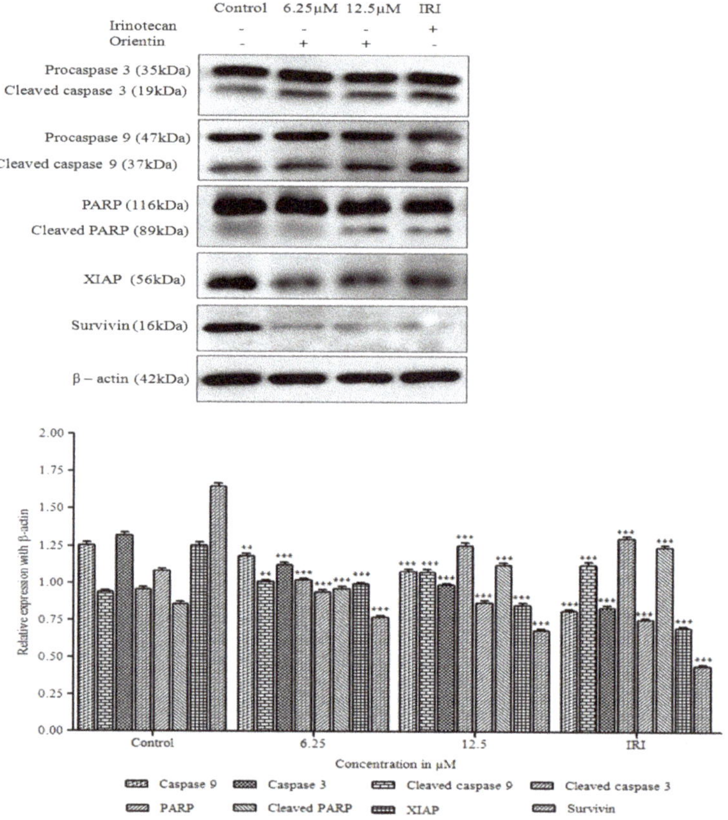

Figure 10. Effect of orientin on caspase cascade and PARP cleavage. HT29 cells treated with orientin activates caspase cascade and subsequently induces PARP cleavage. A marked reduction was observed in the expression of X-linked inhibitor of apoptotic proteins (XIAP) and survivin after treatment with orientin. β-actin was used as an internal control. Quantitative expression of proteins has shown after normalization to β-actin. Values are represented as the mean ± SD of three independent experiments (* $p < 0.05$, ** $p < 0.01$ and *** $p < 0.001$ vs. control).

3.10. Orientin Blocks the Inhibitor of Apoptotic Proteins (IAP)

To determine the effect of orientin on inhibitor of apoptotic proteins (IAP), the analysis of expression of X-linked (X) IAP and survivin, which regulates apoptosis by the inhibition of caspases, were performed. A marked reduction was observed in the expression of XIAP and survivin after treatment with orientin (Figure 11).

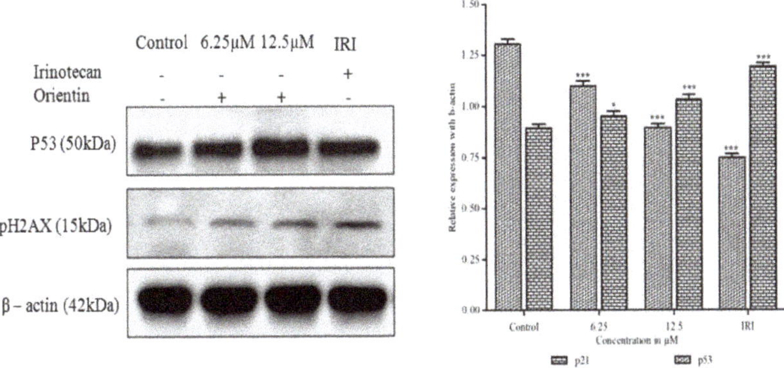

Figure 11. Effect of orientin on p53 and DNA damage. HT29 cells have shown a decrease in the expression of p53 and increase in the expression of pH2AX protein involved in DNA repair. β-actin was used as an internal control. Quantitative expression of proteins has shown after normalization to β-actin. Values are represented as the mean ± SD of three independent experiments (* $p < 0.05$, ** $p < 0.01$ and *** $p < 0.001$ vs. control).

These results suggested that orientin induced apoptosis by blocking the IAP and thereby activating mitochondria mediated caspase dependent intrinsic apoptosis.

3.11. Orientin Induces p53 Expression and DNA Damage

To examine whether orientin could inhibit p53 activation and DNA damage, the expression of p53 and γ-H2AX were assessed by western blotting analysis. The tumor suppressor, p53, was found to be significantly increased after treatment with orientin in a dose dependent manner (* $p < 0.05$ and ** $p < 0.001$) as shown in Figure 11. The increased expression of tumor suppressor p53 after treatment with orientin suggests that p53 could induce the over expression of p21$^{WAF1/CIP1}$, which further inhibited cyclin-CDK complexes activation. The increased level of γ-H2AX with increasing concentration serves as a hallmark of DNA damage after treating with orientin confirmed the DNA damage in HT29 cells.

4. Discussion

Orientin has been reported to exhibit potent anti-proliferative effect against EC109 and MCF-7 cells [24], however, the antiproliferative mechanism is not precisely understood. In this regard, investigated the inhibitory effect of orientin on HT29 cells and the possible molecular mechanism was explicated. The tetrazolium based antiproliferative study revealed that orientin significantly inhibited HT29 cell viability in a dose dependent manner. The light microscopic observations showed typical variations in cell morphology after 24 h exposure. This included the cell shrinkage from its polyhedral origin, rounded off, membrane blebbing and detachment of cells making substantiation for apoptosis [19]. The delimited cell cycle progression and evasion of apoptosis are the common events in the development of colon cancer [25]. Song et al. [26] suggested that the induction of cell cycle arrest at a specific checkpoint and initiation of apoptosis are often used mechanisms for treating cancer with cytotoxic agents. Cell cycle checkpoints fortify the dividing cells from DNA damage and ensure the maintenance of genomic integrity. The increased proportion of G0/G1 cells after treatment illustrates that orientin induced G0/G1 arrest in a dose dependent manner. G0/G1 arrest suggests that the initiation of cell cycle arrest may be accountable for the anti-proliferative potential of orientin.

Cell cycle is firmly driven by activated cyclin dependent serine/threonine kinases and their regulatory cyclin subunits. These cyclin/CDK complexes serve to be a biomarker for proliferation and attractive therapeutic targets for development of anticancer drugs [27]. Cyclin D and cyclin E, together with CDK2 and CDK4 regulate mitotic division and progress the cell cycle through G1 phase. Our results showed that orientin notably decreased the expression of cyclin D1, cyclin E, CDK2 and CDK4. This is in corroboration with our earlier findings where pelargonidin exhibited such a reduction in the expression of CDKs and cyclins in HT29 cells [21]. p21$^{WAF1/CIP1}$, the major inhibitor of cyclin D/CDK complex was observed to be elevated dose dependently. The increased level of p21$^{WAF1/CIP1}$ suggesting that orientin induced p21$^{WAF1/CIP1}$ mediated G0/G1phase arrest in colon HT29 cells, which is in corroboration with earlier findings [28]. Phosphorylation of Rb by CDK4 initiates an intricate process of phosphorylation-mediated disruption of tumor suppressor function that releases E2F and instigates subsequent G1 to S phase progression [29]. Our experimental results indicated that orientin decreased pRb expression and thereby inhibited the subsequent progression of the cell cycle. The enhancement of oxidative stress by triggering intracellular ROS generation could also be a potential therapeutic strategy for treating colon cancer. Although, ROS are involved in multiple signaling cascades of tumor development, its excessive generation could lead to causing severe harm to DNA and proteins, leading to apoptosis [23]. Our experimental findings suggest that, orientin triggered intracellular ROS production extensively in a concentration dependent manner, which could result in the loss of mitochondrial membrane potential and intrinsic apoptosis.

Normally, phosphatidyl serine predominantly exists on the cytosolic side of the membrane, whereas during apoptosis, it gets translocated towards the outer side of the membrane. Such translocation results in the loss of plasma membrane asymmetry, an early marker for apoptosis. Annexin V, the extracellular Ca^{2+} dependent phospholipid binding protein binds to phosphatidyl serine on the cell surface and the permeability of PI in apoptotic cells detects the heterogeneous distribution of cells in early, late apoptotic phases and dead cells [30]. Annexin V-FITC/PI staining revealed that orientin treated HT29 cells significantly underwent apoptosis compared to untreated cells ($p < 0.05$) and the percentage of both early and late apoptotic cells also significantly increased in a concentration dependent manner. Thus, Annexin V-FITC/PI double staining of HT29 cells treated with orientin substantiates the apoptotic effect of orientin. Proteins of Bcl-2 family play a significant part in apoptosis and targeting those proteins have become an effective strategy for treating cancer. These intracellular proteins control pro-apoptotic, anti-apoptotic signals and mediate the membrane potential of mitochondria [24]. Bax induces apoptosis by the loss of mitochondrial permeability in response to cellular stresses. In contrast, Bcl-2 hinders cell death by inhibiting Bax [31]. The decreased level of anti-apoptotic Bcl-2 and Bcl-XL expression accompanied by the increased pro-apoptotic Bax and Bid levels after being treated with orientin demonstrates the significant therapeutic potential of orientin in modulating Bcl-2 family proteins and thereby inducing apoptosis by disruption of outer mitochondrial membrane in HT29 cells.

Cytochrome C release from mitochondria to the cytosol occurs during the early apoptosis. Multiple evidences suggest that active Bax protein ensures apoptosis via pore formation in mitochondrial outer membrane leads to cytochrome C release [32–34]. Orientin activates Bax and translocates cytochrome C to cytosol. Our results are consistent with previous findings where the pro-apoptotic protein Bid mediated Bax activation, which in turn facilitates cytochrome C release and mitochondrial membrane permeabilization [21]. Orientin induces the release of Smac/DIABLO, a mitochondrial membrane regulatory protein along with cytochrome C. This may promote a cytochrome C-dependent caspase cascade by neutralizing IAP [35–37]. The permeabilization of mitochondria and the cytochrome C release substantiates that orientin may induce apoptosis through intrinsic pathway. The cytosolic cytochrome C interacts with apoptotic protease activating factors (Apaf-1) and pro caspase-9 to make apoptosome and initiates the activation cascade of caspases. The binding of Apaf-1 induces conformational change in pro caspase-9 to its active proteolytic caspase-9 [38]. After treatment, an obvious increase in caspases (caspase -9 and caspase-3) activity and cleavage of PARP, an early marker of chemotherapy induced apoptosis, suggesting that orientin induces apoptosis mainly in the intrinsic pathway [39,40]. The decrease in the expression of IAP family members, XIAP and survivin may be attributed to the release of cytochrome C and the depolarization of mitochondrial membrane to elicit the activation of the caspase cascade by orientin [37]. The increased expression in tumor suppressor p53 after treatment with orientin suggests that p53 could induce the over expression of p21$^{WAF1/CIP1}$, which inhibits the cyclin-CDK complexes. The increased level of γ-H2AX serves as a hallmark of DNA damage after treating with orientin confirms the DNA damage induced in HT29 cells. The overall schematic representation of orientin triggered ROS mediated mitochondrial mediated intrinsic apoptosis was shown in Figure 12.

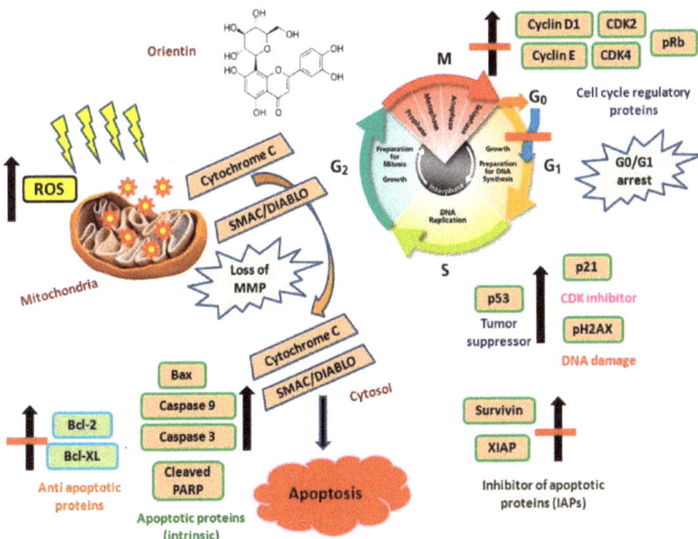

Figure 12. Schematic representation of orientin triggered ROS mediated mitochondrial mediated intrinsic apoptosis.

5. Conclusions

The study demonstrated the anti-proliferative potential of dietary C-glycosyl flavonoid orientin against human CRC HT29 cells by rendering cytotoxicity, morphological changes, chromatin condensation, membrane blebbing and nuclear fragmentation. The overall findings deliberately state that orientin arrested the cell cycle in the G0/G1 phase, induced intracellular ROS and simultaneously depolarized the mitochondrial membrane to activate caspase dependent intrinsic apoptosis. Therefore, the dietary orientin rich in rooibos tea might be a promising chemotherapeutic agent against CRC in human.

Author Contributions: Conceptualization, K.T. and V.M.; methodology, validation and formal analysis, K.T. and K.N.; writing—original draft preparation, K.T., B.B., and V.M.; writing—review and editing, B.B., S.P. and W.L.; supervision and project administration, V.M.

Funding: This research received no external funding. The authors were grateful to the authorities of Periyar University for the necessary facilities to carry out this research work and thank the Sejong University, Republic of Korea for their support.

Conflicts of Interest: The authors declare no conflict of interest.

References

1. Okada, K.; Hakata, S.; Terashima, J.; Gamou, T.; Habano, W.; Ozawa, S. Combination of the histone deacetylase inhibitor depsipeptide and 5-fluorouracil upregulates major histocompatibility complex class II and p21 genes and activates caspase-3/7 in human colon cancer HCT-116 cells. *Oncol. Rep.* **2016**, *36*, 1875–1885. [CrossRef] [PubMed]

2. Thelin, C.; Sikka, S. Epidemiology of colorectal cancer—Incidence, lifetime risk factors statistics and temporal trends. *Lung* **2015**, *1*, 13–20.

3. Soo, H.C.; Chung, F.F.; Lim, K.H.; Yap, V.A.; Bradshaw, T.D.; Hii, L.W.; Tan, S.H.; See, S.J.; Tan, Y.F.; Leong, C.O.; et al. Cudraflavone C induces tumor-specific apoptosis in colorectal cancer cells through inhibition of the phosphoinositide 3-kinase (PI3K)-AKT pathway. *PLoS ONE* **2017**, *12*, 1–20. [CrossRef] [PubMed]

4. Chang, H.; Lei, L.; Zhou, Y.; Ye, F.; Zhao, G. Dietary Flavonoids and the Risk of Colorectal Cancer: An Updated Meta-Analysis of Epidemiological Studies. *Nutrients* **2018**, *10*, 950. [CrossRef] [PubMed]

5. Li, Y.; Zhang, T.; Chen, G.Y. Flavonoids and Colorectal Cancer Prevention. *Antioxidants* **2018**, *7*, 187. [CrossRef] [PubMed]

6. Khan, R.; Khan, A.Q.; Qamar, W.; Lateef, A.; Tahir, M.; Rehman, M.U.; Ali, F.; Sultana, S. Chrysin protects against cisplatin-induced colon toxicity via amelioration of oxidative stress and apoptosis: Probable role of p38MAPK and p53. *Toxicol. Appl. Pharmacol.* **2012**, *258*, 315–329. [CrossRef] [PubMed]

7. Clawson, G.A. Histone deacetylase inhibitors as cancer therapeutics. *Ann. Transl. Med.* **2016**, *4*, 287–291. [CrossRef] [PubMed]

8. Liou, G.Y.; Storz, P. Reactive oxygen species in cancer. *Free Radic. Res.* **2010**, *44*, 479–496. [CrossRef]

9. Yi, J.; Wang, Z.; Bai, H.; Li, L.; Zhao, H.; Cheng, C.; Zhang, H.; Li, J. Polyphenols from pinecones of *Pinus koraiensis* induce apoptosis in colon cancer cells through the activation of caspase in vitro. *RSC Adv.* **2016**, *6*, 5278–5287. [CrossRef]

10. Sergeeva, T.F.; Shirmanova, M.V.; Zlobovskaya, O.A.; Gavrina, A.I.; Dudenkova, V.V.; Lukina, M.M.; Lukyanov, K.A.; Zagaynova, E.V. Relationship between intracellular pH, metabolic co-factors and caspase-3 activation in cancer cells during apoptosis. *BBA-Mol. Cell. Res.* **2017**, *864*, 604–611. [CrossRef]

11. Baig, S.; Seevasant, I.; Mohamad, J.; Mukheem, A.; Huri, H.Z.; Kamarul, T. Potential of apoptotic pathway-targeted cancer therapeutic research: Where do we stand? *Cell Death Dis.* **2016**, *7*, 1–11. [CrossRef] [PubMed]

12. Kim, H.; Jo, A.; Baek, S.; Lim, D.; Park, S.Y.; Cho, S.K.; Chung, J.W.; Yoon, J. Synergistically enhanced selective intracellular uptake of anticancer drug carrier comprising folic acid-conjugated hydrogels containing magnetite nanoparticles. *Sci. Rep.* **2017**, *7*, 1–10. [CrossRef] [PubMed]

13. Yang, F.; Ye, T.; Liu, Z.; Fang, A.; Luo, Y.; Wei, W.; Li, Y.; Li, Y.; Zeng, A.; Deng, Y.; et al. Niclosamide induces colorectal cancer apoptosis, impairs metastasis and reduces immunosuppressive cells in vivo. *RSC Adv.* **2016**, *6*, 106019–106030. [CrossRef]

14. Hata, A.N.; Engelman, J.A.; Faber, A.C. The BCL2 family: Key mediators of the apoptotic response to targeted anticancer therapeutics. *Cancer Discov.* **2015**, *5*, 475–487. [CrossRef] [PubMed]

15. Besbes, S.; Mirshahi, M.; Pocard, M.; Billard, C. New dimension in therapeutic targeting of BCL-2 family proteins. *Oncotarget* **2015**, *6*, 12862–12871. [CrossRef] [PubMed]

16. Lam, K.Y.; Ling, A.P.; Koh, R.Y.; Wong, Y.P.; Say, Y.H. A review on medicinal properties of orientin. *Adv. Pharmacol. Sci.* **2016**, *4104595*, 1–9.

17. Ku, S.K.; Kwak, S.; Bae, J.S. Orientin inhibits high glucose-induced vascular inflammation in vitro and in vivo. *Inflammation* **2014**, *37*, 2164–21673. [CrossRef] [PubMed]

18. Law, B.N.; Ling, A.P.; Koh, R.Y.; Chye, S.M.; Wong, Y.P. Neuroprotective effects of orientin on hydrogen peroxide-induced apoptosis in SH-SY5Y cells. *Mol. Med. Rep.* **2014**, *9*, 947–954. [CrossRef]

19. An, F.; Yang, G.; Tian, J.; Wang, S. Antioxidant effects of the orientin and vitexin in *Trollius chinensis* Bunge in D-galactose-aged mice. *Neural Regen. Res.* **2012**, *7*, 2565–2575.

20. Czemplik, M.; Mierziak, J.; Szopa, J.; Kulma, A. Flavonoid C-glucosides derived from flax straw extracts reduce human breast cancer cell growth in vitro and induce apoptosis. *Front. Pharmacol.* **2016**, *7*, 282. [CrossRef]

21. Karthi, N.; Kalaiyarasu, T.; Kandakumar, S.; Mariyappan, P.; Manju, V. Pelargonidin induces apoptosis and cell cycle arrest via a mitochondria mediated intrinsic apoptotic pathway in HT29 cells. *RSC Adv.* **2016**, *6*, 45064–45076. [CrossRef]

22. Phang, C.W.; Karsani, S.A.; Sethi, G.; Malek, S.N. Flavokawain C inhibits cell cycle and promotes apoptosis, associated with endoplasmic reticulum stress and regulation of MAPKs and Akt signaling pathways in HCT 116 human colon carcinoma cells. *PLoS ONE* **2016**, *11*, e0148775. [CrossRef] [PubMed]

23. Han, M.H.; Park, C.; Jin, C.Y.; Kim, G.Y.; Chang, Y.C.; Moon, S.K.; Kim, W.J.; Choi, Y.H. Apoptosis induction of human bladder cancer cells by sanguinarine through reactive oxygen species-mediated up-regulation of early growth response gene-1. *PLoS ONE* **2013**, *8*, e63425. [CrossRef] [PubMed]

24. Thangaraj, K.; Vaiyapuri, M. Orientin, a C-glycosyl dietary flavone, suppresses colonic cell proliferation and mitigates NF-κB mediated inflammatory response in 1, 2-dimethylhydrazine induced colorectal carcinogenesis. *Biomed. Pharmacother.* **2017**, *96*, 1253–1266. [CrossRef] [PubMed]

25. Wu, X.; Song, M.; Qiu, P.; Li, F.; Wang, M.; Zheng, J.; Wang, Q.; Xu, F.; Xiao, H. A metabolite of nobiletin, 4′-demethylnobiletin and atorvastatin synergistically inhibits human colon cancer cell growth by inducing G0/G1 cell cycle arrest and apoptosis. *Food Funct.* **2018**, *9*, 87–95. [CrossRef] [PubMed]

26. Song, X.L.; Zhang, Y.J.; Wang, X.F.; Zhang, W.J.; Wang, Z.; Zhang, F.; Zhang, Y.J.; Lu, J.H.; Mei, J.W.; Hu, Y.P.; et al. Casticin induces apoptosis and G0/G1 cell cycle arrest in gallbladder cancer cells. *Cancer Cell Int.* **2017**, *17*, 1–10. [CrossRef] [PubMed]

27. Peyressatre, M.; Prevel, C.; Pellerano, M.; Morris, M.C. Targeting cyclin-dependent kinases in human cancers: From small molecules to peptide inhibitors. *Cancers* **2015**, *7*, 179–237. [CrossRef]

28. Kan, W.L.; Yin, C.; Xu, H.X.; Xu, G.; To, K.K.; Cho, C.H.; Rudd, J.A.; Lin, G. Antitumor effects of novel compound, guttiferone K., on colon cancer by p21Waf1/Cip1-mediated G0/G1 cell cycle arrest and apoptosis. *Int. J. Cancer* **2013**, *132*, 707–716. [CrossRef]

29. Asghar, U.; Witkiewicz, A.K.; Turner, N.C.; Knudsen, E.S. The history and future of targeting cyclin-dependent kinases in cancer therapy. *Nat. Rev. Drug Discov.* **2015**, *14*, 130–146. [CrossRef]

30. Hajiaghaalipour, F.; Faraj, F.L.; Bagheri, E.; Ali, H.M.; Abdulla, M.A.; Majid, N.A. Synthesis and Characterization of a New Benzoindole Derivative with Apoptotic Activity Against Colon Cancer Cells. *Curr. Pharma Des.* **2017**, *23*, 6358–6365. [CrossRef]

31. Khodapasand, E.; Jafarzadeh, N.; Farrokhi, F.; Kamalidehghan, B.; Houshmand, M. Is Bax/Bcl-2 ratio considered as a prognostic marker with age and tumor location in colorectal cancer? *Iran. Biomed. J.* **2015**, *19*, 69–75. [PubMed]

32. Bleicken, S.; Zeth, K. Conformational changes and protein stability of the pro-apoptotic protein Bax. *J. Bioenerg. Biomembr.* **2009**, *41*, 29–40. [CrossRef] [PubMed]

33. Dewson, G.K.; Kluck, R.M. Bcl-2 family-regulated apoptosis in health and disease. *Cell Health Cytoskel.* **2010**, *2*, 9–22.

34. Zhang, M.; Zheng, J.; Nussinov, R.; Ma, B. Release of cytochrome C from Bax pores at the mitochondrial membrane. *Sci. Rep.* **2017**, *7*, 2635. [CrossRef] [PubMed]

35. Srinivasula, S.M.; Datta, P.; Fan, X.J.; Fernandes-Alnemri, T.; Huang, Z.; Alnemri, E.S. Molecular determinants of the caspase-promoting activity of Smac/DIABLO and its role in the death receptor pathway. *J. Biol. Chem.* **2000**, *275*, 36152–36157. [CrossRef] [PubMed]

36. Endo, K.; Kohnoe, S.; Watanabe, A.; Tashiro, H.; Sakata, H.; Morita, M.; Kakeji, Y.; Maehara, Y. Clinical significance of Smac/DIABLO expression in colorectal cancer. *Oncol. Rep.* **2009**, *21*, 351–355. [CrossRef] [PubMed]

37. Abdel-Magid, A.F. Modulation of the Inhibitors of Apoptosis Proteins (IAPs) Activities for Cancer Treatment. *ACS Med. Chem. Lett.* **2017**, *8*, 471–473. [CrossRef] [PubMed]

38. Omer, F.A.A.; Hashim, N.B.M.; Ibrahim, M.Y.; Dehghan, F.; Yahayu, M.; Karimian, H.; Salim, L.Z.A.; Mohan, S. Beta-mangostin from *Cratoxylum arborescens* activates the intrinsic apoptosis pathway through reactive oxygen species with downregulation of the HSP70 gene in the HL60 cells associated with a G0/G1 cell-cycle arrest. *Tumor Biol.* **2017**, *39*, 1–12. [CrossRef]

39. Li, M.; Song, L.H.; Yue, G.G.; Lee, J.K.; Zhao, L.M.; Li, L.; Zhou, X.; Tsui, S.K.; Ng, S.S.; Fung, K.P.; et al. Bigelovin triggered apoptosis in colorectal cancer in vitro and in vivo via upregulating death receptor 5 and reactive oxidative species. *Sci. Rep.* **2017**, *7*, 42176. [CrossRef]

40. Jiang, Y.; Wang, X.; Hu, D. Furanodienone induces G0/G1 arrest and causes apoptosis via the ROS/MAPKs-mediated caspase-dependent pathway in human colorectal cancer cells: A study in vitro and in vivo. *Cell Death Dis.* **2017**, *8*, 2815. [CrossRef]

Article

Identification and Isolation of Active Compounds from *Astragalus membranaceus* that Improve Insulin Secretion by Regulating Pancreatic B-Cell Metabolism

Dahae Lee [1], **Da Hye Lee** [2], **Sungyoul Choi** [1], **Jin Su Lee** [2], **Dae Sik Jang** [2,*] and **Ki Sung Kang** [1,*]

[1] College of Korean Medicine, Gachon University, Seongnam 13120, Korea; pjsldh@naver.com (D.L.); pc1075@gachon.ac.kr (S.C.)
[2] Department of Life and Nanopharmaceutical Sciences, Graduate School, Kyung Hee University, Seoul 02447, Korea; marylee93@naver.com (D.H.L.); lee2649318@naver.com (J.S.L.)
* Correspondence: dsjang@khu.ac.kr (D.S.J.); kkang@gachon.ac.kr (K.S.K.); Tel.: +82-2-961-0719 (D.S.J.); +82-31-750-5402 (K.S.K.)

Received: 4 October 2019; Accepted: 17 October 2019; Published: 17 October 2019

Abstract: In type 2 diabetes (T2D), insufficient secretion of insulin from the pancreatic β-cells contributes to high blood glucose levels, associated with metabolic dysregulation. Interest in natural products to complement or replace existing antidiabetic medications has increased. In this study, we examined the effect of *Astragalus membranaceus* extract (ASME) and its compounds **1–9** on glucose-stimulated insulin secretion (GSIS) from pancreatic β-cells. ASME and compounds **1–9** isolated from *A. membranaceus* stimulated insulin secretion in INS-1 cells without inducing cytotoxicity. A further experiment showed that compounds **2**, **3**, and **5** enhanced the phosphorylation of total insulin receptor substrate-2 (IRS-2), phosphatidylinositol 3-kinase (PI3K), and Akt, and activated pancreatic and duodenal homeobox-1 (PDX-1) and peroxisome proliferator-activated receptor-γ (PPAR-γ), which are associated with β-cell function and insulin secretion. The data suggest that two isoflavonoids (**2** and **3**) and a nucleoside (compound **5**), isolated from the roots of *A. membranaceus*, have the potential to improve insulin secretion in β-cells, representing the first step towards the development of potent antidiabetic drugs.

Keywords: *Astragalus membranaceus*; insulin; PI3K; AKT; PPARγ; PDX-1

1. Introduction

Diabetes is characterized by high blood glucose levels (hyperglycemia) and is a common health problem that affects 387 million people worldwide [1]. It is an important risk factor for eye, kidney, nerve, and cardiovascular damage [2]. Type 2 diabetes (T2D) accounts for approximately 90% of all diabetes cases, in which insulin resistance is the primary pathogenic condition. It results in the failure of insulin action in metabolic target tissues, such as muscle, liver, and adipose tissues, due to the insufficient secretion of insulin from the pancreatic β-cells located within the islets of Langerhans [3,4]. A decline in the mass of the pancreatic β-cells, rate of insulin secretion from these cells, or a combination of both, leads to insufficient insulin secretion [5]. Thus, a study focused on preserving the secretory function and mass of the pancreatic β-cells is a strategic approach for the treatment of diabetes.

Interest in natural products to complement or replace existing antidiabetic medications has increased [6]. Approximately 50% of the drugs approved by the US Food and Drug Administration (FDA) are natural compounds or their derivatives, because currently available antidiabetic drugs, including insulin, metformin, and sulfonylureas, are often associated with adverse effects [7]. Some antidiabetic drugs, such as pycnogenol, a trademarked supplement for diabetes derived from the bark extract of the French maritime pine, *Pinus pinaster* Ait. [8], are obtained from natural products [9]. In addition, studies focusing on natural products and traditional medicines for new effective antidiabetic agents have been published [10,11]. Metformin leads to vitamin B12 or folate deficiency and anemia, especially in elderly patients [12,13]. Insulin and sulfonylureas are associated with hypoglycemia [6]. Although insulin lispro was developed in an effort to minimize the risk of hypoglycemia, the number of people with diabetes is predicted to increase to 592 million by 2035 [1]. This emphasizes the need to focus on agents that prevent and treat diabetes using natural products. *Astragalus membranaceus* extract and its constituents are derived from a plant known for its antidiabetic effect [14]. *A. membranaceus* is a Leguminosae flowering plant species used for diabetes treatment in Chinese herbal medicine [15]. It contains saponins, flavonoids, isoflavonoids, sterols, amino acids, polysaccharides, and volatile oils [14,15].

Recent research on T2D focuses on endogenous β-cell function, associated with the rate of insulin secretion from these cells [5]. In T2D models, saponins [16], flavonoids [17], and polysaccharides [18] regulate pancreatic β-cell proliferation and insulin-signaling pathways by enhancing the activation of peroxisome proliferator-activated receptor-γ (PPAR-γ), which is a transcription factor belonging to the nuclear receptor family and expressed in a variety of tissues [19]. The actions of PPAR-γ within pancreatic β-cells are directly involved in β-cell development and function, by regulating gene expression [20]. PPAR-γ directly regulates the nuclear translocation of pancreatic and duodenal homeobox 1 (PDX-1) [21]. PDX-1, a key transcription factor, then binds to the insulin promoter and regulates glucose-stimulated insulin transcription [22]. Additionally, insulin receptor substrate-2 (IRS-2) has important implications for the regulation of the maintenance of β-cell mass and function [23]. Increased expression of IRS-2 stimulates the phosphatidylinositol 3-kinase/protein kinase B (PI3K/Akt) pathway in pancreatic β-cells and promotes β-cell growth and survival.

Here, we report for the first time, to our knowledge, the potential to improve insulin secretion in the INS-1 rat insulin-secreting β-cell line using isoflavonoids isolated from *A. membranaceus*, evaluated as a treatment strategy for T2D. Nine compounds (calycosin, calycosin-7-O-β-D-glucoside, formononetin, formononetin-7-O-β-D-glucoside, adenosine, 3-(β-D-ribofuranosyl)-2,3-dihydro-6H-1,3-oxazine-2,6-dione, acetylastragaloside I, astragaloside I, and astragaloside II) were isolated from the roots of *A. membranaceus* and identified. Additionally, the activities that improved glucose-stimulated insulin secretion (GSIS) and the underlying mechanisms of action were identified. This study demonstrated that two isoflavonoids and a nucleoside isolated from *A. membranaceus* result in the overexpression of IRS-2, PI3K, Akt, PDX-1, and PPAR-γ, which are associated with β-cell function and enhanced insulin secretion.

2. Materials and Methods

2.1. General Experimental Procedures

Column chromatography was performed on silica gel (70–230 and 230–400 mesh ASTM, Merck, Kenilworth, NJ, USA) and Diaion HP-20 (Mitsubishi Chemical Co., Tokyo, Japan). Flash chromatography was performed using the flash purification system (Combi Flash Rf, Teledyne Isco, Lincoln, NE, USA) with Redi Sep-C18 (26 g, 43 g Teledyne Isco). High-performance liquid chromatography (HPLC) was performed using the Gilson purification system, with a J'sphere ODS column (250 × 20.0 mm i.d., 4.0 μm, YMC Co., Tokyo, Japan). Thin-layer chromatography (TLC) analysis was performed on silica gel 60 F_{254} and RP-18 F_{254S} plates (Merck). Nuclear magnetic resonance (NMR) spectra were obtained using a JEOL 500 MHz, using tetramethylsilane as an internal standard, and chemical shifts were expressed as δ values. Organic solvents for the chromatographic separations and extractions were distilled before use.

2.2. Plant Material

The roots of *A. membranaceus* (Leguminosae) were purchased from Hyunjin Pharmaceutical Co. (Seoul, Korea), in January 2015. The origin of the plant material was identified by one of authors D.S.J., and a representative specimen (ASME1-2015) was deposited at the Laboratory of Natural Product Medicine, College of Pharmacy, Kyung Hee University, Seoul, Korea.

2.3. Extraction and Isolation

The dried roots of *A. membranaceus* (6.0 kg) were refluxed once with 60 L of 70% aqueous ethanol (EtOH) for 3 h, and the solvent was evaporated in vacuo at 45 °C to give a 70% EtOH extract (ASME). ASME (1.7 kg) was separated using Diaion HP-20 as the stationary phase and eluted with a MeOH-H_2O gradient (from 0:1 to 1:0 *v/v*), ethyl acetate (EtOAc), and *n*-hexane, to produce seven fractions (K1–7). Fraction K2 (16.4 g) was fractionated using column chromatography (CC) over silica gel (70–230 mesh) and eluted with a gradient of CH_2Cl_2-MeOH-H_2O (from 8:1.8:0.2 to 5:4.5:0.5 *v/v*) to produce five fractions (K2-1–5). Fraction K2-2 (2.6 g) was further separated into nine fractions (K2-2-1–9), using silica gel CC (230–400 mesh) with a stepwise gradient of CH_2Cl_2-MeOH-H_2O (from 9:0.9:0.1 to 5:4.5:0.5 *v/v*). Adenosine (**5**, 20.7 mg) and 3-(β-D-ribofuranosyl)-2,3-dihydro-6H-1,3-oxazine-2,6-dione (**6**, 11.7 mg) were obtained from fraction K2-2-6 (240.0 mg) using a reversed-phase HPLC system with a J'sphere ODS column. Fraction K4 (13.0 g) was separated by silica gel CC (70–230 mesh) with a gradient of CH_2Cl_2-MeOH-H_2O (from 8:1.8:0.2 to 5:4.5:0.5 *v/v*) to yield nine sub-fractions (K4-1–9). Calycosin-7-*O*-β-D-glucoside (**2**, 277.5 mg) was obtained by recrystallization of fraction K4-4 (1.5 g). Fraction K6 (22.3 g) was fractionated using silica gel CC (70–230 mesh) and eluted with a gradient of CH_2Cl_2-MeOH-H_2O (from 8:1.8:0.2 to 7:2.7:0.3 *v/v*) to produce 14 fractions (K6-1–14). Formononetin (**3**, 5.1 g) was obtained from fraction K6-2. Formononetin-7-*O*-β-D-glucoside (**4**, 36.5 mg) and acetylastragaloside I (**7**, 25.5 mg) were obtained from fraction K6-8 (365.8 mg), with a flash chromatography system using a Redi Sep-C18 cartridge (26 g, MeOH-H_2O, gradient from 0:10 to 9:1 *v/v*). Calycosin (**1**, 35.9 mg), astragaloside I (**8**, 249.1 mg), and astragaloside II (**9**, 71.9 mg) were isolated by repeated flash chromatography, using a Redi Sep-C18 cartridge (43 g, MeOH-H_2O, 0:10 to 9:1 *v/v*, gradient), from fractions K6-3 (852. 8 mg), K6-9 (566.8 mg), and K6-12 (536.6 mg), respectively.

2.4. Cell Culture

INS-1 cells (Biohermes, Shanghai, China), a rat insulin-secreting β-cell line, were cultured in a humidified atmosphere at 37 °C containing 5% CO_2, in an RPMI-1640 medium (Cellgro, Manassas, VA, USA), supplemented with 2 mM L-glutamine, 0.05 mM 2-mercaptoethanol, 11 mM D-glucose, 1% penicillin/streptomycin (Invitrogen Co., Grand Island, NY, USA), 10% fetal bovine serum (FBS), 10 mM HEPES, and 1 mM sodium pyruvate.

2.5. Cell Viability

An Ez-Cytox cell viability assay kit (Daeil Lab Service Co., Seoul, Korea) was used to measure cell viability [24]. To determine non-toxic dose ranges of ASME and compounds **1–9** isolated from *A. membranaceus*, INS-1 cells were cultured in 96-well plates for 24 h and, subsequently, treated with ASME and compounds **1–9** for 24 h. Ez-Cytox reagent (10 μL) was added and incubated for 2 h. Absorbance values at 450 nm were measured using a microplate reader (PowerWave XS, Bio-Tek Instruments, Winooski, VT, USA).

2.6. GSIS Assay

An insulin secretion assay was used to determine the effect of ASME and compounds **1–9** on GSIS in INS-1 cells, using gliclazide as the positive control. To assess GSIS after treatment with ASME and compounds **1–9**, INS-1 cells were cultured in 12-well plates for 24 h, washed twice with Krebs–Ringer bicarbonate HEPES buffer and 2.8 mM glucose, and starved in fresh KRBB. After starvation for 2 h, the cells were treated with ASME, compounds **1–9**, and gliclazide. After 2 h, glucose (2.8 and 16.7 mM as basal and stimulant, respectively) was added to each well and incubated for 1 h. GSIS was measured using a rat insulin ELISA kit, as reported previously.

2.7. Western Blot Analysis

INS-1 cells were cultured in 6-well plates for 24 h and treated with compounds **2**, **3**, and **5** for 24 h to assess the levels of protein expression of PPAR-γ, PI3K, Akt, P-IRS-2 (Ser731), IRS-2, P-PI3K, P-Akt (Ser473), and PDX-1 after treatment. Western blot analysis was carried out to evaluate the expression of proteins related to pancreatic β-cell metabolism, as reported previously [25].

2.8. Statistical Analysis

Statistical significance was assessed using one-way analysis of variance (ANOVA) and multiple comparisons with a Bonferroni correction. *P* values less than 0.05 were considered statistically significant. All analyses were determined using SPSS Statistics ver. 19.0 (SPSS Inc., Chicago, IL, USA).

3. Results

3.1. Identification of Compounds 1–9

Nine compounds (**1–9**) were isolated from the roots of *A. membranaceus* in this study. The purity of the isolates (>95%) was determined by NMR. The chemical structures of compounds **1–9** were identified as calycosin (**1**) [26], calycosin-7-O-β-D-glucoside (**2**) [27], formononetin (**3**) [28], formononetin-7-O-β-D-glucoside (**4**) [29], adenosine (**5**) [30], 3-(β-D-ribofuranosyl)-2, 3-dihydro-6H-1,3-oxazine-2,6-dione (**6**) [31], acetylastragaloside I (**7**) [32], astragaloside I (**8**) [32], and astragaloside II (**9**) [33], by 1D and 2D NMR spectroscopic data and by comparison with published values (Figure 1).

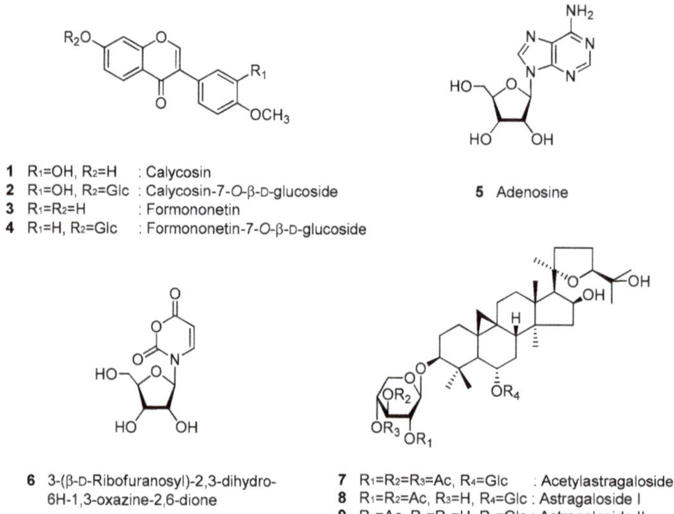

Figure 1. Chemical structures of compounds **1–9**.

3.2. Effect of ASME and Compounds 1–9 on GSIS

The non-toxic dose of ASME and compounds **1–9** was determined using a cell viability assay on INS-1 cells. Although ASME did not show any toxic effect at 12.5, 25, and 50 µM (Figure 2A), some compounds isolated from ASME were slightly cytotoxic at concentrations of 25 µM and above, as cell viability decreased to below 80% (Figure 2B–E,J). ASME at 2.5, 5, and 10 µg/mL, and compounds **1–9** at 2.5, 5, and 10 µM, used in the insulin secretion assay, were based on the results of the cell viability assay. As shown in Figure 3A, ASME led to an increase in GSI in a dose-dependent manner. The GSI levels were 1.37 ± 0.01, 1.43 ± 0.17, and 3.92 ± 0.31 for ASME at 2.5 µg/mL, 5 µg/mL, and 10 µg/mL, respectively. Among compounds isolated from ASME, two isoflavonoids calycosin-7-*O*-β-D-glucoside (**2**), formononetin (**3**), and nucleoside adenosine (**5**) led to a significant increase in GSI in a dose-dependent manner (Figure 3C,D,F). The GSI levels were 6.03 ± 0.41, 6.77 ± 0.43, and 5.97 ± 0.10 for compounds **2**, **3**, and **5** at 10 µM, respectively. These GSI levels were similar to the GSI level of gliclazide at the same concentration (Figure 3K). The GSI levels were 2.03 ± 0.03, 3.16 ± 0.10, and 5.81 ± 0.16 for gliclazide at 2.5 µM, 5 µM, and 10 µM, respectively. Compounds **2**, **3**, and **5** stimulated insulin secretion in INS-1 cells without inducing cytotoxicity.

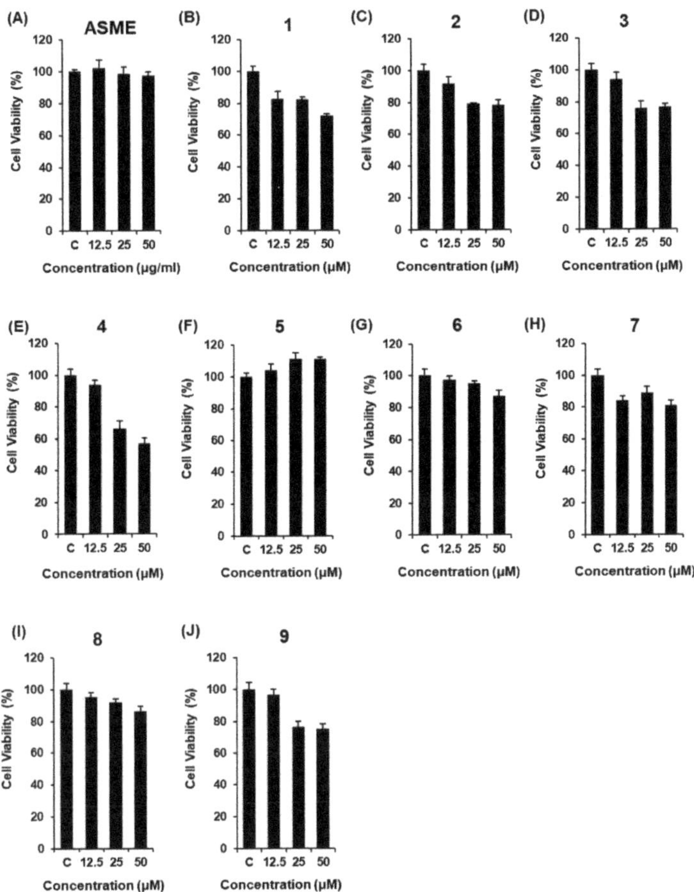

Figure 2. Effect of *A. membranaceus* extract (ASME) and compounds **1–9** isolated from *A. membranaceus*, on the viability of INS-1 cells. Effect of (**A**) ASME and (**B–J**) compounds **1–9**, compared with the control (0 µg/mL), on the viability of INS-1 cells for 24 h by MTT assay.

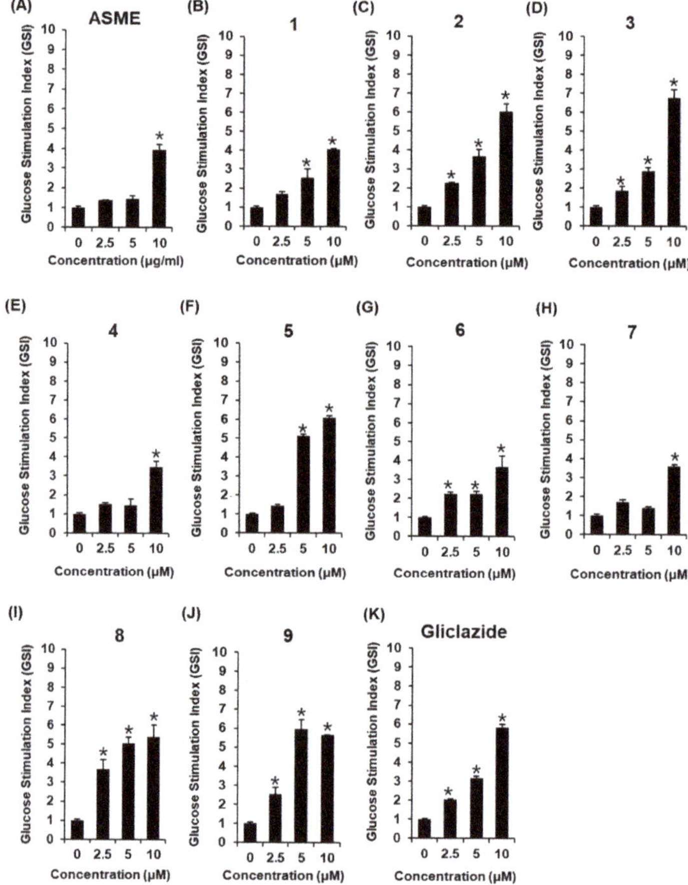

Figure 3. Effect of *A. membranaceus* extract (ASME) and compounds **1–9** isolated from *A. membranaceus* on glucose-stimulated insulin secretion (GSIS) in INS-1 cells. Effect of (**A**) ASME, (**B–J**) compounds **1–9**, and (**K**) gliclazide (positive control) on GSIS in INS-1 cells for 1 h by insulin secretion assay. * $p < 0.05$ compared to the control (0 µM).

3.3. Effect of Compounds 2, 3, and 5 Isolated from A. membranaceus on the Protein Expression of PPARγ, P-IRS-2, IRS-2 (Ser731), P-PI3K, PI3K, P-Akt (Ser473), Akt, and PDX-1

To show the role of PPAR-γ, IRS-2, PI3K, Akt, and PDX-1 in the effect of compounds **2**, **3**, and **5** on GSIS, we measured these protein levels in pancreatic β-cell metabolism, and demonstrated that the protein expression levels of PPAR-γ, P-IRS-2 (Ser731), P-PI3K, P-Akt (Ser473), and PDX-1 increase upon treatment with compounds **2**, **3**, and **5** at 10 µM, compared to untreated controls (Figure 4). A schematic illustration of the proposed mechanism of the effects of compounds **2**, **3**, and **5** in pancreatic β-cell metabolism was shown in Figure 5.

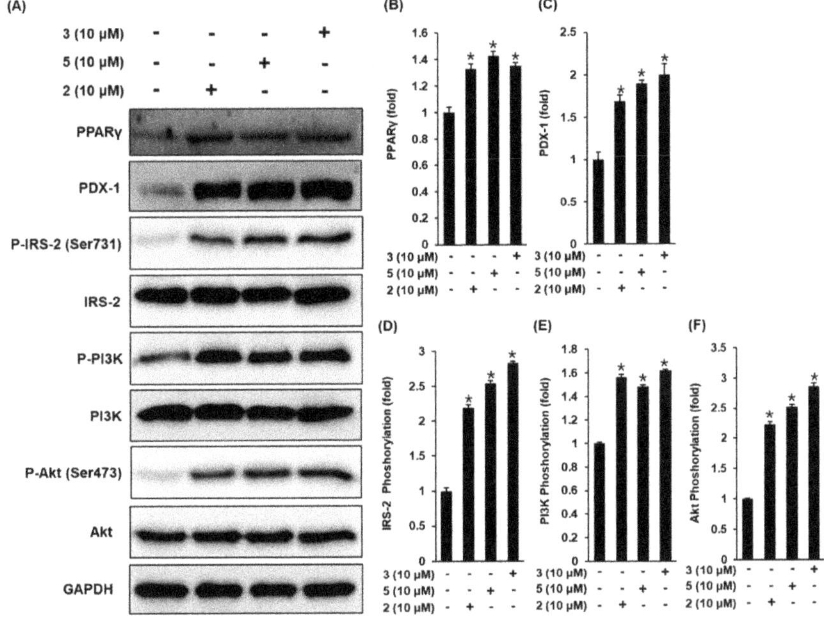

Figure 4. Effect of compounds **2**, **3**, and **5** isolated from *A. membranaceus* on the protein expression levels of peroxisome proliferator-activated receptor-γ (PPAR-γ), pancreatic and duodenal homeobox-1 (PDX-1), P-insulin receptor substrate-2 (IRS-2) (Ser731), IRS-2, P-phosphatidylinositol 3-kinase (PI3K), PI3K, P-Akt (Ser473), and Akt in INS-1 cells. (**A**) Protein expression levels of PPAR-γ, PDX-1, P-IRS-2 (Ser731), IRS-2, P-PI3K, PI3K, P-Akt (Ser473), Akt, and glyceraldehyde 3-phosphate dehydrogenase (GAPDH) in INS-1 cells treated or untreated with 10 μM compounds **2**, **3**, and **5** for 24 h. (**B–F**) Each bar graph presents the densitometric quantification of Western blot bands. * $p < 0.05$ compared to the control (0 μM).

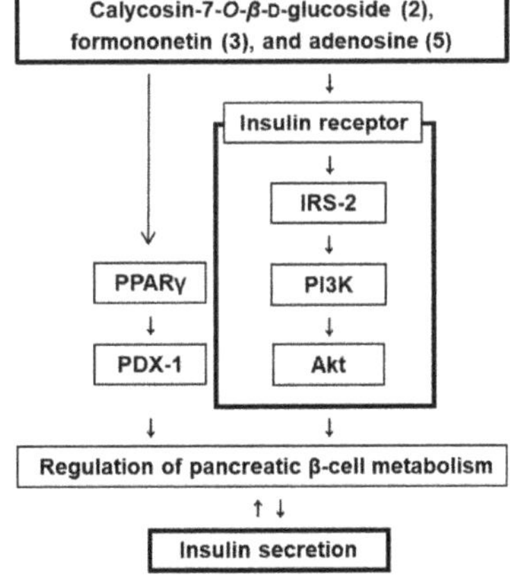

Figure 5. Schematic illustration of the effects of compounds **2**, **3**, and **5** isolated from *A. membranaceus* on the protein expression levels of peroxisome proliferator-activated receptor-γ (PPAR-γ), pancreatic and duodenal homeobox-1 (PDX-1), insulin receptor substrate-2 (IRS-2), phosphatidylinositol 3-kinase (PI3K), P-Akt (Ser473), and Akt in INS-1 cells.

4. Discussion

This study demonstrated that ASME and the compounds identified in this extract exerted insulin secretory effects. ASME and some active compounds isolated from ASME have been reported for their antidiabetic properties [14]. In an STZ-induced rat model of T1D, after treatment with astragalus polysaccharides [34–36] and soy isoflavones [36] isolated from ASME, glucose homeostasis was improved, but not by enhancing the capacity of insulin secretion levels [34,35]. In our study, ASME enhanced GSIS without inducing cytotoxicity in INS-1 cells. From this extract, nine compounds (**1–9**) including calycosin, calycosin-7-*O*-β-D-glucoside, formononetin, formononetin-7-*O*-β-D-glucoside, adenosine, 3-(β-D-ribofuranosyl)-2,3-dihydro-6H-1,3-oxazine-2,6-dione, acetylastragaloside I, astragaloside I, and astragaloside II, were isolated and identified. Among the isolated compounds, calycosin-7-*O*-β-D-glucoside (**2**), formononetin (**3**), and adenosine (**5**) led to a significant increase in GSI in a dose-dependent manner, without inducing cytotoxicity in INS-1 cells. The effects of these three compounds were similar to the effect of gliclazide, a medicine used to treat T2D and classified in the sulfonylurea class of insulin secretagogues [37], which improved pancreatic β-cell sensitivity to glucose and enhanced insulin secretion in clinical studies [38]. These results suggest that sufficient secretion of insulin after treatment with ASME and its bioactive compounds, in response to an increase of glucose, may inhibit characteristic diabetic hyperglycemia and improve the sensitivity of pancreatic β-cells to glucose.

The most effective isoflavonoid identified in our study, formononetin (**3**), has been reported for its antidiabetic effect. In an experimental model of T2D, treatment with formononetin improves insulin sensitivity and reduces hyperglycemia by activating sirtuin 1 (SIRT1), an important regulator of energy metabolism that is involved in the regulation of insulin production and sensitivity and controls co-regulators, such as nuclear factor-kappa B (NF-κB), FOXO proteins, and PPAR-γ in pancreatic β cells [39]. In our previous study, the mechanisms of action mediating insulin secretion by compounds **2**, **3**, and **5** were evaluated. PPAR-γ is an important regulator of glucose metabolism by regulating gene expression [40], and is activated by compounds **2**, **3**, and **5**. Thiazolidinediones (TZDs), PPARγ agonists, are widely used antidiabetic drugs, but have side effects, including weight gain and hepatic dysfunction. Because of these side effects, a variety of natural compounds, including stilbenes, flavonoids, neolignans, sesquiterpenes, amorfrutins, and coumarins have been identified as PPAR-γ agonists in an attempt to increase the effectiveness of PPAR-γ, while limiting its side effects [41].

PPAR-γ binds to the PDX-1 promoter to upregulate PDX-1 expression, which is associated with pancreatic development and the capacity of β-cells [42,43]. In human pancreatic β-cells, PDX-1 mRNA levels are increased in the presence of gliclazide [44], which has a similar insulin secretory capacity to compounds **2**, **3**, and **5**. These three active compounds also increase the protein expression of PDX-1. Pancreatic β-cell malfunction is characterized by the lack of insulin production and secretion to regulate glucose metabolism, which results in hyperglycemia in PDX-1 knockout mice [45] and IRS-2 knockout mice [46]. IRS-2 is a member of a family of large adaptor proteins, linking insulin receptors to the activation of the PI3K/Akt pathway, which plays an important role in β-cell function [47,48]. Upregulation of IRS-2, PI3K, and Akt leads to the proliferation of these proteins in pancreatic β-cells, maintaining functional β-cell mass and enhancing insulin secretion [49]. In pancreatic islets isolated from patients with T2D, these expression levels are reduced [50,51]. We investigated the role of IRS-2 in the presence of compounds **2**, **3**, and **5**. IRS-2 phosphorylation at Ser731 was increased by compounds **2**, **3**, and **5**. In addition, PI3K-dependent Akt phosphorylation at Ser473 was observed after treatment with compounds **2**, **3**, and **5**.

Consequently, the three active compounds (**2**, **3**, and **5**) facilitated insulin secretion by enhancing the expression of IRS-2, PI3K, Akt, PPAR-γ, and PDX-1. Despite these findings, further investigation is needed to determine how inhibitors of IRS-2, PI3K, Akt, PPAR-γ, and PDX-1 affect GSIS. Studies on the solubility, membrane permeability, absorption, distribution, and metabolism of the active compounds in vivo are also required, because these factors limit the oral bioavailability of the compounds [52,53]. In addition, in vivo studies of T2D are required to assess the antidiabetic potential of the active compounds, because their effects on insulin secretion from the pancreatic β cells may not be as significant upon oral administration. Clarifying the underlying mechanisms of action through further investigation may lead to the development of new drugs to prevent or delay the development of diabetes in patients who do not adequately respond to currently available antidiabetic drugs.

5. Conclusions

Thus, calycosin-7-*O*-β-D-glucoside (**2**), formononetin (**3**), and adenosine (**5**), isolated from the roots of *A. membranaceus*, were observed to potentiate GSIS from the pancreatic β-cells. Moreover, our results suggested that IRS-2, PI3K, Akt, PPAR-γ, and PDX-1 played important roles in these effects, highlighting the potential of these 3 active compounds in antidiabetic research.

Author Contributions: K.S.K. and D.S.J. conceived and designed the experiments; D.L., D.H.L., S.C., and J.S.L. performed the experiments; K.S.K. and S.C. analyzed the data; D.L. and K.S.K. interpreted the data and contributed to manuscript structure and flow; and D.L. and J.S.L. wrote the paper. All authors reviewed and confirmed the manuscript.

Funding: This work was funded by the National Research Foundation of Korea (NRF) grant funded by Ministry of Science and ICT (MSIT), Republic of Korea (NRF-2019R1A2C1083945). The present study was also funded by the Basic Science Research Program through the National Research Foundation of Korea (NRF) (2019R1F1A1059173). This research was also funded by the Nano Convergence Industrial Strategic Technology Development Program (No. 20000105) funded by the Ministry of Trade, Industry & Energy (MOTIE, Korea).

Conflicts of Interest: The authors declare no competing financial interest.

References

1. Shrestha, P.; Ghimire, L. A Review about the Effect of Life style Modification on Diabetes and Quality of Life. *Glob. J. Health Sci.* **2012**, *4*, 185–190. [CrossRef] [PubMed]
2. Sami, W.; Ansari, T.; Butt, N.S.; Ab Hamid, M.R. Effect of diet on type 2 diabetes mellitus: A review. *Int. J. Health Sci.* **2017**, *11*, 65.
3. Hameed, I.; Masoodi, S.R.; Mir, S.A.; Nabi, M.; Ghazanfar, K.; A Ganai, B. Type 2 diabetes mellitus: From a metabolic disorder to an inflammatory condition. *World J. Diabetes* **2015**, *6*, 598–612. [CrossRef] [PubMed]
4. Skyler, J.S.; Bakris, G.L.; Bonifacio, E.; Darsow, T.; Eckel, R.H.; Groop, L.; Groop, P.-H.; Handelsman, Y.; Insel, R.A.; Mathieu, C. Differentiation of diabetes by pathophysiology, natural history, and prognosis. *Diabetes* **2017**, *66*, 241–255. [CrossRef]
5. Cantley, J.; Ashcroft, F.M. Q&A: Insulin secretion and type 2 diabetes: Why do β-cells fail? *BMC Boil.* **2015**, *13*, 33.
6. Tabatabaei-Malazy, O.; Larijani, B.; Abdollahi, M. Targeting metabolic disorders by natural products. *J. Diabetes Metab. Disord.* **2015**, *14*, 57. [CrossRef]
7. Chaudhury, A.; Duvoor, C.; Dendi, V.S.R.; Kraleti, S.; Chada, A.; Ravilla, R.; Marco, A.; Shekhawat, N.S.; Montales, M.T.; Kuriakose, K.; et al. Clinical Review of Antidiabetic Drugs: Implications for Type 2 Diabetes Mellitus Management. *Front. Endocrinol.* **2017**, *8*, 28. [CrossRef]
8. Liu, X.; Wei, J.; Tan, F.; Zhou, S.; Würthwein, G.; Rohdewald, P. Antidiabetic effect of Pycnogenol®French maritime pine bark extract in patients with diabetes type II. *Life Sci.* **2004**, *75*, 2505–2513. [CrossRef]
9. Ríos, J.; Francini, F.; Schinella, G. Natural Products for the Treatment of Type 2 Diabetes Mellitus. *Planta Medica* **2015**, *81*, 975–994. [CrossRef]
10. Xu, L.; Li, Y.; Dai, Y.; Peng, J. Natural products for the treatment of type 2 diabetes mellitus: Pharmacology and mechanisms. *Pharmacol. Res.* **2018**, *130*, 451–465. [CrossRef]
11. Li, W.; Yuan, G.; Pan, Y.; Wang, C.; Chen, H. Network Pharmacology Studies on the Bioactive Compounds and Action Mechanisms of Natural Products for the Treatment of Diabetes Mellitus: A Review. *Front. Pharmacol.* **2017**, *8*, 74. [CrossRef] [PubMed]
12. Bailey, C.J.; Day, C. Metformin: Its botanical background. *Pract. Diabetes Int.* **2004**, *21*, 115–117. [CrossRef]
13. Group, D.P.P.R. Long-term safety, tolerability, and weight loss associated with metformin in the Diabetes Prevention Program Outcomes Study. *Diabetes Care* **2012**, *35*, 731–737.

14. Agyemang, K.; Han, L.; Liu, E.; Zhang, Y.; Wang, T.; Gao, X. Recent Advances in *Astragalus membranaceus* Anti-Diabetic Research: Pharmacological Effects of Its Phytochemical Constituents. *Evidence-Based Complement. Altern. Med.* **2013**, *2013*, 654643. [CrossRef] [PubMed]

15. Kim, J.; Moon, E.; Kwon, S. Effect of *Astragalus membranaceus* extract on diabetic nephropathy. *Endocrinol. Diabetes Metab. Case Rep.* **2014**, *2014*, 140063. [CrossRef] [PubMed]

16. Kai, Z.; Michela, P.; Antonio, P.; Annamaria, P. Biological Active Ingredients of Traditional Chinese Herb *Astragalus membranaceus* on Treatment of Diabetes: A Systematic Review. *Mini-Rev. Med. Chem.* **2015**, *15*, 315–329. [CrossRef] [PubMed]

17. Lavle, N.; Shukla, P.; Panchal, A. Role of flavonoids and saponins in the treatment of diabetes mellitus. *J. Pharm.* **2016**, *6*, 41–53.

18. Wang, J.; Jia, J.; Song, L.; Gong, X.; Xu, J.; Yang, M.; Li, M. Extraction, Structure, and Pharmacological Activities of Astragalus Polysaccharides. *Appl. Sci.* **2019**, *9*, 122. [CrossRef]

19. Gupta, D.; Kono, T.; Evans-Molina, C. The role of peroxisome proliferator-activated receptor γ in pancreatic β cell function and survival: therapeutic implications for the treatment of type 2 diabetes mellitus. *Diabetes Obes. Metab.* **2010**, *12*, 1036–1047. [CrossRef]

20. Kim, H.-S.; Hwang, Y.-C.; Koo, S.-H.; Park, K.S.; Lee, M.-S.; Kim, K.-W.; Lee, M.-K. PPAR-γ Activation Increases Insulin Secretion through the Up-regulation of the Free Fatty Acid Receptor GPR40 in Pancreatic β-Cells. *PLoS ONE* **2013**, *8*, e50128. [CrossRef]

21. Houseknecht, K.L.; Cole, B.M.; Steele, P.J. Peroxisome proliferator-activated receptor gamma (PPARγ) and its ligands: A review. *Domest. Anim. Endocrinol.* **2002**, *22*, 1–23. [CrossRef]

22. Keane, K.N.; Cruzat, V.F.; Carlessi, R.; De Bittencourt, P.I.H.; Newsholme, P. Molecular Events Linking Oxidative Stress and Inflammation to Insulin Resistance and β-Cell Dysfunction. *Oxidative Med. Cell. Longev.* **2015**, *2015*, 181643. [CrossRef] [PubMed]

23. Morales, N.B.; De Plata, C.A. Role of AKT/mTORC1 pathway in pancreatic β-cell proliferation. *Colomb. Medica* **2012**, *43*, 235–243.

24. Trinh, T.A.; Park, E.-J.; Lee, D.; Song, J.H.; Lee, H.L.; Kim, K.H.; Kim, Y.; Jung, K.; Kang, K.S.; Yoo, J.-E. Estrogenic Activity of Sanguiin H-6 through Activation of Estrogen Receptor α Coactivator-binding Site. *Nat. Prod. Sci.* **2019**, *25*, 28–33. [CrossRef]

25. Roy, A.; Park, H.-J.; Jung, H.A.; Choi, J.S. Estragole Exhibits Anti-inflammatory Activity with the Regulation of NF-κB and Nrf-2 Signaling Pathways in LPS-induced RAW 264.7 cells. *Nat. Prod. Sci.* **2018**, *24*, 13–20. [CrossRef]

26. Chun-qing, S.; Zhi-ren, Z.; Di, L.; Zhi-bi, H. Isoflavones from *Astragalus membranaceus*. *Acta. Bot. Sin.* **1997**, *39*, 764–768.

27. Yu, D.-H.; Bao, Y.-M.; Wei, C.-L.; An, L.-J. Studies of chemical constituents and their antioxidant activities from Astragalus mongholicus Bunge. *Biomed. Environ. Sci.* **2005**, *18*, 297–301.

28. Han, T.; Li, H.; Zhang, Q.; Zheng, H.; Qin, L. New thiazinediones and other components from Xanthium strumarium. *Chem. Nat. Compd.* **2006**, *42*, 567–570. [CrossRef]

29. Choi, C.W.; Choi, Y.H.; Cha, M.-R.; Yoo, D.S.; Kim, Y.S.; Yon, G.H.; Hong, K.S.; Kim, Y.H.; Ryu, S.Y. Yeast α-Glucosidase Inhibition by Isoflavones from Plants of Leguminosae as an in Vitro Alternative to Acarbose. *J. Agric. Food Chem.* **2010**, *58*, 9988–9993. [CrossRef]

30. Patching, S.G.; Baldwin, S.A.; Baldwin, A.D.; Young, J.D.; Gallagher, M.P.; Henderson, P.J.F.; Herbert, R.B. The nucleoside transport proteins, NupC and NupG, from Escherichia coli: specific structural motifs necessary for the binding of ligands. *Org. Biomol. Chem.* **2005**, *3*, 462–470. [CrossRef]

31. Jiang, Y.; Choi, H.G.; Li, Y.; Park, Y.M.; Lee, J.H.; Kim, D.H.; Lee, J.-H.; Son, J.K.; Na, M.K.; Lee, S.H. Chemical constituents of Cynanchum wilfordii and the chemotaxonomy of two species of the family Asclepiadacease, C. wilfordii and C. auriculatum. *Arch. Pharm. Res.* **2011**, *34*, 2021–2027. [CrossRef] [PubMed]

32. Hirotani, M.; Zhou, Y.; Lui, H.; Furuya, T. Astragalosides from hairy root cultures of *Astragalus membranaceus*. *Phytochemistry* **1994**, *36*, 665–670. [CrossRef]

33. Kim, J.S.; Yean, M.H.; Lee, S.Y.; Kang, S.S. Phytochemical studies on Astragalus root (1)-Saponins. *Nat. Prod. Sci.* **2008**, *14*, 14–37.

34. Zou, F.; Mao, X.-Q.; Wang, N.; Liu, J.; Ou-Yang, J.-P. Astragalus polysaccharides alleviates glucose toxicity and restores glucose homeostasis in diabetic states via activation of AMPK. *Acta Pharmacol. Sin.* **2009**, *30*, 1607–1615. [CrossRef] [PubMed]

35. Wang, N.; Zhang, D.; Mao, X.; Zou, F.; Jin, H.; Ouyang, J. Astragalus polysaccharides decreased the expression of PTP1B through relieving ER stress induced activation of ATF6 in a rat model of type 2 diabetes. *Mol. Cell. Endocrinol.* **2009**, *307*, 89–98. [CrossRef]

36. Liao, W.; Shi, Y. Effect of astragalus polysaccharides and soy isoflavones on glucose metabolism in diabetic rats. *Acta Acad. Med. Mil. Tertiae* **2007**, *29*, 416–418.

37. Ida, S.; Kaneko, R.; Murata, K. Effects of oral antidiabetic drugs on left ventricular mass in patients with type 2 diabetes mellitus: a network meta-analysis. *Cardiovasc. Diabetol.* **2018**, *17*, 129. [CrossRef]

38. Van Haeften, T.; Veneman, T.; Gerich, J.; Van Der Veen, E. Influence of gliclazide on glucose-stimulated insulin release in man. *Metabolism* **1991**, *40*, 751–755. [CrossRef]

39. Oza, M.J.; Kulkarni, Y.A. Formononetin Treatment in Type 2 Diabetic Rats Reduces Insulin Resistance and Hyperglycemia. *Front. Pharmacol.* **2018**, *9*, 9. [CrossRef]

40. Leonardini, A.; Laviola, L.; Perrini, S.; Natalicchio, A.; Giorgino, F. Cross-talk between PPAR and insulin signaling and modulation of insulin sensitivity. *PPAR Res.* **2009**. [CrossRef]

41. Ammazzalorso, A.; Amoroso, R. Inhibition of PPARγ by Natural Compounds as a Promising Strategy in Obesity and Diabetes. *Open Med. Chem. J.* **2019**, *13*, 7–15. [CrossRef]

42. Remedi, M.S.; Emfinger, C. Pancreatic β-cell identity in diabetes. *Diabetes Obes. Metab.* **2016**, *18*, 110–116. [CrossRef] [PubMed]

43. Hsu, W.-H.; Pan, T.-M. A novel PPARgamma agonist monascin's potential application in diabetes prevention. *Food Funct.* **2014**, *5*, 1334–1340. [CrossRef] [PubMed]

44. Del Guerra, S.; D'Aleo, V.; Lupi, R.; Masini, M.; Bugliani, M.; Boggi, U.; Filipponi, F.; Marchetti, P. Effects of exposure of human islet beta-cells to normal and high glucose levels with or without gliclazide or glibenclamide. *Diabetes Metab.* **2009**, *35*, 293–298. [CrossRef] [PubMed]

45. Gao, T.; McKenna, B.; Li, C.; Reichert, M.; Nguyen, J.; Singh, T.; Yang, C.; Pannikar, A.; Doliba, N.; Zhang, T. Pdx1 maintains β cell identity and function by repressing an α cell program. *Cell Metab.* **2014**, *19*, 259–271. [CrossRef] [PubMed]

46. Withers, D.J.; Gutierrez, J.S.; Towery, H.; Burks, D.J.; Ren, J.-M.; Previs, S.; Zhang, Y.; Bernal, D.; Pons, S.; Shulman, G.I. Disruption of IRS-2 causes type 2 diabetes in mice. *Nature* **1998**, *391*, 900. [CrossRef]

47. Keegan, A.D.; Zamorano, J.; Keselman, A.; Heller, N.M. IL-4 and IL-13 Receptor Signaling From 4PS to Insulin Receptor Substrate 2: There and Back Again, a Historical View. *Front. Immunol.* **2018**, *9*, 9. [CrossRef]

48. Lin, X.; Taguchi, A.; Park, S.; Kushner, J.A.; Li, F.; Li, Y.; White, M.F. Dysregulation of insulin receptor substrate 2 in β cells and brain causes obesity and diabetes. *J. Clin. Investig.* **2004**, *114*, 908–916. [CrossRef]

49. Mohanty, S.; Spinas, G.; Maedler, K.; Zuellig, R.; Lehmann, R.; Donath, M.; Trüb, T.; Niessen, M. Overexpression of IRS2 in isolated pancreatic islets causes proliferation and protects human β-cells from hyperglycemia-induced apoptosis. *Exp. Cell Res.* **2005**, *303*, 68–78. [CrossRef]

50. Gunton, J.E.; Kulkarni, R.N.; Yim, S.; Okada, T.; Hawthorne, W.J.; Tseng, Y.-H.; Roberson, R.S.; Ricordi, C.; O'Connell, P.J.; Gonzalez, F.J.; et al. Loss of ARNT/HIF1β Mediates Altered Gene Expression and Pancreatic-Islet Dysfunction in Human Type 2 Diabetes. *Cell* **2005**, *122*, 337–349. [CrossRef]

51. Choi, M.R.; Kwak, S.M.; Bang, S.H.; Jeong, J.E.; Kim, D.J. Chronic saponin treatment attenuates damage to the pancreas in chronic alcohol-treated diabetic rats. *J. Ginseng Res.* **2017**, *41*, 503–512. [CrossRef] [PubMed]

52. Kim, H.; Lee, J.H.; Kim, J.E.; Kim, Y.S.; Ryu, C.H.; Lee, H.J.; Kim, H.M.; Jeon, H.; Won, H.-J.; Lee, J.-Y.; et al. Micro-/nano-sized delivery systems of ginsenosides for improved systemic bioavailability. *J. Ginseng Res.* **2018**, *42*, 361–369. [CrossRef] [PubMed]

53. Mohammed, A.; Islam, M.S. Spice-Derived Bioactive Ingredients: Potential Agents or Food Adjuvant in the Management of Diabetes Mellitus. *Front. Pharmacol.* **2018**, *9*, 893. [CrossRef] [PubMed]

 *biomolecules*

Article

Chemical Constituents of the Leaves of Butterbur (*Petasites japonicus*) and Their Anti-Inflammatory Effects

Jin Su Lee [1], Miran Jeong [2], Sangsu Park [3], Seung Mok Ryu [4], Jun Lee [4], Ziteng Song [5], Yuanqiang Guo [5], Jung-Hye Choi [1,2], Dongho Lee [6,*] and Dae Sik Jang [1,2,*]

[1] Department of Life and Nanopharmaceutical Sciences, Graduate School, Kyung Hee University, Seoul 02447, Korea; lee2649318@naver.com (J.S.L.); jchoi@khu.ac.kr (J.-H.C.)
[2] College of Pharmacy, Kyung Hee University, Seoul 02447, Korea; jeongmiran@hanmail.net
[3] Department of Fundamental Pharmaceutical Sciences, Graduate School, Kyung Hee University, Seoul 02447, Korea; x-zara@nate.com
[4] Herbal Medicine Resources Research Center, Korea Institute of Oriental Medicine, Jeollanam-do 58245, Korea; smryu@kiom.re.kr (S.M.R.); junlee@kiom.re.kr (J.L.)
[5] State Key Laboratory of Medicinal Chemical Biology, College of Pharmacy, Tianjin Key Laboratory of Molecular Drug Research, and Drug Discovery Center for Infectious Disease, Nankai University, Tianjin 300350, China; kdszt152@163.com (Z.S.); victgyq@nankai.edu.cn (Y.G.)
[6] Department of Biosystems and Biotechnology, College of Life Sciences and Biotechnology, Korea University, Seoul 02841, Korea
* Correspondence: donghoLee@korea.ac.kr (D.L.); dsjang@khu.ac.kr (D.S.J.); Tel.: +82-2-3290-3017 (L.D.); +82-2-961-0719 (D.S.J.)

Received: 30 October 2019; Accepted: 27 November 2019; Published: 29 November 2019

Abstract: Two new aryltetralin lactone lignans, petasitesins A and B were isolated from the hot water extract of the leaves of butterbur (*Petasites japonicus*) along with six known compounds. The chemical structures of lignans **1** and **2** were elucidated on the basis of 1D and 2D nuclear magnetic resonance (NMR) spectroscopic data, electronic circular dichroism (ECD) and vibrational circular dichroism (VCD) spectra. Petasitesin A and cimicifugic acid D showed significant inhibitory effects on the production of both prostaglandin E2 (PGE_2) and NO in RAW264.7 macrophages. The expressions of inducible nitric oxide synthase (iNOS) and cyclooxygenase-2 (COX-2) were inhibited by compound **1** in RAW264.7 cells. Furthermore, compounds **1** and **3** exhibited strong affinities with both iNOS and COX-2 enzymes in molecular docking studies.

Keywords: *Petasites japonicus*; Asteraceae; lignan; anti-inflammation; NO; PGE_2; iNOS; COX-2; molecular docking

1. Introduction

Petasites japonicus Maxim (Asteraceae), known as butterbur, Japanese butterbur, and giant butterbur, is used as a botanical dietary supplement in the USA. The aerial parts of *P. japonicus* have been used in traditional Japanese folk medicine as an antipyretic, antitussive, or wound healing agent [1]. The constituents of *P. japonicus* have been reported and include flavonoids [2], sesquiterpenes [3–5], triterpenes [6], and various types of phenolic compounds [7]. Moreover, the leaves or stalks of *P. japonicus* are commonly consumed as vegetables in Korea and Japan. In the course of searching for active compounds from higher plants [8,9], the leaves of *P. japonicus* were selected for a detailed study since a hot water extract of the leaves of *P. japonicus* have shown inhibitory activity against nitric oxide (NO) production in RAW 264.7 cells [half maximal inhibitory concentration (IC_{50}) value: 19 ± 4.9 µg/mL]. Phytochemical study on the hot water extract resulted in the isolation of two new aryltetralin lactone lignans (**1** and **2**) along with six previously known compounds (Figure 1). The chemical structures of the new lignans **1** and **2** were determined by interpretation of 1D and 2D nuclear magnetic resonance (NMR) spectroscopic data, and by electronic circular dichroism (ECD) and vibrational circular dichroism (VCD) studies.

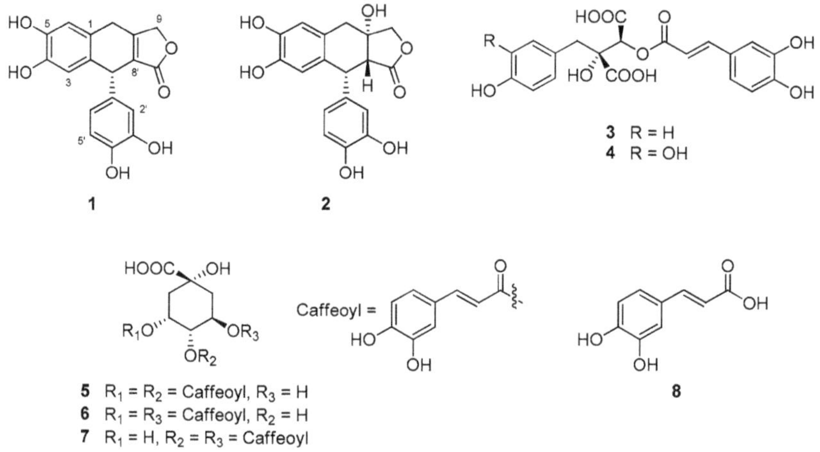

Figure 1. Chemical structures of the isolates **1–8**.

All the isolates from the leaves of *P. japonicus* were evaluated for their inhibitory effects on lipopolysaccharide (LPS)-induced production of pro-inflammatory mediators NO and prostaglandin E_2 (PGE_2) in the LPS-stimulated RAW 264.7 macrophages. We describe isolation of the secondary metabolites from the leaves of *P. japonicus*, structure elucidation of the two new lignans (**1** and **2**), and anti-inflammatory effects of the isolates as well as the possible mechanism.

2. Materials and Methods

2.1. General Experimental Procedures

General experimental procedures are described in the Supplementary Materials.

2.2. Plant Material

The leaves of *Petasites japonicus* (Asteraceae) were obtained from Nature Bio Co. (Seoul, Republic of Korea), in October 2016. The plant material was identified by one of the authors (D.S.J.) and the plant specimen (PEJA-2016) has been deposited in the Laboratory of Natural Product Medicine, College of Pharmacy, Kyung Hee University.

2.3. Isolation of Compounds

The dried leaves (500 g) were extracted once with 10 L of boiled water for 4 h and the solvent was evaporated with freeze drying. The extract (100.0 g) was separated over Diaion HP-20 (Mitsubishi, Tokyo, Japan) column eluted with an H_2O-acetone gradient (from 1:0 to 0:1 *v/v*, gradient) to give 15 fractions (K1–K15). A part of fraction K3 was fractionated with medium pressure liquid chromatography (MPLC) using Redi Sep (Teledyne Isco, Lincoln, NE, USA)-C18 cartridge (13 g, acetonitrile–H_2O, 0:1 to 3:7 *v/v*, gradient) and purified by high performance liquid chromatography (HPLC) using YMC Pack ODS-A column (Phenomenex, Torrance, CA, USA), yielding compound **8** (7.7 mg). Fraction K6 was separated over Sephadex LH-20 (Amersham Pharmacia Biotech, Buckinghamshire, United Kingdom) column with an acetone–H_2O mixture (6:4 *v/v*) as solvent to give three fractions (K6-1–K6-3). Fraction K6-2 was fractionated further using Sephadex LH-20 with an acetone–H_2O mixture (2:8 *v/v*) to yield six subfractions (K6-2-1–K6-2-6). Compound **4** (0.5 g) and **3** (27.1 mg) were obtained from fraction K6-2-4 using LiChroprep RP-18 (Merck, Kenilworth, NJ, USA) CC. Fraction K8 was loaded to Sephadex LH-20 as stationary phase eluting with EtOH–H_2O mixture (1:1 *v/v*) to afford 12 pooled fractions (K8-1–K8-12). Compound **5** (40.3 mg), **6** (4.2 mg), and **7** (4.7 mg) were purified from fraction K8-6 by HPLC with a Luna 10 µm C18(2) 100 Å column. Subfraction K8-9 was purified using Luna 10 µm C18 (2) 100 Å column to obtain compound **2** (61.6 mg). Fraction K9 was fractionated further using Sephadex LH-20 column eluted with the EtOH–H_2O mixture (1:1 *v/v*) to generate ten fractions (K9-1–K9-10). Fraction K9-6 was purified by HPLC using YMC Pack ODS-A column, yielding compound **1** (4.2 mg).

2.3.1. Petasitesin A (**1**)

Dark brownish powder; $[\alpha]_D^{23}$: −11.5° (*c* 0.1, MeOH); ultraviolet (UV) (MeOH) λ_{max} (log ε) 204 nm (3.85), 262 nm (3.57); CD (CH_3CN) λ_{max} 214 (−10.4), 239 (2.1), 254 (−5.3); infrared (IR) (ATR) ν_{max} 3333, 2915, 2847, 1718, 1524, 1240 cm^{-1}; High resolution electrospray ionization mass spectrometry (HRESIMS) (HRESIMS) (negative mode) *m/z* 325.0714 [M−H]$^-$ (calculated for $C_{18}H_{13}O_6$, 325.0712) (Figure S1); ^{1}H and ^{13}C NMR data (Table 1) (Figures S2 and S3); 2D NMR data (Figures S4–S7).

Table 1. ^{1}H (500 MHz) and ^{13}C NMR (125 MHz) data of compounds **1** and **2**.

Position	1 (in Acetone-d_6)		2 (in CD$_3$OD)	
	δ_C	δ_H Multi (*J* in Hz)	δ_C	δ_H Multi (*J* in Hz)
1	129.8		127.0	
2	136.9		130.1	
3	116.9	6.60 s	117.1	6.55 s
4	145.1		145.5	
5	145.3		145.7	
6	115.7	6.72 s	117.0	6.64 s
7	29.0	3.86 d (23.0) 3.62 overlapped	39.6	2.83 d (14.5) 2.67 d (14.5)
8	160.9		78.0	
9	72.4	4.97 d (17.0) 4.89 d (17.0)	80.1	4.17 d (10.0) 3.93 d (10.0)
1'	123.0		135.5	
2'	116.2	6.59 d (2.0)	116.0	6.58 d (2.0)
3'	145.9		146.3	
4'	144.6		144.8	
5'	116.1	6.63 d (8.0)	116.2	6.68 d (8.0)
6'	120.3	6.45 dd (8.0, 2.0)	120.2	6.52 dd (8.0, 2.0)
7'	42.3	4.53 s	47.4	4.18 d (3.0)
8'	128.2		56.2	3.24 d (3.0)
9'	173.9		181.0	

2.3.2. Petasitesin B (**2**)

Dark brownish powder; $[\alpha]_D^{23}$: −20.6° (*c* 0.1, MeOH); UV (MeOH) λ_{max} (log ε) 206 nm (4.34), 287 nm (3.48); CD (CH_3CN) λ_{max} 203 (4.0), 210 (−5.4), 222 (−4.4), and 229 (1.9); IR (ATR) ν_{max} 3305, 1768, 1606, 1514 cm^{-1}; HRESIMS (negative mode) *m/z* 343.0810 [M−H]$^-$ (calculated for $C_{18}H_{15}O_7$, 343.0818) (Figure S8); ^{1}H and ^{13}C NMR data (Table 1) (Figures S9 and S10); 2D NMR data (Figures S11–S14).

2.4. Computational Methods

ECD and VCD calculations of compounds **1** and **2** were conducted as described previously [10,11]. In brief, their 3D models were built from Chem3D modeling. Conformational analysis was performed by the MMFF force field as implemented in Spartan'14 software (Wavefunction, Inc., Irvine, CA, USA; 2014). Geometrical optimization of the selected conformers was performed at the B3LYP/6–31 + G (d,p) level by Gaussian 09 software (Revision E.01; Gaussian, Inc., Wallingford, CT, USA; 2009).

The theoretical ECD and VCD spectra were calculated at the CAM-B3LYP/SVP level with a CPCM solvent model (acetonitrile) and at the DFT [B3LYP/6–31 + G(d,p)] basis set level by the Gaussian 09 software, respectively.

2.5. Measurement of NO Production

The 3-[4"C-dimethylthiazol-2-yl]-2,5-dipheyl tetrazolium bromide (MTT) and Griess reaction assays were used for cell viability studies and measuring nitrite levels, respectively, as reported previously [12].

2.6. Measurement of PGE_2

The RAW 264.7 macrophage cell lines were pretreated with various concentrations of the extract and isolates **1–8** for 1 h and then stimulated with or without LPS (1 µg/mL) for 24 h. A selective COX-2 inhibitor, NS-398 (*N*-[2-(cyclohexyloxy)-4-nitrophenyl]methanesulfonamide; Sigma Aldrich, St. Louis, MO, USA) was used as a positive control for blocking PGE_2 production. PGE_2 levels in cell culture mediums were measured using the same methods as described in the previous paper [12].

2.7. Measurement of iNOS and COX-2 Expression

Quantitative polymerase chain reaction (qPCR) using Thermal Cycler Dice Real Time PCR System (Takara Bio Inc., Shiga, Japan) was used to determine the steady-state mRNA levels of inducible nitric oxide synthase (iNOS) and cyclooxygenase-2 (COX-2) as reported previously [12].

2.8. Molecular Docking Studies

The software AutoDock Vina with AutoDock Tools (The Scripps Research Institute, La Jolla, CA, USA: ADT 1.5.6) using the hybrid Lamarckian Genetic Algorithm (LGA) was used for performing molecular docking simulations as reported in the literature [13,14]. In short, the 3D crystal structures (resolution: 2.5 Å) of iNOS (PDB code: 3E6T) and COX-2 (PDB code: 1PXX) were obtained from the RCSB Protein Data Bank. The configurations of compounds **1** and **3** were determined by their nuclear overhauser effect spectroscopy (NOESY) spectra and time-dependent density functional theory (TDDFT) ECD calculations. Chem3D Pro 14.0 software (CambridgeSoft, Waltham, MA, USA) was used for construction of the standard 3D structures (PDB format) of compounds **1** and **3**.

3. Results

3.1. Structure Elucidation of Compounds *1* and *2*

Compound **1** was obtained as a dark brownish powder, and its molecular formula was identified as $C_{18}H_{14}O_6$ by HRESIMS (*m/z* 325.0714 [M−H]⁻; calculated for $C_{18}H_{13}O_6$, 325.0712). It exhibited UV maxima at 262 nm and IR maxima at 3333, 1718, and 1524 cm⁻¹, suggesting the presence of a hydroxyl, ester group, and aromatic ring. The ^{13}C NMR spectral data of compound **1** (Table 1) exhibited 18 carbon signals including a carbonyl carbon (δ_C 173.9), 12 aromatic carbons (from δ_C 115.7 to 145.9), an oxygenated methylene carbon (δ_C 72.4), a methine carbon (δ_C 42.3), and a methylene carbon (δ_C 29.0).

The remaining two quaternary carbons (δ_C 160.9 and 128.2) were derived from a double bond. The ^1H NMR spectrum revealed one 1,2,4,5-tetrasubstituted aromatic ring [δ_H 6.72 (s, H-6) and 6.60 (s, H-3)], and the one 1,3,4-trisubstituted aromatic ring [δ_H 6.63 (d, *J* = 8.0, H-5′), 6.59 (d, *J* = 2.0, H-2′), and 6.45 (dd, *J* = 8.0 and 2.0, H-6′)], an oxygenated methylene [δ_H 4.97 (d, *J* = 17.0) and 4.89 (d, *J* = 17.5), H-9], a methine [δ_H 4.53 (s, H-7′)], and a methylene [δ_H 3.86 (d, *J* = 23.0) and 3.62 (overlapped), H-7]. The heteronuclear multiple bond correlation spectroscopy (HMBC) correlations of **1** (Figure 2) suggest aryltetralin lactone type lignan with a double bond at C-8 and C-8′.

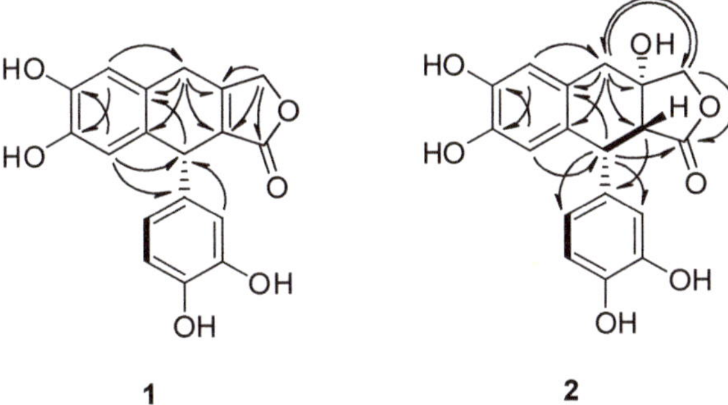

Figure 2. Selected correlations of compounds **1** and **2**: correlation spectroscopy (COSY, ▬) and heteronuclear multiple bond correlation spectroscopy (HMBC, →) (in acetone-d_6 and methanol-d_4).

The absolute configuration at C-7′ of compound **1** was established by comparing its experimental ECD spectrum with those calculated spectra of (7′R) and (7′S) models using the time-dependent density functional theory (TDDFT) method. The experimental ECD spectrum of compound **1** exhibited a positive Cotton effect (CE) at 239 nm (Δε +2.1) and negative CEs at 214 nm (Δε −10.4) and 254 nm (Δε −5.3). The experimental data (Figure 3) was in agreement with the calculated ECD spectrum of the (7′R) model, suggesting the absolute configuration of compound **1** as (7′R). Thus, the structure of **1** was elucidated as (R)-9-(3, 4-dihydroxyphenyl)-6,7-dihydroxy-4,9-dihydronaphtho[2′′C-c]furan-1(3H)-one, and was named as petasitesin A.

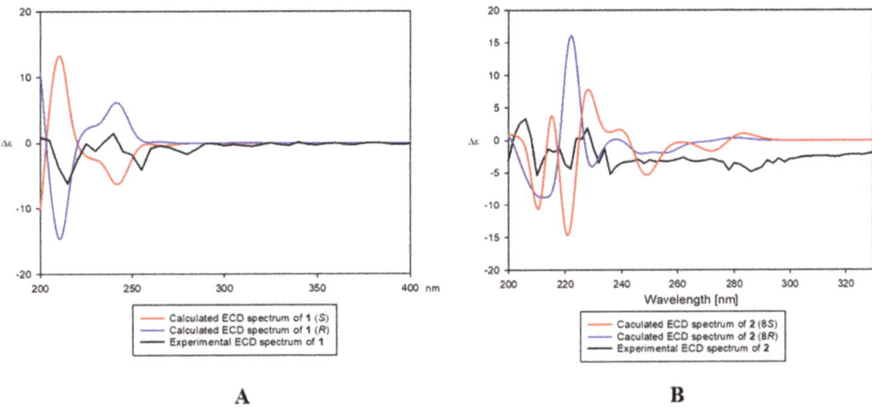

Figure 3. Comparison of experimental and calculated electronic circular dichroism (ECD) spectra of compounds **1** (**A**) and **2** (**B**).

The compound **2** was isolated as a dark brownish powder. Its molecular formula was established as $C_{18}H_{16}O_7$ by HRESIMS (m/z 343.0810 [M−H]⁻; calculated for $C_{18}H_{15}O_7$, 343.0818). The ¹H and ¹³C NMR data of **2** were similar to those of **1** (Table 1), although the NMR solvents were different from each other due to the different solubility of the compounds. Comparison of the ¹³C NMR data and molecular weights from **1** and **2** suggested that the carbons C-8 and C-8′ of **1** with a double bonded linkage (δ_C 160.9 and 128.2) were replaced by an oxygenated quaternary (δ_C 78.0) and methine (δ_C 56.2) carbon atoms. The correlation spectroscopy (COSY) correlation between H-7′ (δ_H 4.18) and H-8′ (δ_H 3.24), and the HMBC experiment revealed aryltetralin lactone type lignan (Figure 4). Considering a biogenetic relationship with **1**, the absolute configuration at C-7′ of **2** was suggested to be (R)-configuration [15].

The coupling constant of 3.0 Hz between H-7′ and H-8′ suggested the *cis*-geometry of C-7′ and C-8′. It was further confirmed by the NOESY interaction of H-7′ and H-8′ [16].

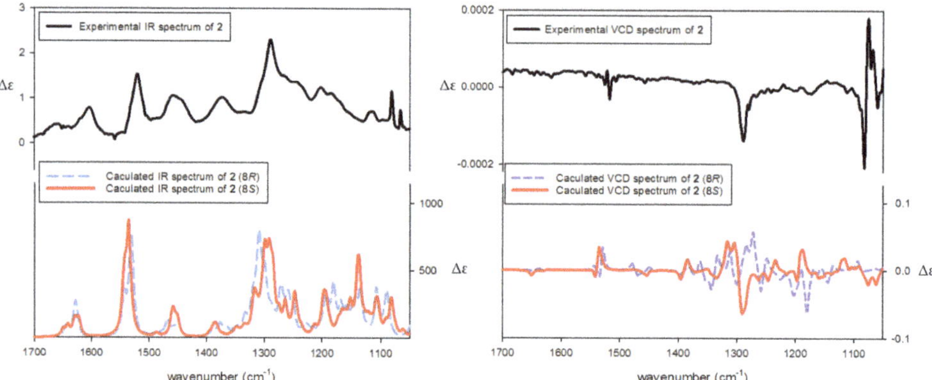

Figure 4. Comparison of experimental and calculated vibrational circular dichroism (VCD) spectra of compound **2** (*c* 0.5 M, DMSO-d_6).

To determine the absolute configuration C-8 of **2**, experimental ECD spectrum of **2** was compared with the calculated spectra of (8*R*,7′*R*,8′*R*) and (8*S*,7′*R*,8′*R*) models using the TDDFT method. The experimental ECD spectrum of **2** showed positive CEs at 203 nm (Δε +4.0) and 229 nm (Δε +1.9), and negative CEs at 210 nm (Δε −5.4) and 222 nm (Δε −4.4). The experimental spectrum (Figure 3) was also in agreement with the calculated ECD spectrum of (8*S*,7′*R*,8′*R*) model. Moreover, the VCD spectrum of **2** was measured additionally to establish the configuration at C-8. The conformity of the experimental IR and VCD spectra and theoretical spectra of **2** suggested the absolute configuration of **2** as (8*S*,7′*R*,8′*R*) (Figure 4). Therefore, the structure of **2** was proposed as (9*R*,3a*S*,9a*R*)-9-(3,4-dihydroxyphenyl)-6,7,3a-trihydroxy-4,9,3a,9a-tetrahydronaphtho[2,3-*c*]furan-1(3*H*)-one, and was named as petasitesin B.

Compounds **3–8** were identified as cimicifugic acid D (**3**) [17], fukinolic acid (**4**) [7], 3,4-dicaffeoylquinic acid (**5**) [18], 3,5-dicaffeoylquinic acid (**6**) [18], 4,5-dicaffeoylquinic acid (**7**) [18], and caffeic acid (**8**) [19] by comparison of their NMR data with those reported.

3.2. Anti-inflammatory Effects of the Isolates

As shown in Table 2, cimicifugic acid D (**3**) and the new compound **1** (petasitesin A) exhibited significant inhibitory activities against NO production with IC_{50} values of 12 ± 1.1 and 15 ± 1.4 μM, respectively, without affecting the cell viability (Figure S15). 4,5-Dicaffeoylquinic acid showed mild activity with an observed IC_{50} value of 38.9 ± 0.72 μM. On the other hand, compound **1** showed the most potent inhibitory effect on PGE_2 production with an IC_{50} value of 17 ± 3.2 μM (Table 2) in a dose-dependent manner (Figure 5). These results suggest that compound **1** might have an anti-inflammatory effect due to inhibition of the production of NO and PGE_2 which are the key inflammatory mediators of macrophages. It is worth noting that compound **1** significantly suppressed the expression of NO and PGE_2 synthesis enzymes, inducible nitric oxide synthase (iNOS) and cyclooxygenase-2 (COX-2), respectively (Figure 6), in a concentration-dependent manner. The data indicate that the inhibitory effect of compound **1** on NO and PGE_2 production in macrophages is related to the regulation of iNOS and COX-2 expression.

Table 2. Inhibitory effects of the compounds from *P. japonicus* on NO and PGE$_2$ production in lipopolysaccharide (LPS)-induced RAW 264.7 cells.

Compound	IC$_{50}$ (μM)[a]	
	NO	PGE$_2$
1	15 ± 1.4	17 ± 3.2
2	>50	>50
Cimicifugic acid D (3)	12 ± 1.1	43 ± 7.9
4,5-Dicaffeoylquinic acid	38.9 ± 0.72	>50
Caffeic acid	>50	45.7 ± 0.87

[a] The values represent the means of the results from three independent experiments with similar patterns. l-N^6-(1-Iminoethyl)lysine (l-NIL) and *N*-[2-(cyclohexyloxy)-4-nitrophenyl]methanesulfonamide (NS-398) were used as a positive control substance for NO [half maximal inhibitory concentration IC$_{50}$) value = 1.62 ± 0.08 μM] and prostaglandin E$_2$ (PGE$_2$) productions (IC$_{50}$ value = 3.3 ± 0.15 μM), respectively. Three known compounds, fukinolic acid, 3,4-dicaffeoylquinic acid, and 3,5-dicaffeoylquinic acid were inactive (IC$_{50}$ value > 50 μM) in this assay system.

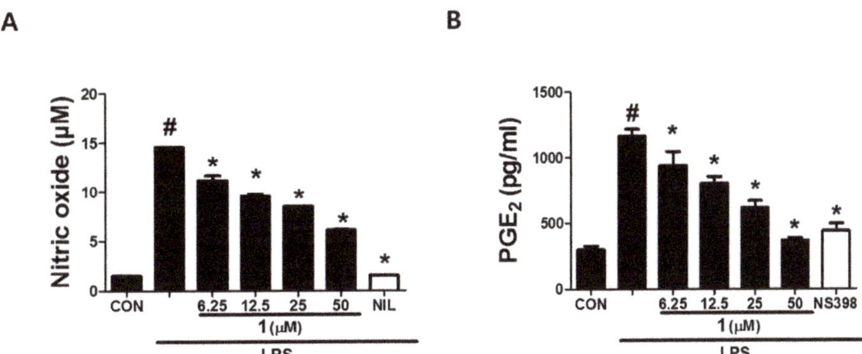

Figure 5. Effects of compound **1** (6.25, 12.5, 25 or 50 μM) on LPS-stimulated NO (**A**) and PGE$_2$ (**B**) productions in RAW 264.7 macrophages. #: $p < 0.05$ as compared with the untreated group, *: $p < 0.05$ as compared with the LPS only-treated group.

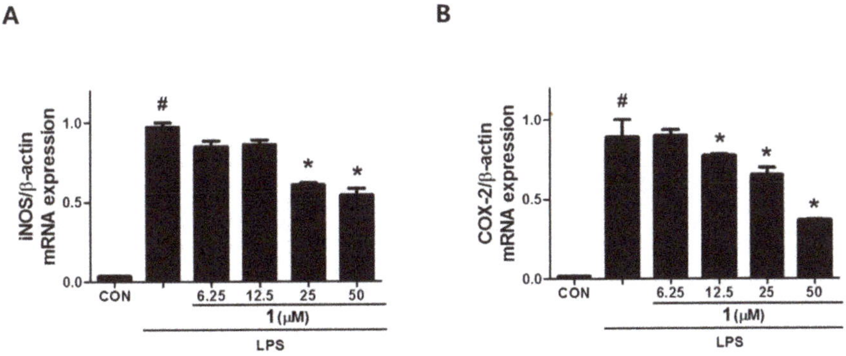

Figure 6. Effect of compound **1** on the expression of iNOS (**A**) and COX-2 (**B**) in RAW 264.7 macrophages. #: $p < 0.05$ as compared with the untreated group, *: $p < 0.05$ as compared with the LPS only-treated group.

To better understand the molecular mechanism of inhibitory activities against NO and PGE$_2$ production, the most active compounds **1** and **3** were subjected to molecular docking studies. The results showed that **1** and **3** had strong affinities with both NO and PGE$_2$ synthesis enzymes, iNOS and COX-2 (Figure 7, Table 3). The binding residues and logarithms of free binding energy are given in Table 3. These results implicated that **1** and **3** may directly interact with the cavity residues of iNOS and COX-2, leading to the activity reduction of free iNOS and COX-2 enzymes. Taken together, these results indicate petasitesin A (**1**), a novel lignan isolated from butterbur leaves extract, exhibits anti-inflammatory properties by suppressing NO and PGE$_2$ production via inhibiting the expression of iNOS and COX-2 and binding to the free iNOS and COX-2 enzymes.

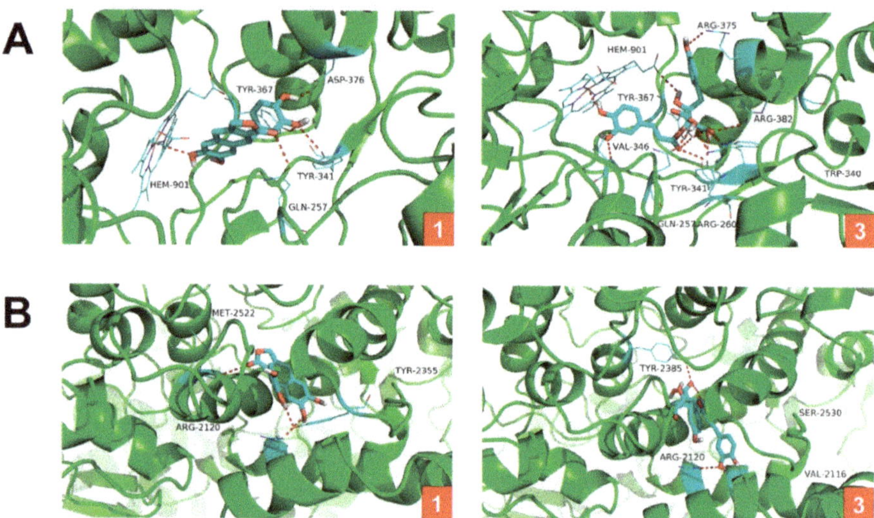

Figure 7. Molecular docking results of compounds **1** and **3** with iNOS (**A**) and COX-2 (**B**) enzymes. Molecular docking simulations obtained at the lowest energy conformation, highlighting potential hydrogen contacts of **1** and **3**, respectively (nitrogen is blue; oxygen is red; carbon is cyan; hydrogen is gray). For clarity, only interacting residues are labeled. Hydrogen bonding interactions are shown by dashes. These figures were created by PyMOL (Schrödinger, LLC, New York, NY, USA: version 1.3).

Table 3. Logarithms of free binding energies (FBE, kcal/mol) of NO inhibitors to the active cavities of iNOS (PDB Code: 3E6T) and COX-2 (PDB code: 1PXX) and targeting residues of the binding site located on the mobile flap.

Compound	−Log (FBE)		Targeting Residues	
	iNOS	COX-2	iNOS	COX-2
1	−8.8	−7.5	ASP-376, TYR-367, TYR-341, GLN-257, HEM-901	ARG-2120, TYR-3355, MET-2522
3	−10.0	−8.3	ARG-260, ARG-375, ARG-382 TYR-341, TYR-367, TRP-340, GLN-257, HEM-901, VAL-346	ARG-2120, TYR-2385, VAL-2116, SER-2530

4. Discussion

In the present study, we isolated two new aryltetralin lactone lignans, petasitesin A and B (**1** and **2**) from the leaves of *P. japonicus*. To the best of our knowledge, this is the first report on the isolation of the aryltetralin lactone type lignans from the leaves of *P. japonicas*. Although cimicifugic acid D (**3**) has been isolated from *Cimicifuga* spp. including black cohosh (*Cimicifuga racemosa*) and possesses vasoactive effect and hyaluronidase inhibitory activity [20,21], this is the first finding that it presents in *P. japonicus* and inhibits pro-inflammatory mediators, NO and PGE$_2$.

A new lignan petasitesin A (**1**) showed a potent inhibitory effect on the production of both NO and PGE$_2$ in LPS-stimulated macrophages (IC$_{50}$ values < 20 μM). Our molecular docking studies reveal that petasitesin A (**1**) can interact with the cavity residues of both iNOS and COX-2. Interestingly, petasitesin A (**1**) also inhibited the mRNA expression of iNOS and COX-2 induced by LPS in macrophages. However, the molecular mechanism of action underlying the gene expression regulation by petasitesin A remains to be investigated. Considering that LPS binds to toll-like receptor 4 (TLR4), the TLR4-mediated NF-κB pathway is likely associated with the inhibition of iNOS and COX-2 expression by petasitesin A. In fact, NF-κB is a key transcriptional factor to regulate the iNOS and COX-2 gene in macrophages under the inflammatory condition. In this regard, the effect of petasitesin A on the NF-κB pathway can be further elucidated.

5. Conclusions

New lignans (compounds **1** and **2**) and six known compounds were isolated and identified from the leaves of *P. japonicus*. Petasitesin A (**1**) and cimicifugic acid D (**3**) inhibit production of inflammatory mediators NO and PGE$_2$. PetasitesinA (**1**) inhibits iNOS and COX-2 expression, and petasitesin A (**1**) and cimicifugic acid D (**3**) have strong affinities with both iNOS and COX-2 enzymes in molecular docking studies. Thus, petasitesin A (**1**) and cimicifugic acid D (**3**) are worthy of further pharmacological evaluation for their potential as anti-inflammatory drugs.

Supplementary Materials: The following are available online at http://www.mdpi.com/2218-273X/9/12/806/s1, Figure S1: HR-ESI-MS spectrum of compound **1**, Figure S2: The ^{1}H-NMR (500 MHz, CD$_3$COCD$_3$) spectrum of compound **1**, Figure S3: The ^{13}C-NMR (125 MHz, CD$_3$COCD$_3$) spectrum of compound **1**, Figure S4: The HSQC spectrum of compound **1** in CD$_3$COCD$_3$, Figure S5: The COSY spectrum of compound **1** in CD$_3$COCD$_3$, Figure S6: The HMBC spectrum of compound **1** in CD$_3$COCD$_3$, Figure S7: The NOESY spectrum of compound **1** in CD$_3$COCD$_3$, Figure S8: HR-ESI-MS spectrum of compound **2**, Figure S9: The ^{1}H-NMR (500 MHz, CD$_3$OD) spectrum of compound **2**, Figure S10: The ^{13}C-NMR (125 MHz, CD$_3$OD) spectrum of compound **2**, Figure S11: The HSQC spectrum of compound **2** in CD$_3$OD, Figure S12: The COSY spectrum of compound **2** in CD$_3$OD, Figure S13: The HMBC spectrum of compound **2** in CD$_3$OD, Figure S14: The NOESY spectrum of compound **2** in CD$_3$OD, Figure S15: Cell viability of the isolates from *P. japonicus*.

Author Contributions: D.L. and D.S.J. conceived and designed the experiments; J.S.L., M.J., S.P., S.M.R., and Z.S. performed the experiments; J.S.L. and S.M.R. analyzed the data; J.L., Y.G., and J.-H.C. interpreted the data and contributed to manuscript structure and flow; and J.S.L., M.J., and Z.S. wrote the paper. All authors reviewed and confirmed the manuscript.

Funding: This research was supported by Korea Institute of Planning and Evaluation for Technology in Food, Agriculture and Forestry (IPET) through Agri-Bio Industry Technology Development Program, funded by Ministry of Agriculture, Food and Rural Affairs (MAFRA), Republic of Korea (116044-1, 118046-03). This work was also supported by the National Research Foundation of Korea (NRF) grant funded by Ministry of Science and ICT (MSIT), Republic of Korea (NRF-2019R1A2C1083945).

References

1. Kusano, A.; Seyama, Y.; Nagai, M.; Shibano, M.; Kusano, G. Effects of fukinolic acid and cimicifugic acids from *Cimicifuga* species on collagenolytic activity. *Biol. Pharm. Bull.* **2001**, *24*, 1198–1201. [CrossRef] [PubMed]

2. Mizushina, Y.; Ishidoh, T.; Kamisuki, S.; Nakazawa, S.; Takemura, M.; Sugawara, F.; Yoshida, H.; Sakaguchi, K. Flavonoid glycoside: A new inhibitor of eukaryotic DNA polymerase α and a new carrier for inhibitor-affinity chromatography. *Biochem. Biophys. Res. Commun.* **2003**, *301*, 480–487. [CrossRef]

3. Naya, K.; Hayashi, M.; Takagi, I.; Nakamura, S.; Kobayashi, M. The structural elucidation of sesquiterpene lactones from *Petasites japonicus* Maxim. *Bull. Chem. Soc. Jpn.* **1972**, *45*, 3673–3685. [CrossRef]

4. Tori, M.; Kawahara, M.; Sono, M. Eremophilane-type sesquiterpenes from fresh rhizomes of Petasites japonicus. *Phytochemistry* **1997**, *47*, 401–409. [CrossRef]

5. Wang, S.; Jin, D.Q.; Xie, C.; Wang, H.; Wang, M.; Xu, J.; Guo, Y. Isolation, characterization, and neuroprotective activities of sesquiterpenes from Petasites japonicus. *Food Chem.* **2013**, *141*, 2075–2082. [CrossRef] [PubMed]

6. Shimoda, H.; Tanaka, J.; Yamada, E.; Morikawa, T.; Kasajima, N.; Yoshikawa, M. Anti type I allergic property of Japanese butterbur extract and its mast cell degranulation inhibitory ingredients. *J. Agric. Food Chem.* **2006**, *54*, 2915–2920. [CrossRef] [PubMed]

Biomolecules **2019**, *9*, 806

7. Sakamura, S.; Yoshihara, T.; Toyoda, K. The constituents of Petasites japonicus: Structures of fukiic acid and fukinolic acid. *Agric. Biol. Chem.* **1973**, *37*, 1915–1921. [CrossRef]

8. Jeong, M.; Kim, H.M.; Kim, H.J.; Choi, J.H.; Jang, D.S. Kudsuphilactone B, a nortriterpenoid isolated from *Schisandra chinensis* fruit, induces caspase-dependent apoptosis in human ovarian cancer A2780 cells. *Arch. Pharm. Res.* **2017**, *40*, 500–508. [CrossRef]

9. Du, Y.E.; Lee, J.S.; Kim, H.M.; Ahn, J.H.; Jung, I.H.; Ryu, J.H.; Choi, J.H.; Jang, D.S. Chemical constituents of the roots of *Codonopsis lanceolata*. *Arch. Pharm. Res.* **2018**, *41*, 1082–1091. [CrossRef]

10. Kwon, J.; Lee, H.; Ko, W.; Kim, D.C.; Kim, K.W.; Kwon, H.C.; Guo, Y.; Sohn, J.H.; Yim, J.H.; Kim, Y.C.; et al. Chemical constituents isolated from Antarctic marine-derived *Aspergillus* sp. SF-5976 and their anti-inflammatory effects in LPS-stimulated RAW 264.7 and BV2 cells. *Tetrahedron* **2017**, *73*, 3905–3912. [CrossRef]

11. Ryu, S.M.; Lee, H.M.; Song, E.G.; Seo, Y.H.; Lee, J.; Guo, Y.; Kim, B.S.; Kim, J.J.; Hong, J.S.; Ryu, K.H.; et al. Antiviral activities of trichothecenes isolated from *Trichoderma albolutescens* against pepper mottle virus. *J. Agric. Food. Chem.* **2017**, *65*, 4273–4279. [CrossRef] [PubMed]

12. Shin, J.S.; Lee, K.G.; Lee, H.H.; Lee, H.J.; An, H.J.; Nam, J.H.; Jang, D.S.; Lee, K.T. α-Solanine Isolated from *Solanum Tuberosum*, L. cv Jayoung Abrogates LPS-Induced Inflammatory Responses Via NF-κB Inactivation in RAW 264.7 Macrophages and Endotoxin-Induced Shock Model in Mice. *J. Cell. Biochem.* **2016**, *117*, 2327–2339. [CrossRef] [PubMed]

13. Ma, J.; Ren, Q.; Dong, B.; Shi, Z.; Zhang, J.; Jin, D.Q.; Xu, J.; Ohizumi, Y.; Lee, D.; Guo, Y. NO inhibitory constituents as potential anti-neuroinflammatory agents for AD from *Blumea balsamifera*. *Bioorg. Chem.* **2018**, *76*, 449–457. [CrossRef] [PubMed]

14. Liu, F.; Yang, X.; Ma, J.; Yang, Y.; Xie, C.; Tuerhong, M.; Jin, D.Q.; Xu, J.; Lee, D.; Ohizumi, Y.; et al. Nitric oxide inhibitory daphnane diterpenoids as potential anti-neuroinflammatory agents for AD from the twigs of *Trigonostemon thyrsoideus*. *Bioorg. Chem.* **2017**, *75*, 149–156. [CrossRef]

15. Won, T.H.; Song, I.H.; Kim, K.H.; Yang, W.Y.; Lee, S.K.; Oh, D.C.; Oh, W.K.; Oh, K.B.; Shin, J. Bioactive metabolites from the fruits of *Psoralea corylifolia*. *J. Nat. Prod.* **2015**, *78*, 666–673. [CrossRef]

16. Liu, Y.; Young, K.; Rakotondraibe, L.H.; Brodie, P.J.; Wiley, J.D.; Cassera, M.B.; Callmander, M.W.; Rakotondrajaona, R.; Rakotobe, E.; Rasamison, V.E.; et al. Antiproliferative compounds from *Cleistanthus boivinianus* from the Madagascar dry forest1. *J. Nat. Prod.* **2015**, *78*, 1543–1547. [CrossRef]

17. Takahira, M.; Kusano, A.; Shibano, M.; Kusano, G.; Miyase, T. Piscidic acid and fukiic acid esters from Cimicifuga simplex. *Phytochemistry* **1998**, *49*, 2115–2119. [CrossRef]

18. Wu, Q.Z.; Zhao, D.X.; Xiang, J.; Zhang, M.; Zhang, C.F.; Xu, X.H. Antitussive, expectorant, and anti-inflammatory activities of four caffeoylquinic acids isolated from *Tussilago farfara*. *Pharm. Biol.* **2016**, *54*, 1117–1124. [CrossRef]

19. Park, H.Y.; Nam, M.H.; Lee, H.S.; Jun, W.; Hendrich, S.; Lee, K.W. Isolation of caffeic acid from Perilla frutescens and its role in enhancing γ-glutamylcysteine synthetase activity and glutathione level. *Food Chem.* **2010**, *119*, 724–730. [CrossRef]

20. Noguchi, M.; Nagai, M.; Koeda, M.; Nakayama, S.; Sakurai, N.; Takahira, M.; Kusano, G. Vasoactive effects of cimicifugic acids C and D, and fukinolic acid in cimicifuga rhizome. *Biol. Pharm. Bull.* **1998**, *21*, 1163–1168. [CrossRef]

21. Iwanaga, A.; Kusano, G.; Warashina, T.; Miyase, T. Hyaluronidase inhibitors from "*Cimicifugae Rhizoma*" (a mixture of the rhizomes of *Cimicifuga dahurica* and *C. heracleifolia*). *J. Nat. Prod.* **2010**, *73*, 573–578. [CrossRef] [PubMed]

 *biomolecules*

Article

Discovery of Kynurenines Containing Oligopeptides as Potent Opioid Receptor Agonists

Edina Szűcs [1,2], Azzurra Stefanucci [3,*], Marilisa Pia Dimmito [3], Ferenc Zádor [1], Stefano Pieretti [4], Gokhan Zengin [5], László Vécsei [6], Sándor Benyhe [1], Marianna Nalli [7] and Adriano Mollica [3]

[1] Institute of Biochemistry, Biological Research Center, Hungarian Academy of Sciences, Temesvári krt. 62., H-6726 Szeged, Hungary; szucs.edina@brc.hu (E.S.); zador.ferenc@brc.hu (F.Z.); benyhe.sandor@brc.hu (S.B.)
[2] Doctoral School of Theoretical Medicine, Faculty of Medicine, University of Szeged, Dómtér 10, H-6720 Szeged, Hungary
[3] Department of Pharmacy, University of Chieti-Pescara "G. d'Annunzio", Via dei Vestini 31, 66100 Chieti, Italy; marilisa.dimmito@unich.it (M.P.D.); a.mollica@unich.it (A.M.)
[4] National Center for Drug Research and Evaluation, Istituto Superiore di Sanità, Viale Regina Elena 299, 00161 Rome, Italy; stefano.pieretti@iss.it
[5] Department of Biology, Science Faculty, Selcuk University, 42250 Konya, Turkey; gokhanzengin@selcuk.edu.tr
[6] MTA-SZTE Neuroscience Research Group, Department of Neurology, Interdisciplinary Excellence Centre, Faculty of Medicine, University of Szeged, H-6725 Szeged, Hungary; vecsei.laszlo@med.u-szeged.hu
[7] Laboratory affiliated with the Institute Pasteur Italy-Cenci Bolognetti Foundation, Department of Drug Chemistry and Technologies, Sapienza University of Rome, Piazzale Aldo Moro 5, I-00185 Roma, Italy; marianna.nalli@uniroma1.it
* Correspondence: a.stefanucci@unich.it

Received: 27 December 2019; Accepted: 6 February 2020; Published: 12 February 2020

Abstract: Kynurenine (kyn) and kynurenic acid (kyna) are well-defined metabolites of tryptophan catabolism collectively known as "kynurenines", which exert regulatory functions in host-microbiome signaling, immune cell response, and neuronal excitability. Kynurenine containing peptides endowed with opioid receptor activity have been isolated from natural organisms; thus, in this work, novel opioid peptide analogs incorporating L-kynurenine (L-kyn) and kynurenic acid (kyna) in place of native amino acids have been designed and synthesized with the aim to investigate the biological effect of these modifications. The kyna-containing peptide (**KA1**) binds selectively the μ-opioid receptor with a Ki = 1.08 ± 0.26 (selectivity ratio μ/δ/κ = 1:514:10,000), while the L-kyn-containing peptide (**K6**) shows a mixed binding affinity for μ, δ, and κ-opioid receptors, with efficacy and potency (E_{max} = 209.7 + 3.4%; $LogEC_{50}$ = −5.984 + 0.054) higher than those of the reference compound DAMGO. This novel oligopeptide exhibits a strong antinociceptive effect after i.c.v. and s.c. administrations in in vivo tests, according to good stability in human plasma ($t_{1/2}$ = 47 min).

Keywords: peptides; kynurenines; binding affinity; μ-opioid receptor; pharmacophore; G-protein activation

---

1. Introduction

The kynurenine pathway (KP) is an essential part of the tryptophan metabolism in mammalian tissues, where it is responsible for the formation of two principal metabolites, namely, L-kynurenine (kyn) and kynurenic acid (kyna). Kyn can arise in peptides and proteins by post/translational modifications or direct oxidation of tryptophan. It is present in lens crystallins, human Cu^{2+}/Zn^{2+} dismutase, milk proteins, actin oxidized in vivo, and several bioactive compounds produced by bacteria and marine organisms [1]. Daptomycin is a cyclic kyn-containing lipodepsipeptide approved by the Food and Drug Administration (FDA), isolated from *Streptomyces roseoporus* used in the treatment of

Gram-positive pathogen skin infections [2]. Cyclomontanin B isolated from *Annona montana* exhibits promising anti-inflammatory activity [3]. The kyn-containing peptide FP-Kyn-L-NH$_2$ is the minor component of Australian red tree frog skin *Litoria rubella* collected in central Australia, endowed with opioid activity at 10^{-7} M (Figure 1) [4].

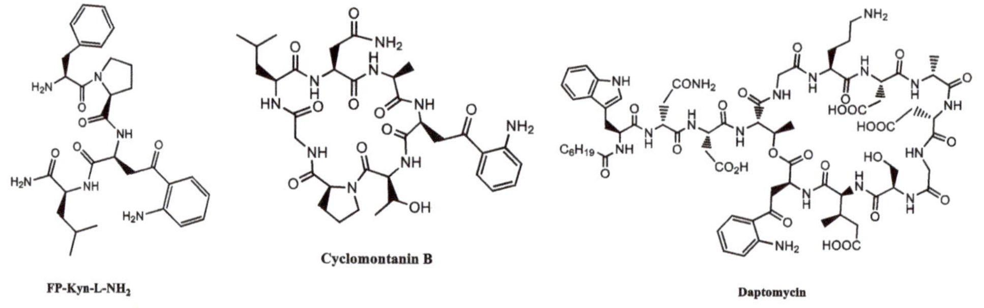

Figure 1. Natural bioactive compounds containing kyn residue.

The occurrence of kyn in natural products suggests a possible specificity towards their biological targets. The enzymes of the human kynurenine pathway are expressed in different tissues and cell types throughout the body [1]. In humans, the majority of kyn is excreted by urine; thus, its bioavailability increases according to the tryptophan flux downstream of the KP [5]. Kyn is able to penetrate the central nervous system (CNS) by transport across the blood–brain barrier (BBB), but it is also produced locally [6].

Kyna has been originally discovered in canine urine, but a huge amount has been measured in the gut, bile, human saliva, synovial and amniotic fluid; it has also been detected in food products such as broccoli, some potatoes, and honeybee products [1]. Kyna possesses an antagonistic effect on the N-methyl-D-aspartate (NMDA) receptor and other glutamate receptors such as AMPA and kainate receptors [7,8]. Kyna is also found to have an agonistic effect on the G protein coupled receptor GPR35 [9,10], which can be found in various tissues and organs such as gastrointestinal tract, liver, immune system, central nervous system, and cardiovascular system [11]. NMDA receptors are essential for the control of the glutamatergic work at the CNS; in contrast to the kainate and AMPA receptors, the NMDA mediates the influx of Ca^{2+} ions into neurons, playing an important role in synaptic plasticity, memory, and learning [7,8]. Overactivation of NMDA receptors can lead to excitotoxicity, severe cell damage, and apoptosis of neurons, which are strongly related to neurodegenerative and CNS disorders such as depression, stroke, ischemia, and neuropathic pain [10–12]. Different therapeutic approaches based on thekynurenine pathway have been postulated to circumvent this problem, such as the use of kynurenic acid prodrugs or analogs able to penetrate more readily than the parent compounds or the involvement of ascorbate conjugation to promote the interaction of kyna with SVCT2 transport protein [13–15]. Intracisternal kyna attenuates formalin-induced nociception in animals together with antagonist activity at the glycine binding site of NMDA, which is associated with analgesic properties in rats [16]. At the peripheral sites, kyna decreases the nociceptive behavior in the tail flick and hot plate tests [16]. Administration of L-kyn and probenecid together with kyna analogs inhibits NMDA receptors in animal models of trigeminal activation and sensitization [17]. Noteworthy, kyna and its analogs are able to act on second-order neurons, decreasing mechanical allodynia and pain sensitivity in different animal pain models [18]. Considering the presence of kyn residue in natural peptide sequences and the important role exerted by both kynurenines at the CNS [19–21], we plan to investigate the biological consequences of the insertion of these residues in opioid pharmacophoric sequences. Kyn could be used in place of phenylalanine, considering its aromatic side chain, whereas kyna could be used as C-terminus to mimic an additional aromatic residue. In this preliminary work, we performed the synthesis and biological screening of six novel kynurenines containing peptides, aiming to investigate the modifications imposed by the presence of kyn and kyna on the biological properties of known endogenous and synthetic opioid peptides in vivo and in vitro. Peptide **KA1** retains the DAMGO primary sequence, but the OH terminal group is esterified by kynurenic acid. Peptides **K2** and **K3** are EM-2 analogs in which the Phe residues in positions 3 and 4 have been replaced with kyn and kyn *C*-terminal amides, respectively. Peptides **K4–K6** are enkephalin-like peptides containing kyn in position 5, bearing as *C*-terminal the methyl ester, acid, and amide group, respectively. The novel chemical entities were prepared following solution phase peptide synthesis and were obtained as TFA salts in good overall yields and excellent purities.

2. Materials and Methods

2.1. Chemistry

All reagents and solvents were acquired from Sigma-Aldrich (Milano, Italy). Solution phase peptide synthesis was applied to prepare the final products **KA1, K2–K6** as TFA salts, following the procedures reported below. Boc-protected intermediates were purified by silica gel column chromatography where necessary, or by trituration in Et_2O. Final products **KA1, K2–K6** were purified by RP–HPLC on a Waters XBridge BEH130 (C18 5.0 μm, 250 × 10 mm column; flow rate of 7 mL/min; Waters Binary pump 1525; eluent: linear gradient of H_2O/ACN 0.1% TFA, ranging from 5% to 95% ACN in 32 min). The purity of the N^α-Boc-protected products was confirmed by NMR analysis on a Varian Mercury 300 MHz. The purity of all final compounds was assessed by NMR analysis, ESI–LRMS, and by analytical RP–HPLC (C18-bonded 4.6 × 150 mm; flow rate of 1 mL/min; eluent: gradient of H_2O/ACN 0.1% TFA, ranging from 5% to 95% ACN in 26 min, recorded at 254, 275, and 213 nm and was found to be ≥95%). The mass spectrometry (MS) equipment was composed as follows: LCQ Thermo Finnigan ion trap mass spectrometer (San Jose, CA, USA) with an electrospray ionization (ESI) source; capillary temperature: 300 °C; spray voltage: 4.00 kV; nitrogen (N_2) as the sheath and auxiliary gas.

2.2. General Procedures

2.2.1. Formation of Ethanolamine-Kynurenic Acid Ester

Kynurenic acid (2 mmol, 1 equiv.) was dissolved in DMF (5 mL) stirring in agitation at 0 °C, then a mixture of Boc-*N*-aminoethanol (2 mmol, 1 equiv.) and DMAP (0.6 mmol, 0.3 equiv.) in DMF (3 mL) was transferred in the round bottom flask. After 10 min, EDC.HCl (2.2 mmol, 1.1 equiv.) was added to the reaction mixture in agitation at 0 °C for 10 min, then at r.t. overnight. The solvent was removed in a rotary evaporator and the oily residue was taken up with EtOAc and washed with 5% citric acid solution (3 times), $NaHCO_3$ s.s. (3 times), and NaCl s.s. (3 times). Organic phases were collected and dried on Na_2SO_4 anhydrous, filtered and dried in rotavapor and high vacuum to give a yellow oily product. The crude product was triturated with Et_2O (2 times), the aqueous layer filtered up, and the white solid product dried in a rotary evaporator and high vacuum.

2.2.2. Coupling Reaction

Boc-protected compound (1.1 equiv.) was dissolved in DMF (5 mL) in an iced-cooled bottom flask, then EDC.HCl (1.1 equiv.) and HOBt hydrate (1.1 equiv.) were added stirring for 10 min. A solution of *N*-terminal free intermediate (1 equiv.) and DIPEA (3.3 equiv.) in DMF (5 mL) was transferred in the ice-cooled bottom flask at 0 °C and allowed to react at r.t. overnight. The solvent was removed in a rotary evaporator; the oily residue was taken up with EtOAc and washed with 5% citric acid solution (3 times), $NaHCO_3$ s.s. (3 times), and NaCl s.s. (3 times). Organic phases were collected and dried on Na_2SO_4 anhydrous, filtered and dried in a rotavapor and high vacuum to give a yellow oily product. The crude product was triturated with Et_2O (2 times), the aqueous layer filtered up and the white solid product dried in a rotary evaporator and high vacuum.

2.2.3. Amidation

The Boc-protected or free *N*-terminal compound (1 equiv.) was dissolved in THF (7 mL) stirring at −15 °C, then NMM (2.5 equiv.) and *i*BCF (2.1 equiv.) were added to the solution allowing to react for 30 min. Then NH_4OH aq. solution (0.21 mL) was added to the reaction mixture at −15 °C for 30 min. The reaction mixture was allowed to react at r.t. for 2 h. The solvent was removed in rotavapor and the solid residue was dissolved in EtOAc washing with 5% citric acid solution (3 times), $NaHCO_3$ s.s. (3 times), and NaCl s.s. (3 times). Organic phases were collected and dried on Na_2SO_4 anhydrous, filtered and dried in a rotavapor and high vacuum to give a yellow oily product. The crude product was triturated with Et_2O (2 times), the aqueous layer filtered up, and the product dried in a rotary evaporator and high vacuum to give a white solid product.

2.2.4. Saponification

Boc-protected compound (0.1 mmol, 1 equiv.) was dissolved in THF (5 mL) stirring at r.t. then NaOH 1M (4 equiv.) was added dropwise, allowing it to react for 3 h. The solvent was removed in a rotavapor; the oily residue was taken up with water and washed with Et_2O (2 times). The aqueous solution was acidified with HCl 1 M until complete precipitation of the solid residue, which was extracted with EtOAc 3 times. The organic layers were collected, dried on Na_2SO_4 anhydrous, filtered and dried in a rotavapor and high vacuum to give a white solid product.

2.3. Synthesis and Characterization

Description of the reaction procedures [22–24] and compounds characterization are reported in detail in the Supplementary Materials.

2.4. In Vitro Biological Assays

2.4.1. Chemicals

Tris-HCl, EGTA, NaCl, $MgCl_2 \cdot 6H_2O$, GDP, the GTP analog GTPγS, and the L-tryptophan metabolite kynurenic acid were from Sigma-Aldrich (Budapest, Hungary); Tyr-D-Ala-Gly-(NMe)Phe-Gly-ol (DAMGO) was purchased from Bachem Holding AG (Bubendorf, Switzerland); endomorphin-2 (EM-2) was kindly provided by MTA-ELTE Research Group of Peptide Chemistry (Budapest, Hungary); $Ile^{5,6}$-deltorphine II ($Ile^{5,6}$Delt II) was purchased from Isotope Laboratory of BRC (Szeged, Hungary); and the highly selective KOR agonist diphenethylamine derivative, HS665 [25], was offered by Dr. Helmut Schmidhammer (University of Innsbruck, Austria). Naloxone was provided by Endo Laboratories DuPont de Nemours (Wilmington, DE, USA). The non-competitive NMDA antagonist, (+)-MK 801 maleate (MK-801) and L-kynurenine (L-kyn) were obtained from Tocris Bioscience (Bristol, UK). A solution of each ligand in water was stored in 1 mM stock solution at −20 °C. The radiolabelled GTP analog, $[^{35}S]$GTPγS (specific activity: 1000 Ci/mmol), was acquired from Hartmann Analytic (Braunschweig, Germany). $[^3H]$DAMGO [26] (specific activity: 38.8 Ci/mmol), $[^3H]Ile^{5,6}$Delt II (specific activity: 19.6 Ci/mmol), and $[^3H]$HS665 [27] (specific activity: 13.1 Ci/mmol) were radiolabelled in the Isotope Laboratory of BRC (Szeged, Hungary). $[^3H]$MK-801 [28] (specific activity: 30 Ci/mmol) was purchased from PerkinElmer (Boston, MA, USA) and the UltimaGold™ MV aqueous scintillation cocktail was from PerkinElmer (Boston, MA, USA).

2.4.2. Animal

Male and female Wistar rats and guinea pigs were used for membrane preparations. The animals were guarded in a temperature-controlled room, ranging from 21 to 24 °C, under a 12:12 light and dark cycle with water and food ad libitum. All housing and experiments were conducted in accordance with the European Communities Council Directives (86/606/ECC) and the Hungarian Act for the Protection of Animals in Research (XXVIII.tv. 32.§). The total number of animals, as well as their suffering, was minimized whenever possible.

2.4.3. Preparation of Brain Samples for Binding Assays

Rats and guinea pigs were decapitated, and their brains were quickly removed. The brains were used for membrane preparation following the procedure reported by Benyhe [29] for binding and [^{35}S]GTPγS binding experiments, in agreement with the protocol of Zádor et al. [30].

Homogenization of brains was performed in 30 volumes (*v/w*) of ice-cold 50 mM Tris-HCl pH 7.4 buffer with a Teflon-glass Braun homogenizer at 1500 rpm. The centrifuge was settled at 18,000 rpm for 20 min at 4 °C, the resulting supernatant discarded, and the pellet taken up in the original volume of Tris-HCl buffer. Incubation of homogenate at 37 °C for 30 min was performed in a shaking water-bath. Five volumes of 50 mM Tris-HCl pH 7.4 buffer were used to suspend the final pellet at −80 °C. For the [^{35}S]GTPγS binding experiments, the brains were homogenized with a Dounce in 5 volumes (*v/w*) of ice-cold TEM (Tris-HCl, EGTA, MgCl$_2$) at −80 °C. The protein content of the membrane preparation was determined by the method of Bradford, BSA being used as a standard [31].

2.4.4. Receptor Binding Assays

Functional [^{35}S]GTPγS Binding Experiments

The functional [^{35}S]GTPγS binding experiments were performed as previously described [32,33]. Briefly the membrane homogenates were incubated at 30 °C for 60 min in Tris-EGTA buffer (pH 7.4) composed of 50 mM Tris-HCl, 1 mM EGTA, 3 mM MgCl$_2$, 100 mM NaCl, containing 20 MBq/0.05 cm^3 [^{35}S]GTPγS (0.05 nM) and increasing concentrations (10^{-10} to 10^{-5} M) of ligands. The experiments were performed in the presence of excess GDP (30 µM) in a final volume of 1 mL. Total binding was measured in the absence of test compounds, non-specific binding was determined in the presence of 10 µM unlabeled GTPγS. The reaction was terminated by rapid filtration under vacuum and washed three times with 5 mL ice-cold 50 mM Tris-HCl (pH 7.4) buffer through Whatman GF/B glass fibers. The radioactivity of the filters was detected in an UltimaGold^{TM} MV aqueous scintillation cocktail with a Packard Tricarb 2300TR liquid scintillation counter.

Binding Experiments

In MOR, DOR, and KOR displacement, aliquots of frozen rat and guinea pig brain membrane homogenates were thawed and suspended in 50 mM Tris-HCl buffer (pH 7.4); in NMDA displacement, the Tris-HCl buffer (pH 7.4) contained 100 µM glycine and 100 µM L-glutamic acid. Samples were incubated in the presence of the unlabeled ligands in increasing concentrations (10^{-10} to 10^{-5} M) for at 35 °C for 45 min with [^{3}H]DAMGO, for at 35 °C for 45 min with [^{3}H]Ile^{5,6}Delt II, at 25 °C for 30 min with [^{3}H]HS665, and at 25 °C for 120 min with [^{3}H]MK-801. The non-specific and total binding was determined in the presence and absence of 10 µM unlabelled naloxone (MOR and DOR), HS665 (KOR), and MK-801 (NMDA). Radioactivity of the filter disks was measured, as written above.

Data Analysis

Experimental data are presented as means ± S.E.M. Sigmoid dose-response curves were fitted with GraphPad Prism 5.0 (GraphPad Prism Software Inc., San Diego, CA, USA). An unpaired *t*-test with two-tailed *p* value was performed to determine the significance level. In the competition binding assays, the 'One site competition' fitting was used to establish the equilibrium binding affinity (K$_i$ value).

2.5. In Vivo Tests

2.5.1. Animals

In our experiments, we used CD-1 male mice (Harlan, Italy, 25–30 g) maintained in colony, housed in cages (7 mice per cage) under standard light/dark cycle (from 7:00 a.m. to 7:00 p.m.), temperature (21 ± 1 °C) and relative humidity (60% ± 10%) for at least 1 week. Food and water were available ad libitum. The Service for Biotechnology and Animal Welfare of the Istituto Superiore di Sanità and the Italian Ministry of Health authorized the experimental protocol according to Legislative Decree 26/14.

2.5.2. Treatment Procedure

DMSO was purchased from Merck (Rome, Italy). Peptides solutions were freshly prepared using saline containing 0.9% NaCl and DMSO in the ratio DMSO/saline 1:5 *v/v* every experimental day. These solutions were injected at a volume of 10 µL/mouse for intracerebroventricular (i.c.v.) administrations or at a volume of 20 µL/mouse for subcutaneous administrations.

2.5.3. Surgery for i.c.v. Injections

For i.c.v. injections, mice were lightly anesthetized with isoflurane, and an incision was made in the scalp, and the bregma was located. Injections were performed using a 10 µL Hamilton microsyringe equipped with a 26-gauge needle, 2 mm caudal and 2 mm lateral from the bregma at a depth of 3 mm.

2.5.4. Tail Flick Test

The tail flick latency was obtained using a commercial unit (Ugo Basile, Gemonio, Italy), consisting of an infrared radiant light source (100 W, 15 V bulb) focused onto a photocell utilizing an aluminum parabolic mirror. During the trials, the mice were gently hand-restrained using leather gloves. Radiant heat was focused 3–4 cm from the tip of the tail, and the latency (s) of the tail withdrawal to the thermal stimulus was recorded. The measurement was interrupted if the latency exceeded the cut off time (15 s at 15 V). The baseline latency was calculated as mean of three readings recorded before testing at intervals of 15 min and the time course of latency determined at 15, 30, 45, 60, 90, and 120 min after treatment. Data were expressed as time course of the maximum percentage effect (%MPE) = (post-drug latency − baseline latency)/(cut-off time − baseline latency) × 100.

2.5.5. Formalin Test

In the formalin test, the injection of a dilute solution of formalin (1%, 20 µL/paw) into the dorsal surface of the mouse hind paw evoked biphasic nociceptive behavioral responses, such as licking, biting the injected paw, or both, occurring from 0 to 10 min after formalin injection (the early phase) and a prolonged phase, occurring from 15 to 40 min (the late phase). Before the test, mice were individually placed in a Plexiglas observation cage (30 × 14 × 12 cm) for one hour, to acclimatize to the testing environment. The total time the animal spent licking or biting its paw during the early and late phase of formalin-induced nociception was recorded.

2.5.6. Data Analysis and Statistics

Experimental data were expressed as mean ± s.e.m. In the tail flick test, significant differences among the groups were evaluated with two-way ANOVA followed by Sidak's multiple comparisons test. Formalin test data were analyzed by using one-way ANOVA, followed by Holm–Sidak's multiple comparisons test. GraphPad Prism 6.03 software was used for all the analyses. Statistical significance was set at $p < 0.05$. The data and statistical analysis comply with the recommendations on experimental design and analysis in pharmacology.

2.6. Stability in Human Plasma Sample Preparation

Five microliters of **K6** (500 µg in 250 µL of water) were added to 45 µL of fresh human plasma, then incubated at 37 ± 1 °C. Prepared samples were removed at several designated time points and incubation was stopped by adding an equal volume of the sequencing mixture (5% aqueous $ZnSO_4$ solution, MeOH, and ACN; 5:3:2), which precipitated proteins, to achieve a final concentration of 100 µg/mL. The mixture was vortexed and centrifuged at $12,000\times g$ for 5 min, then 20 µL of clear supernatant was directly injected into the HPLC system (Waters model 600 solvent pump and 2996 photodiode array detector, with XBridge BEH 130 C18, 4.6×250 mm, 5 µm). The samples were tested in three independents experiments (n = 3) and reported values represent the mean ± standard error (SEM). Data were analysed using simple regression analysis (significant deviation from zero: $F_{1,13} =$ 171.9, $p < 0.0001$; Y = $-0.9705 \times$ X + 95.98).

3. Results and Discussion

3.1. Chemistry

Simple modifications of kyna scaffold through the insertion of aromatic substituents [34], *C*-terminal derivatization as methyl ester [35], or amide bearing a water-soluble side chain [36] have been abundantly documented in the literature, rather than the insertion of kyna into a peptide sequence. On the contrary, different papers highlight the scarce propensity of the kyn carboxylic group to react with *N*-terminus free amino acids involved in the coupling reaction, which renders this unusual amino acid difficult to manage as a building block for peptide synthesis [22,23].

Taking this in mind, we focused our attention on the functionalization of kyna via a variant of Steglich esterification with the previously prepared Boc-protected 2-aminoethanol [23] so as to obtain intermediate **2** in 60% yield after purification by column chromatography (see General Procedure) [24]. Compound **2** was deprotected with a mixture of TFA:DCM = 1:1 at r.t. for 1 h and the so obtained intermediate was coupled with Boc-*N*(Me)Phe-OH, following the standard procedure for coupling reaction [36].

Peptide elongation/deprotection steps were repeated to reach the complete Boc-protected peptide **6** in 66% yield, starting from intermediate compound **5**, after silica gel column chromatography (Scheme 1). The final peptide **KA1** has been obtained in 71% yield after RP–HPLC purification. The purity of the sample was assessed by analytical RP–HPLC recorded at 254 nm and was found to be ≥95%.

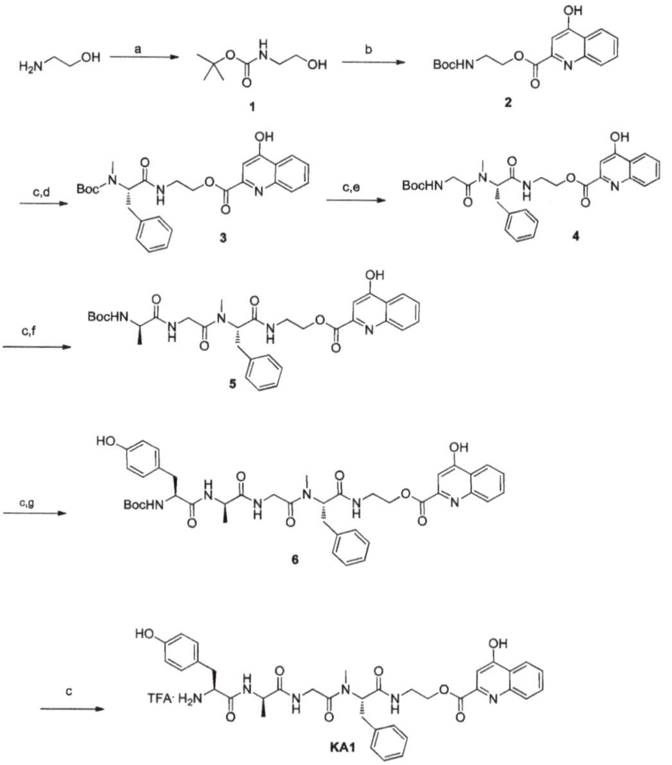

Scheme 1. Synthesis of peptide **KA1**. Reagents and conditions: (**a**) Boc$_2$O, NaOH 1 M, THF, 10 min at 0 °C, then 16 h at r.t. quantitative; (**b**) kynurenic acid, EDC·HCl, DMAP, DMF, 10 min at 0 °C, then 36 h at r.t., 60% yield after silica gel column chromatography; (**c**) TFA/DCM 1:1 1 h at r.t. quantitative; (**d**) Boc-*N*-Me-Phe-OH, EDC·HCl, HOBt, DIPEA, DMF, 10 min at 0 °C, then 16 h at r.t., 75% yield after reaction work-up; (**e**) Boc-Gly-OH, EDC·HCl, HOBt, DIPEA, DMF, 10 min at 0 °C, then 16 h at r.t., 70% yield after reaction work-up; (**f**) Boc-DAla-OH, EDC·HCl, HOBt, DIPEA, DMF, 10 min at 0 °C, then 16 h at r.t., 80% yield after silica gel column chromatography; (**g**) Boc-Tyr-OH, EDC·HCl, HOBt, DIPEA, DMF, 10 min at 0 °C, then 16 h at r.t., 66% yield from **5**, after silica gel column chromatography.

Then L-kyn was converted in its Boc-derivative **7** following the procedure reported by Tsentalovich et al. [23] to prepare the EM-2 analogs **K2** and **K3** (Scheme 2). Firstly intermediate **7** was coupled with H-Phe-NH$_2$, previously prepared following the general procedure of amidation, to obtain intermediate **8** in 61% yield. The Boc-protecting group was removed from compound **8**, and the so obtained TFA salt was coupled with Boc-Pro-OH to give intermediate **9** in good yield (81%). Repeated steps of coupling/purification/deprotection afforded the final compound **K2** in 72% yield from **10**, and excellent purity after RP–HPLC purification of the crude product. Conversely, Boc-Kynurenine **7** was transformed in the amide Boc-derivative **11** in a 71% yield. Then it was deprotected with a mixture of TFA:DCM = 1:1 at r.t. for 1 h and the so obtained TFA salt was coupled with BocPhe-OH to afford intermediate **12** quantitatively. Then repeated steps of coupling/purification/deprotection were performed to reach peptide **K3** in high yield (83% from **14**) and excellent purity after RP–HPLC purification of the crude compound.

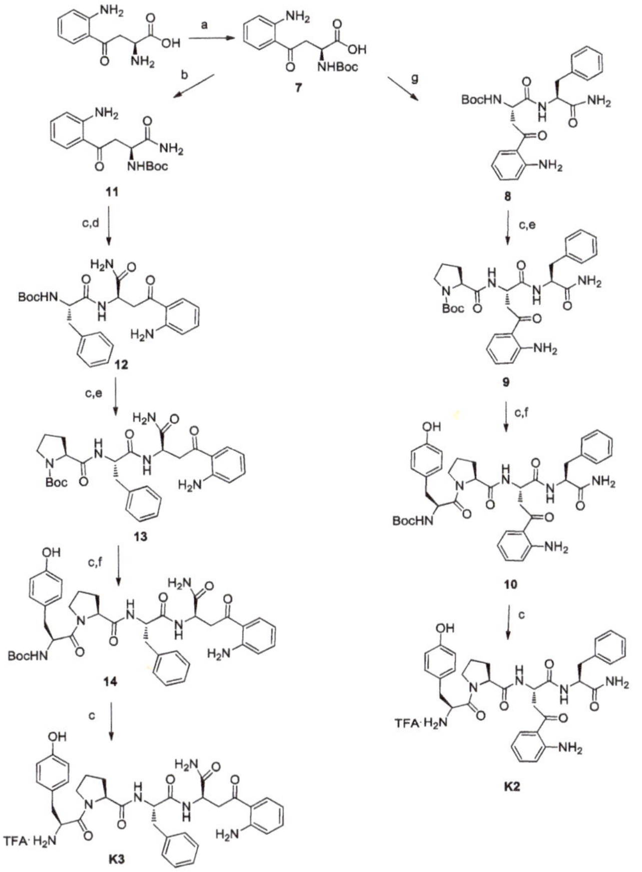

Scheme 2. Synthesis of peptides **K2, K3**. Reagents and conditions: (**a**) Boc$_2$O, NaHCO$_3$, NaOH 1 M, dioxane/H$_2$O (2:1), 10 min at 0 °C, then 30 min at r.t. quantitative; (**b**) *i*BCF, NMM, NH$_4$OH, THF, 30 min at −10 °C, then 16 h at r.t., 71% yield after trituration; (**c**) TFA/DCM 1:1, 1 h at r.t. quantitative; (**d**) Boc-Phe-OH, EDC·HCl, HOBt, DIPEA, DMF, 10 min at 0 °C, then 16 h at r.t. quantitative; (**e**) Boc-Pro-OH, EDC·HCl, HOBt, DIPEA, DMF, 10 min at 0 °C, 16 h at r.t., 72% yield for **13** after trituration, 81% yield for **9**; (**f**) Boc-Tyr-OH, EDC·HCl, HOBt, DIPEA, DMF, 10 min at 0 °C, 16 h at r.t., 95% yield for **14** after trituration, 73% yield for **10**; (**g**) H-Phe-NH$_2$, EDC·HCl, HOBt, DIPEA, DMF, 10 min at 0 °C, then 16 h at r.t., 61% yield from **7** after trituration.

Finally, linear peptides **K4–K6** have been synthesized as methyl ester, acid, and amide derivatives, respectively, starting from L-kyn methyl ester **15** (Scheme 3), prepared following the procedure described in the literature [24,36]. Repeated steps of deprotection/coupling reaction were performed to reach intermediate **19** in high yield after trituration in Et$_2$O. Boc group removal of intermediate **19** gave **K4** in good yield and excellent purity after RP–HPLC purification. Saponification of intermediate compound **19** afforded **20,** following the general procedure; the conversion of the latter in the amide **21** was performed as previously described by Stefanucci et al. [24]. Boc group removal of the linear peptides **20** and **21** afforded products **K5** and **K6** in good yields (72% and 52%, respectively), and excellent purity after RP–HPLC purification (98% and 96%, respectively).

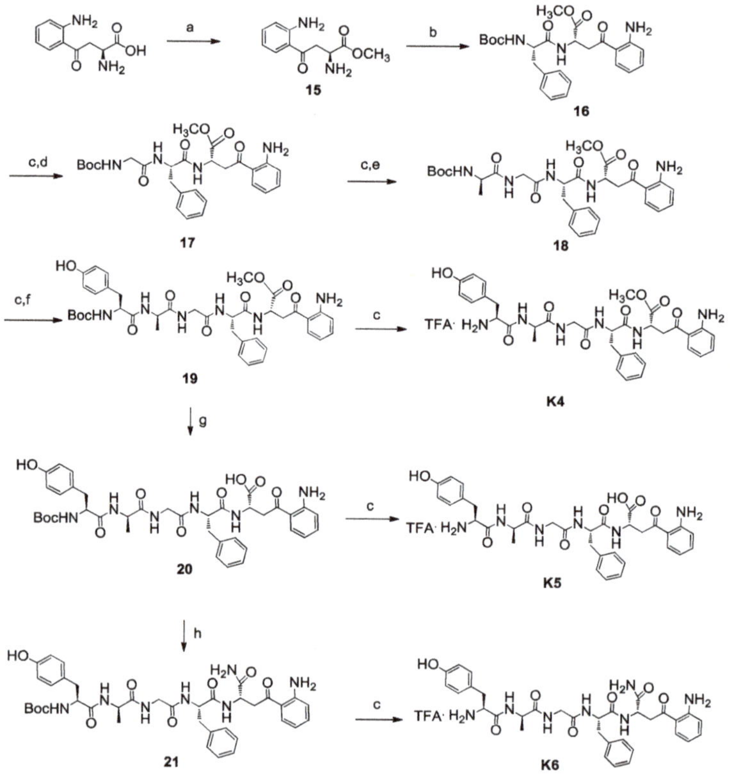

Scheme 3. Synthesis of linear peptides **K4–K6**. Reagents and conditions: (**a**) SOCl$_2$, MeOH, 30 min at 0 °C, 16 h at r.t., quantitative; (**b**) Boc-Phe-OH, EDC·HCl, HOBt, DIPEA, DMF, 10 min at 0 °C, then 16 h at r.t., 60% after silica gel column chromatography; (**c**) TFA/DCM 1:1, 1 h at r.t. quantitative; (**d**) Boc-Gly-OH, EDC·HCl, HOBt, DIPEA, DMF, 10 min at 0 °C, 16 h at r.t., quantitative; (**e**) Boc-DAla-OH, EDC·HCl, HOBt, DIPEA, DMF, 10 min at 0 °C, 16 h at r.t., 84% yield after silica gel column chromatography; (**f**) Boc-Tyr-OH, EDC·HCl, HOBt, DIPEA, DMF, 10 min at 0 °C, 16 h at r.t., 72% yield after trituration; (**g**) NaOH 1 M, THF, 3 h at r.t., quantitative; (**h**) *i*BCF, NMM, NH$_4$OH (aq), THF, 30 min at −15 °C, 2 h at r.t., 52% yield after trituration.

3.2. In Vitro Studies

3.2.1. Binding Assays

Kyna and its analog (KYNA1) previously reported by Zádor et al. did not directly bind μ, δ, κ-receptors in vitro [37]. However, after chronic and acute administration, they altered opioid receptor function in vivo and in vitro through the NMDA receptor co-localized in the cortex and striatum of mice and rats, though the interaction of opioid receptors and NMDA have been deeply discussed in the literature [37,38]. Kyna is able to bind to the NMDA receptor at micromolar affinity [38]. To test if our novel peptides are able to target both of these systems, they were examined in a receptor binding radioassay using highly specific tritium-labelled primary ligands for opioid and NMDA receptor binding sites. [^{3}H]DAMGO, [^{3}H]Ile^{5,6}Delt II, and [^{3}H]MK-801 equilibrium competition (displacement) studies were conducted in rat brain homogenates, while κ-opioid receptor tests were performed with [^{3}H]HS665 in guinea pig brain homogenates. The novel ligands showed similar equilibrium inhibitory affinities (K$_i$ value) in the μ-opioid system as DAMGO except **K3** (Table 1, Figure S1A). In the δ-opioid system, the ligands showed lower binding affinity (higher Ki) compared to the selective δ-opioid receptor selective agonist Ile^{5,6}Delt II (Figure S1B). In the κ-opioid receptor system, the compounds showed higher Ki values than that of the selective κ-opioid agonist HS665 (Figure S1C). In the NMDA receptor binding assays, the peptides did not produce any competing activity (Figure S1D). Peptide **KA1** possesses the best binding affinity, with a Ki value very close to that of the reference compound DAMGO (1.08 ± 0.26 nM vs. 0.90 ± 0.28 nM), suggesting that the insertion of kyna into the DAMGO sequence does not impair its binding potency at the μ-opioid receptor. The peptide **K6**, presenting the enkephalin-like structure linked to L-kyn C-terminal amide, is able to bind all three opioid receptors with significant affinity, showing a moderate preference for the μ-opioid receptor (affinity ratio 1:18:70 for μ, δ, k, respectively), whereas compounds **K4** and **K5** are able to bind only μ- and δ-opioid receptors. It is reasonable to believe that the C-terminal amide derivatization in peptide **K6** confers the ability to bind to the κ-opioid receptor. Concerning the endomorphin-2 (EM-2) analogs **K2** and **K3**, the replacement of Phe^3 with L-kyn improves the binding affinity and selectivity of **K2** for μ-opioid receptors with respect to the standard compound EM-2, with a weak affinity for κ-opioid receptors, while the incorporation of L-Kyn amide in position 4 causes the loss of selectivity for MOR in favor of a modest binding affinity for μ- and δ-opioid receptors. Peptide **K2** shows a Ki value two-folds lower than that of EM-2 on the μ-opioid receptor, which let us suppose the positive influence of L-Kyn in position 3 on **K2** binding ability.

Table 1. Displacement of [^{3}H]DAMGO, [^{3}H]Ile^{5,6}Delt II, [^{3}H]HS665, and [^{3}H]MK-801 by DAMGO, Ile^{5,6}Delt II, HS665, MK-801, and oligopeptides in membranes of rat and guinea pig brains. The IC$_{50}$ values for the MOR, DOR, KOR, and NMDA, according to the competition binding curves (see Figure S1), were converted into equilibrium inhibitory constant (K$_i$) values using the Cheng–Prusoff [39] equation.

Ligand	K$_i$ + S.E.M. (nM) Opioid System			NMDA System
	DAMGO [a]	Ile^{5,6}Delt II [a]	HS665 [b]	MK-801 [a]
DAMGO	0.90 ± 0.28	n.d. [c]	n.d. [c]	n.d. [c]
Ile^{5,6}Delt II	n.d. [c]	8.85 ± 0.77	n.d. [c]	n.d. [c]
HS665	n.d. [c]	n.d. [c]	2.38 ± 0.25	n.d. [c]
MK-801	n.d. [c]	n.d. [c]	n.d. [c]	11.45 ± 1.04
EM-2	3.16 ± 0.3	n.d. [c]	n.d.[c]	n.d. [c]
KYNA	n.d. [c]	n.d. [c]	n.d. [c]	>10000
L-kyn	n.d. [c]	n.d. [c]	n.d. [c]	>10000
KA1	1.08 ± 0.26	554.7 ± 0.8	>10000	>10000
K2	1.39 ± 0.30	>10000	1043 ± 0.3	>10000
K3	197.3 ± 0.36	158.8 ± 1.6	>10000	>10000
K4	2.29 ± 0.28	31.2 ± 0.7	>10000	>10000
K5	9.11 ± 0.32	94.4 ± 0.8	>10000	>10000
K6	1.84 ± 0.27	32.5 ± 0.8	127.7 ± 0.3	>10000

[a] rat brain membrane, [b] guinea pig brain membrane, [c] not determined.

3.2.2. Binding-Protein Activation Assays

The effect of kyn or kyna combined peptides on G-protein activation was investigated in functional [^{35}S]GTPγS binding assays in rat and guinea pig brain membranes. All ligands produced dose-dependent stimulations described by sigmoid curves (Figure S2). **K6** showed higher efficacy (E$_{max}$) than DAMGO (Table 2). Moreover, 10 μM cyprodime and 10 μM naltrindole, which are selective MOR and DOR antagonists, respectively [40,41], significantly reversed the agonist effects of the ligands in rat brain membrane homogenates (Figure S3A). In guinea pig brain membrane homogenates, the ligands did not activate G-protein except **K4** and **K6** (Figure S3B). Additionally, 10 μM norbinaltorphine decreased significantly the agonist effect of **K4** but did not change the effect of **K6** (Table 3). Altogether, these data reveal that peptide **KA1** acts as a selective μ-opioid agonist, being able to decrease the GTPγS binding percentage under the basal level in the presence of 10 μM cyprodime. Peptides **K2** and **K3** possess a mixed μ/δ agonist activity profile, while peptides **K4**–**K6** show a modest mixed

μ/δ-opioid receptor agonism. Probably peptide **K6** is the strongest mixed $\mu/\delta/\kappa$ opioid agonist due to its ability to bind protein G with an efficacy value over the basal level in presence of each selective opioid antagonist at 10 μM.

Table 2. G-protein activation DAMGO, and novel oligopeptides in [^{35}S]GTPγS binding assays using rat brain membrane homogenates. The values were calculated according to dose-response binding curves.

Ligand	Maximal Stimulation (Efficacy)	Potency
	$E_{max} \pm$ S.E.M. (%)	Log $EC_{50} \pm$ S.E.M.
DAMGO	172.0 ± 3.5	-6.384 ± 0.101
KA1	140.9 ± 1.4	-6.504 ± 0.076
K2	121.6 ± 2.5	-7.535 ± 0.354
K3	114.0 ± 2.1	-6.993 ± 0.422
K4	155.5 ± 4.8	-6.073 ± 0.172
K5	149.2 ± 3.5	-5.990 ± 0.111
K6	209.7 ± 3.4	-5.984 ± 0.054

Table 3. The maximal G-protein efficacy (E_{max}) of novel oligopeptides in the absence or presence of the selective MOR antagonist cyprodime and the selective DOR antagonist naltrindole in rat brain membrane homogenates and in the absence or presence of the selective KOR antagonist norbinaltorphine in guinea pig brain membrane homogenates in [^{35}S]GTPγS binding assays. The values were calculated according to bar graphs in Figure S3.

	Ligand	MOR	DOR	Ligand	KOR
		Ligand + Cyp	Ligand + NTI		Ligand + Nor-BNI
			$E_{max} \pm$ S.E.M. (%)		
KA1	139.6 ± 8.2	94.7 ± 3.7 **	102.5 ± 1.3 *	111.7 ± 1.1	100.7 ± 6.2 [ns]
K2	129.9 ± 8.2	88.0 ± 3.3 **	88.8 ± 1.9 **	92.8 ± 2.0	100.7 ± 6.0 [ns]
K3	118.9 ± 3.0	98.5 ± 1.2 **	97.8 ± 1.7 **	95.2 ± 2.2	106.0 ± 6.3 [ns]
K4	160.6 ± 5.7	101.6 ± 2.6 ***	103.6 ± 3.2 ***	125.2 ± 2.0	104.9 ± 9.5 *
K5	153.1 ± 5.9	108.8 ± 1.6 **	103.8 ± 1.9 ***	106.9 ± 0.3	110.5 ± 5.0 [ns]
K6	211.7 ± 3.1	108.5 ± 2.4 ***	113.3 ± 2.9 ***	139.8 ± 2.3	142.1 ± 7.1 [ns]

Experimental data were processed by GraphPad Prism 5.0 using bar graphs. ns: not significant; * $p < 0.05$; ** $p < 0.01$; *** $p < 0.001$ based on unpaired t-tests.

3.3. In Vivo Studies

Data obtained from binding experiments indicate **K6** as the best of the series; thus, tail flick and formalin tests were performed in order to evaluate the effects of this novel peptide in two different pain animal models. The tail flick test measures the time in which the animal withdraws the tail from a thermal nociceptive stimulus; this time is increased by analgesics. Compounds were centrally injected, and the response measured from 15 to 120 min after the administration. Compounds **K6** and DAMGO greatly increase the time of response to thermal nociceptive stimuli and both effects were in the same order to magnitude (Figure 2).

Since it was reported that some opioid peptides such as DAMGO are also peripheral acting [42], we performed a formalin test administering **K6** and DAMGO subcutaneously in mice paws. The formalin test measures the behavioral response to chemical nociceptive stimuli evoked by a formalin diluted solution injected in the mice paw. From 0 to 10 min after the injection, an early phase response occurs with direct stimulation of peripheral nociceptors, while a response to inflammatory pain appears from 15 to 40 min as a late prolonged phase. Compound **K6** was able to reduce the nociceptive response to formalin in the early phase, whereas a light but not significant reduction was induced in the late phase of the formalin test (Figure 3). After the DAMGO injection, we observed a reduction of the nociceptive behavior both early and late in the formalin test (Figure 3). The formalin early phase, which depends upon the direct excitement of sensory neurons through TRPA1 cation channel activation [43] of MORs at the peripheral endings of nociceptors, is responsible for meaningful analgesia [44], and it is not surprising that **K6** and DAMGO reduced formalin-induced nociception in the early phase of the test. The differences observed in the late phase are probably due to a different metabolic fate of **K6** and DAMGO after subcutaneous administration, depending upon the protease activity that might act in different ways on **K6** and DAMGO chemical structures.

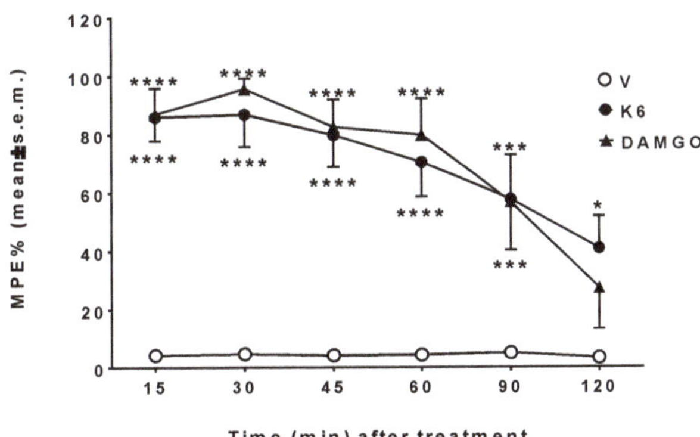

Figure 2. The effect of **K6** and DAMGO in the tail flick test. Compounds were administered in the left cerebral ventricle at the dose of 10 μg/10 μL, and the time to respond to thermal stimuli measured from 15 to 120 min. V is for vehicle-treated animals. Statistical analysis: two-way ANOVA followed by Sidak's multiple comparisons test. * is for $p < 0.05$, *** is for $p < 0.001$, and **** is for $p < 0.0001$ vs. V. n = 7.

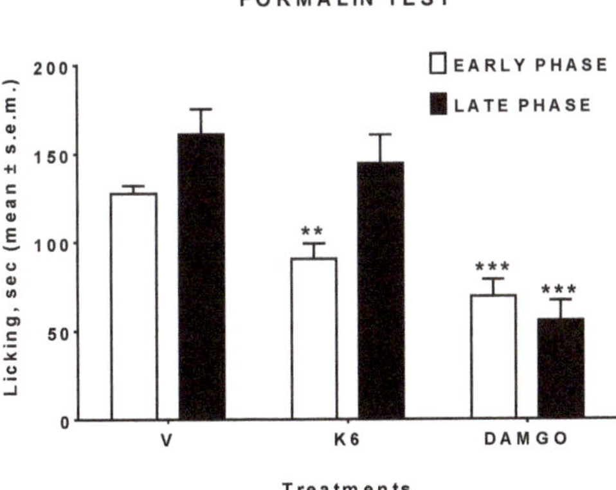

Figure 3. The effect of **K6** and DAMGO in the formalin test. Compounds were administered subcutaneously (s.c.) in the dorsal hind paw of mice at the dose of 100 μg/20 μL, 15 min before a s.c. injection of dilute formalin solution (1% in saline, 20 μL/paw). Early phase represents the formalin-induced nociceptive behavior recorded from 0 to 10 min after formalin injection; late phase is for the formalin-induced nociceptive behavior recorded from 15 to 40 min after formalin injection. V is for vehicle-treated animals. Statistical analysis: one-way ANOVA followed by Holm–Sidak's multiple comparisons test. ** is for $p < 0.01$ and *** is for $p < 0.001$ vs. V. n = 7.

3.4. Plasma Stability Results

The plasma stability of compound **K6** was tested by incubation in human plasma at 37 °C. The degradation curve (Figure 4) was built by plotting the total amount of remaining peptide (expressed as µg/mL) vs. time (minutes). Concentration data were obtained in triplicate and analyzed as simple linear regression using GraphPad 8.3.1. The novel compound exhibits good stability in human plasma, showing a $t_{1/2}$ = 47 min, according to the results obtained from the tail flick test after i.c.v. administration.

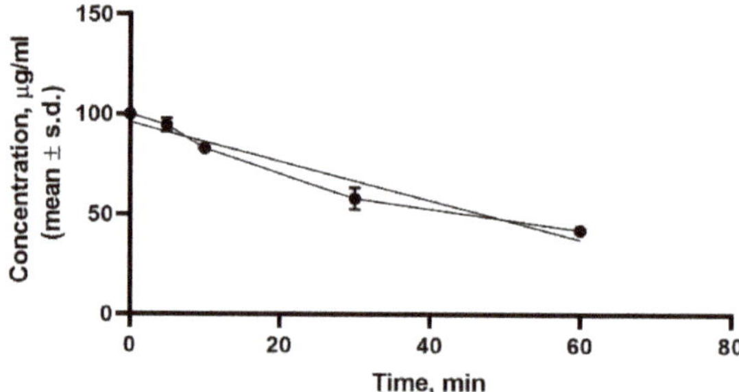

Figure 4. Stability of compound **K6** in human plasma. Concentration of K6 in µg/mL measured from 0 to 60 min in triplicate.

4. Conclusions

Chemical modification of endogenous opioid peptides promotes the development of novel analogs with increased potency and improved pharmacokinetic properties, e.g., the synthetic bivalent peptide biphalin enhanced stroke immunohistochemical and behavioral neuroprotection in comparison to DPDPE and DAMGO, reducing glutamate toxicity and oxidative stress [45–48]. Metabolites of the KP, especially kyna and L-kyn, play crucial roles in maintaining the normal brain function, preventing the over-activation of excitatory amino acid receptors, thus offering novel therapeutical opportunities for brain neuroprotection. In this work, we have synthesized six novel opioid analogs incorporating the L-kyn and kyna residues at different positions of several opioid peptide scaffolds. They were characterized by in vitro and in vivo assays to evaluate their ability to bind the NMDA/opioid receptors and to induce analgesic effects after i.c.v. and s.c. administration. These novel peptides do not bind to the NMDA receptors, and some of them showed good/high affinity for opioid receptors with different selectivity profiles. In particular, **KA1** exhibits the binding constant (Ki = 1.08) very close to that of DAMGO for the µ-opioid receptor and a pronounced selectivity but medium-low efficacy (E_{max} = 139%); thus the esterification of the ethanolamine portion with kyna does not add any particular advantage to the parent peptide DAMGO. On the other hand, the presence of L-kyn residue in place of native Phe in position 3 (**K2**) leads a potent and selective opioid fragment toward MOR (selectivity ration µ/δ/κ = 1:10,000:750), whereas the substitution in position 4 (**K3**) leads to a weak and unselective agonist at MOR and DOR. In the analogs **K4–K6**, the kyna residue was inserted in the fifth position of the YaGP peptide, with different C-terminus, respectively, methyl ester, free carboxylic acid, and amide. **K4** and **K5** show a similar mixed binding affinity for MOR and DOR, with a preference for MOR and none or weak affinity for KOR.

On the contrary, peptide **K6** shows an interesting behavior since it is able to bind all three opioid receptors with binding affinity ranging from high to modest (Ki^{μ} = 1.84 nM, Ki^{δ} = 32.5 nM, Ki^{κ} = 127.7 nM), with a potency ($logEC_{50}$ = −5.598) and efficacy at MOR (E_{max} = 211%) higher than that of DAMGO (E_{max} = 172%). Its activity on the GTPγS binding assay on rat brain membrane and guinea pig ileum is well antagonized by the co-administration of the selective antagonists for MOR and DOR at 10 µM concentration, prompting us to deeply investigate its anti-nociceptive effect in vivo. The formalin test, which is a model of inflammatory pain, revealed that the antinociceptive effect exerted by **K6** after subcutaneous administration is significant only in the early phase, whereas DAMGO is also active in the late phase. In the tail flick test, the **K6** analgesic profile after i.c.v. administration is superimposable to that of DAMGO, according to a good stability profile in human plasma. These data are encouraging to further develop opioid peptides containing kynurenine moieties since the insertion of kyna and kyn in our opioid model improved, in some cases, the binding affinity and was able to modulate the selectivity. Also, the possible role played by the metabolism of these peptides and their possible implication in different neuropathic and chronic pain models is unknown and worth further investigation.

Supplementary Materials: The following are available online at http://www.mdpi.com/2218-273X/10/2/284/s1, Figure S1: MOR (A), DOR (B), KOR (C) and NMDA (D) binding affinity of the novel oligopeptides, Figure S2: The effect of oligopeptides on G-protein activity compared to DAMGO in [^{35}S]GTPγS binding assay in rat brain membrane homogenates, Figure S3: The effect of novel oligopeptides on G-protein activity in [^{35}S]GTPγS binding assays in the absence or presence of the selective MOR antagonist cyprodime (Cyp) and the selective DOR antagonist naltrindole (NTI) in rat brain membrane homogenates (Figure A) and the selective KOR antagonist norbinaltorphine (nor-BNI) in guinea pig brain membrane homogenates (Figure B).

Author Contributions: Conceptualization, A.S. and S.B.; Data curation, E.S., F.Z., S.P. and G.Z.; Formal analysis, F.Z. and L.V.; Investigation, A.S., M.N. and S.P.; Methodology, E.S. and M.P.D.; Validation, S.B.; Writing–original draft, A.S.; Writing–review & editing, A.M., L.V. and S.B. All authors have read and agreed to the published version of the manuscript.

Funding: This research was supported by the project GINOP 2.3.2-15-2016-00034, provided by National Research, Development and Innovation Office (NKFI), Budapest, Hungary, and the Ministry of Human Capacities, Hungary grant 20391-3/2018/FEKUSTRAT.

Conflicts of Interest: The authors declare no conflict of interest.

Abbreviations

KP	kynurenine pathway
kyn	kynurenine
kyna	kynurenic acid
FDA	Food and Drug Administration
CNS	central nervous system
BBB	blood brain barrier
GPR35	G protein-coupled receptor 35
SVCT2	Sodium-dependent vitamin C transporter 2
MOR	μ-opioid receptor; DOR, δ-opioid receptor
DOR	δ-opioid receptor
KOR	k-opioid receptor; GPCRs, G protein coupled receptors
GPCRs	G protein coupled receptors
NMR	Nuclear magnetic resonance
TFA	trifloroacetic acid
ACN	acetonitrile
RP-HPLC	Reverse Phase High performance liquid chromatography
TMS	trimethylsilane
ESI	Electrospray ionization
LRMS	Low Resolution Mass Spectroscopy
HOBt	1-hydroxybenzotriazole
DMAP	4-Dimethylaminopyridine
EDC·HCl	1-Ethyl-3-(3-dimethylaminopropyl)carbodiimide hydrochloride
EtOAc	ethyl acetate
THF	tetrahydrofurane
NMM	N-methylmorpholine

iBCF	isobuthylchloroformiate
DMSO	dimethylsulfoxide
Boc	tert-butyloxycarbonyl
Ar	aryl
EM-2	endomorphine-2
BSA	Bovine serum albumin
i.c.v.	intracerebroventricular
Cyp	cyprodime
NTI	naltrindole
Nor-BNI	norbinaltorphine
DPDPE	[D-Pen,D-Pen5]Enkephalin
DIPEA	N,N-Diisopropylethylamine
DMF	dimethylformamide
DCM	dichloromethane
MeOH	methanol
TIPS	triisopropylsilane
EGTA	ethylene glycol-bis(2-aminoethylether)-N,N,N′,N′-tetraacetic acid
GDP	guanosine 5′-diphosphate
GTP	guanosine-5′-triphosphate
DMSO	dimethylsulfoxide
DAMGO	[D-Ala^2, N-MePhe^4, Gly-ol]-enkephalin
IleDelt II	Ile^{5,6}-deltorphin II
[^{35}S]GTPγS	guanosine-5′-[^{35}S]thiophosphate
NMDA	N-methyl-D-aspartate receptor
Tris-HCl	tris-(hydroxymethyl)-aminomethane hydrochloride

References

1. Cervenka, I.; Agudelo, L.Z.; Ruas, J.L. Kynurenines: Tryptophan's metabolites in exercise, inflammation, and mental health. *Science* **2017**, *357*, 9794. [CrossRef]
2. Ye, Y.; Xia, Z.; Zhang, D.; Sheng, Z.; Zhang, P.; Zhu, H.; Xu, N.; Liang, S. Multifunctional pharmaceutical effects of the antibiotic daptomycin. *BioMed Res. Int.* **2019**, *2019*, 8609218. [CrossRef] [PubMed]
3. Chuang, P.-H.; Hsieh, P.-W.; Yang, Y.-L.; Hua, K.-F.; Chang, F.-R.; Shiea, J.; Wu, S.-H.; Wu, Y.-C. Cyclopeptides with Anti-inflammatory Activity from Seeds of Annona Montana. *J. Nat. Prod.* **2008**, *71*, 1365–1370. [PubMed]
4. Ellis-Steinborner, S.T.; Scanlon, D.; Musgrave, I.F.; Nha Tran, T.T.; Hack, S.; Wang, T.; Abell, A.D.; Tyler, M.J.; Bowie, J.H. An unusual kynurenine-containig opioid tetrapeptide from the skin gland secretion of the Australian red tree frog Litoria rubella. Sequence determination by electrospray mass spectrometry. *Rapid Commun. Mass Spectrom.* **2011**, *25*, 1735–1740. [CrossRef] [PubMed]
5. Guillemin, G.J.; Kerr, S.J.; Smythe, G.A.; Smith, D.G.; Kapoor, V.; Armati, P.J.; Croitoru, J.; Brew, B.J. Kynurenine pathway metabolism in human astrocytes: A paradox for neuronal protection. *J. Neurochem.* **2001**, *78*, 842–853.
6. Schwarcz, R.; Bruno, J.P.; Muchowski, P.J.; Wu, H.-Q. Kynurenines in the mammalian brain: When physiology meets pathology. *Nat. Rev. Neurosci.* **2012**, *13*, 465–477.
7. Vécsei, L.; Szalárdy, L.; Fülöp, F.; Toldi, J. Kynurenines in the CNS: Recent advances and new questions. *Nat. Rev. Drug. Discov.* **2013**, *12*, 64–82. [CrossRef]
8. Yeganeh Salehpour, M.; Mollica, A.; Momtaz, S.; Sanadgol, N.; Farzaei, M.H. Melatonin and multiple sclerosis: From plausible neuropharmacological mechanisms of action to experimental and clinical evidence. *Clin. Drug Investig.* **2019**, *39*, 607–624. [CrossRef]
9. Birch, P.J.; Grossman, C.J.; Hayes, A.G. Kynurenate and FG9041 have both competitive and non-competitive antagonist actions at excitatory amino acid receptors. *Eur. J. Pharmacol.* **1988**, *151*, 313–315. [CrossRef]
10. Perkins, M.N.; Stone, T.W. Actions of kynurenic acid and quinolinic acid in the rat hippocampus in vivo. *Exp. Neurol.* **1985**, *88*, 570–579. [CrossRef]
11. Wang, J.; Simonavicius, N.; Wu, X.; Swaminath, G.; Reagan, J.; Tian, H.; Ling, L. Kynurenic acid as a Ligand for Orphan G Protein-coupled Receptor GPR35. *J. Biol. Chem.* **2006**, *281*, 22021–22028. [CrossRef] [PubMed]

12. Shore, D.M.; Reggio, P.H. The therapeutic potential of orphan GPCRs, GPR35 and GPR55. *Front. Pharmacol.* **2015**, *6*, 69. [CrossRef] [PubMed]

13. Bonina, F.P.; Arenare, L.; Ippolito, R.; Boatto, G.; Battaglia, G.; Bruno, V.; de Caprariis, P. Synthesis, pharmacokinetics and anticonvulsant activity of 7-chlorokynurenic acid prodrugs. *Int. J. Pharm.* **2000**, *202*, 79–88. [CrossRef]

14. Luhavaya, H.; Sigrist, R.; Chekan, J.R.; McKinnie, S.M.K.; Moore, B.S. Biosynthesis of l-4-Chlorokynurenine, an antidepressant prodrug and a non-proteinogenic amino acid found in lipopeptide. *Antibiotics* **2019**, *58*, 8394–8399.

15. Manfredini, S.; Vertuani, S.; Pavan, B.; Vitali, F.; Scaglianti, M.; Bortolotti, F.; Biondi, C.; Scatturin, A.; Prasad, P.; Dalpiaz, A. Design, synthesis and activity of ascorbic acid prodrugs of nipecotic, kynurenic and diclophenamic acids, liable to increase neurotropic activity. *J. Med. Chem.* **2002**, *45*, 559–562. [CrossRef]

16. Vamos, E.; Pardutz, A.; Klivenyi, P.; Toldi, J.; Vecsei, L. The role of kynurenines in disorders of the central nervous system: Possibilities for neuroprotection. *J. Neurol. Sci.* **2009**, *283*, 21–27. [CrossRef]

17. Knyihár-Csillik, E.; Toldi, J.; Mihály, J.A.; Krisztin-Péva, B.; Chadaide, Z.; Németh, H.; Vécsei, L. Kynurenine in combination with probenecid mitigates the stimulation-induced increase of c-fos immunoreactivity of the rat caudal trigeminal nucleus in an experimental migraine model. *J. Neural. Transm.* **2007**, *114*, 417–421. [CrossRef]

18. Gábor, N.-G.; Dvorácskó, S.; Bohár, Z.; Benyhe, S.; Tömböly, C.; Párdutz, A.; Vécsei, L. Interactions between the kynurenine and the endocannabinoid system with special emphasis on migraine. *Int. J. Mol. Sci.* **2017**, *18*, 1617. [CrossRef]

19. Stone, T.W.; Caroline, M.; Forrest, M.; Darlington, L.G. Kynurenine pathway inhibition as a therapeutic strategy for neuroprotection. *FEBS J.* **2012**, *279*, 1386–1397. [CrossRef]

20. Deora, G.S.; Kantham, S.; Chan, S.; Dighe, S.N.; Veliyath, S.K.; McColl, G.; Parat, M.O.; McGeary, R.P.; Ross, B.P. Multifunctional analogs of kynurenic acid for the treatment of Alzheimer's disease: Synthesis, pharmacology, and molecular modeling studies. *ACS Chem. Neurosci.* **2017**, *8*, 2667–2675. [CrossRef]

21. Zádori, D.; Nyiri, G.; Szonyi, A.; Szatmári, I.; Fülöp, F.; Toldi, J.; Freund, T.F.; Vécsei, L.; Klivényi, P. Neuroprotective effects of a novel kynurenic acid analogue in a transgenic mouse model of Huntington's disease. *J. Neurol. Transm.* **2011**, *18*, 865–875. [CrossRef] [PubMed]

22. Ghilardi, A.; Pezzoli, D.; Bellucci, M.C.; Malloggi, C.; Negri, A.; Sganappa, A.; Tedeschi, G.; Candiani, G.; Volonterio, A. Synthesis of multifunctional PAMAM-Aminoglycoside conjugates with enhanced transfection efficiency. *Bioconjug Chem.* **2013**, *24*, 1928–1936. [CrossRef]

23. Tsentalovich, Y.P.; Yanshole, V.V.; Polienko, Y.F.; Morozov, S.V.; Grigor'ev, I.A. Deactivation of excited states of kynurenine covalently linked to nitroxides. *Photochem. Photobiol.* **2011**, *87*, 22–31. [CrossRef] [PubMed]

24. Stefanucci, A.; Novellino, E.; Mirzaie, S.; Macedonio, G.; Pieretti, S.; Minosi, P.; Szűcs, E.; Erdei, A.I.; Zádor, F.; Benyhe, S.; et al. Opioid receptor activity and analgesic potency of DPDPE peptide analogues containing a xylene bridge. *ACS Med. Chem. Lett.* **2017**, *8*, 449–454. [CrossRef] [PubMed]

25. Spetea, M.; Berzetei-Gurske, I.P.; Guerrieri, E.; Schmidhammer, H. Discovery and pharmacological evaluation of a diphenethylamine derivative (HS665), a highly potent and selectivek-opioid receptor agonist. *J. Med. Chem.* **2012**, *55*, 10302–10306. [CrossRef] [PubMed]

26. Oktem, H.A.; Moitra, J.; Benyhe, S.; Tóth, G.; Lajtha, A.; Borsodi, A. Opioid receptor labeling with the chloromethyl ketone derivative of [3H]Tyr-D-Ala-Gly-(Me)Phe-Gly-ol (DAMGO) II: Covalent labeling of mu opioid binding site by 3H-Tyr-D-Ala-Gly-(Me)Phe chloromethyl ketone. *Life Sci.* **1991**, *48*, 1763–1768. [CrossRef]

27. Guerrieri, E.; Mallareddy, J.R.; Tóth, G.; Schmidhammer, H.; Spetea, M. Synthesis and pharmacological evaluation of [(3)H]HS665, a novel, highly selective radioligand for the kappa opioid receptor. *ACS Chem. Neurosci.* **2015**, *6*, 456–463. [CrossRef]

28. Basu, N.; Scheuhammer, A.M.; Rouvinen-Watt, K.; Grochowina, N.; Evans, R.D.; O'Brien, M.; Chan, H.M. Decreased N-methyl-d-aspartic acid (NMDA) receptor levels are associated with mercury exposure in wild and captive mink. *Neurotoxicology* **2007**, *28*, 587–593. [CrossRef]

29. Benyhe, S.; Farkas, J.; Tóth, G.; Wollemann, M. Met5-enkephalin-Arg6-Phe7, an endogenous neuropeptide, binds to multiple opioid and nonopioid sites in rat brain. *J. Neurosci. Res.* **1997**, *48*, 249–258. [CrossRef]

30. Zádor, F.; Kocsis, D.; Borsodi, A.; Benyhe, S. Micromolar concentrations of rimonabant directly inhibits delta opioid receptor specific ligand binding and agonist-induced G-protein activity. *Neurochem. Int.* **2014**, *67*, 14–22. [CrossRef]

31. Bradford, M.M. A rapid and sensitive method for the quantitation of microgram quantities of protein utilizing the principle of protein-dye binding. *Anal. Biochem.* **1976**, *72*, 248–254. [CrossRef]

32. Sim, L.J.; Selley, D.E.; Childers, S.R. In vitro autoradiography of receptor-activated G proteins in rat brain by agonist-stimulated guanylyl 5'-[gamma-[35S]thio]-triphosphate binding. *Proc. Natl. Acad. Sci. USA* **1995**, *92*, 7242–7246. [CrossRef] [PubMed]

33. Traynor, J.R.; Nahorski, S.R. Modulation by mu-opioid agonists of guanosine-5'-O-(3-[35S]thio)triphosphate binding to membranes from human neuroblastoma SH-SY5Y cells. *Mol. Pharmacol.* **1995**, *47*, 848–854. [PubMed]

34. Neises, B.; Steglich, W. Simple Method for the Esterification of Carboxylic Acids. *Angew. Chem. Int. Ed. Engl.* **1978**, *17*, 522–524. [CrossRef]

35. Mollica, A.; Costante, R.; Stefanucci, A.; Pinnen, F.; Luisi, G.; Pieretti, S.; Borsodi, A.; Bojinik, E.; Benyhe, S. Hybrid peptides endomorphin-2/DAMGO: Design, synthesis and biological evaluation. *Eur. J. Med. Chem.* **2013**, *68*, 167–177. [CrossRef]

36. Stefanucci, A.; Lei, W.; Hruby, V.J.; Macedonio, G.; Luisi, G.; Carradori, S.; Streicher, J.M.; Mollica, A. Fluorescent-labeled bioconjugates of the opioid peptides biphalin and DPDPE incorporating fluorescein-maleimide linkers. *Future Med. Chem.* **2017**, *9*, 859–869. [CrossRef]

37. Zádor, F.; Samavati, R.; Szlavicz, E.; Tuka, B.; Bojnik, E.; Fulop, F.; Toldi, J.; Vecsei, L.; Borsodi, A. Inhibition of opioid receptor mediated G-Protein activity after chronic administration of kynurenic acid and its derivative without direct binding to opioid receptors. *CNS Neurol. Disord. Drug Targets* **2014**, *13*, 1520–1529. [CrossRef]

38. Kemp, J.A.; Grimwood, S.; Foster, A.C. Characterization of the antagonism of excitatory amino acid receptors in rat cortex by kynurenic acid. *Br. J. Pharmacol.* **1987**, *91*, 314P.

39. Cheng, Y.-C.; Prusoff, W.H. Relationship between the inhibition constant (K1) and the concentration of inhibitor which causes 50 per cent inhibition (I50) of an enzymatic reaction. *Biochem. Pharmacol.* **1973**, *22*, 3099–3108.

40. Márki, A.; Monory, K.; Otvos, F.; Tóth, G.; Krassnig, R.; Schmidhammer, H.; Traynor, J.R.; Roques, B.P.; Maldonado, R.; Borsodi, A. Mu-opioid receptor specific antagonist cyprodime: Characterization by in vitro radioligand and [^{35}S]GTPγS binding assays. *Eur. J. Pharmacol.* **1999**, *383*, 209–214. [CrossRef]

41. Portoghese, P.S.; Sultana, M.; Takemori, A.E. Naltrindole, a highly selective and potent non-peptide delta opioid receptor antagonist. *Eur. J. Pharmacol.* **1988**, *146*, 185–186. [CrossRef]

42. Smith, H.S. Peripherally-acting opioids. *Pain Physician* **2008**, *11*, 121–132.

43. McNamara, C.R.; Mandel-Brehm, J.; Bautista, D.M.; Siemens, J.; Deranian, K.L.; Zhao, M.; Hayward, N.J.; Chong, J.A.; Julius, D.; Moran, M.M.; et al. TRPA1 mediates formalin-induced pain. *Proc. Natl. Acad. Sci. USA* **2007**, *104*, 13525–13530. [CrossRef] [PubMed]

44. Stein, C. Opioid receptors. *Annu. Rev. Med.* **2016**, *67*, 433–451. [CrossRef]

45. Yang, L.; Wang, H.; Shah, K.; Karamyan, V.T.; Abbruscato, T.J. Opioid receptor agonists reduce brain edema in stroke. *Brain Res.* **2011**, *1383*, 307–316. [CrossRef]

46. Yang, L.; Shah, K.; Wang, H.; Karamyan, V.T.; Abbruscato, T.J. Characterization of neuroprotective effects of biphalin, an opioid receptor agonist, in a model of focal brain ischemia. *J. Pharmacol. Exp. Ther.* **2011**, *339*, 499–508. [CrossRef]

47. Yang, L.; Islam, M.R.; Karamyan, V.T.; Abbruscato, T.J. In vitro and in vivo efficacy of a potent opioid receptor agonist, biphalin, compared to subtype-selective opioid receptor agonists for stroke treatment. *Brain Res.* **2015**, *1609*, 1–11. [CrossRef]

48. Davis, T.P.; Abbruscato, T.J.; Egleton, R.D. Peptides at the blood brain barrier: Knowing me knowing you. *Peptides* **2015**, *72*, 50–56. [CrossRef]

 *biomolecules* 

Article

Development and Characterization of a Fucoidan-Based Drug Delivery System by Using Hydrophilic Anticancer Polysaccharides to Simultaneously Deliver Hydrophobic Anticancer Drugs

Yen-Ho Lai [1], Chih-Sheng Chiang [2], Chin-Hao Hsu [1], Hung-Wei Cheng [1] and San-Yuan Chen [1,3,4,5,*]

[1] Department of Materials Science and Engineering, National Chiao Tung University, Hsinchu 30010, Taiwan; angus80817@gmail.com (Y.-H.L.); eddy610061@hotmail.com (C.-H.H.); mlb14756@gmail.com (H.-W.C.)

[2] Cell Therapy Center, China Medical University Hospital, Taichung 40454, Taiwan; brian78630g@gmail.com

[3] Frontier Research Centre on Fundamental and Applied Sciences of Matters, National Tsing Hua University, Hsinchu 30013, Taiwan

[4] School of Dentistry, College of Dental Medicine, Kaohsiung Medical University, Kaohsiung 80708, Taiwan

[5] Graduate Institute of Biomedical Science, China Medical University, Taichung 40454, Taiwan

* Correspondence: sanyuanchen@mail.nctu.edu.tw; Tel.: +886-3573-1818

Received: 4 June 2020; Accepted: 26 June 2020; Published: 28 June 2020

Abstract: Fucoidan, a natural sulfated polysaccharide, which can activate the immune response and lessen adverse effects, is expected to be an adjuvant agent in combination with chemotherapy. Using natural hydrophilic anticancer polysaccharides to simultaneously encapsulate hydrophobic anticancer drugs is feasible, and a reduced side effect can be achieved to amplify the therapeutic efficacy. In this study, a novel type of fucoidan-PLGA nanocarrier (FPN-DTX) was developed for the encapsulation of the hydrophobic anticancer drug, docetaxel (DTX), as a drug delivery system. From the comparison between FPN-DTX and the PLGA particles without fucoidan (PLGA-DTX), FPNs–DTX with fucoidan were highly stable with smaller sizes and dispersed well without aggregations in an aqueous environment. The drug loading and release can be further modified by modulating relative ratios of Fucoidan (Fu) to PLGA. The (FPN 3-DTX) nanoparticles with a 10:3 ratio of Fu:PLGA displayed uniform particle size with higher encapsulation efficiency than PLGA NPs and sustained drug release ability. The biocompatible fucoidan-PLGA nanoparticles displayed low cytotoxicity without drug loading after incubation with MDA-MB-231 triple-negative breast cancer cells. Despite lower cellular uptake than that of PLGA-DTX due to a higher degree of negative zeta potential and hydrophilicity, FPN 3-DTX effectively exerted better anticancer ability, so FPN 3-DTX can serve as a competent drug delivery system.

Keywords: fucoidan; PLGA; docetaxel; drug delivery system; anticancer therapy/cancer treatment

---

1. Introduction

Cancer is one of the leading causes of death worldwide, and a major reason for the limited therapeutic efficacy against cancer is the difficulty to deliver therapeutic agents precisely and effectively to the tumor sites without inducing severe adverse effects on healthy tissues and organs [1,2]. Drug delivery systems (DDS) such as micelles, vesicles/liposomes, nanocapsules, porous particles, nanospheres, and inorganic nanoparticles have emerged to resolve therapeutic issues by enhancing the accumulation of active pharmaceutical molecules in tumor sites [3]. These nanosystems are capable of accumulating in tumors via passive targeting because leaky blood vessels in tumors induce enhanced permeability and retention (EPR) effects [4–7]. Among these DDSs, liposomes serve as prevalent carriers in biomedical applications, and several liposomal formulations have been clinically approved, e.g., Doxil®, Caelyx®, and Myocet®, for cancer treatment [8,9]. Hydrophilic compounds can be incorporated in the aqueous core of liposomes, whereas hydrophobic ones can be accommodated within the lipid bilayer of liposomes. Unfortunately, limited space offered by the lipid bilayer compromises the payload of hydrophobic compounds. Moreover, rapid release of hydrophobic drugs in the lipid bilayer will cause high drug leakage and limit the applicability of liposomes as an ideal candidate for DDS of hydrophobic drugs to the targeted sites [10–12]. On the other hand, Poly(lactic-co-glycolic acid) (PLGA) is one of the most successfully used biodegradable materials approved by the FDA for drug delivery due to its superior biocompatibility and biodegradability. PLGA with PEG surface modification (PEGylation) to increase the hydrophilicity of the formulation has been widely used to load and deliver a high payload of hydrophobic anticancer bioactive agents, including drugs, proteins/peptides, and small molecules [13–17].

In addition to common anticancer drugs, many natural polysaccharides acquired from natural sources such as plants and algae display anticancer properties. Fucoidan (Fu), a natural product from *Fucus vesiculosus*, has recently drawn considerable attention because of its anti-coagulant, anti-virus, and anti-thrombotic activities [18–20]. Pozharitskaya et al. further elucidates some mechanisms of radical scavenging and the anti-inflammatory, anti-hyperglycemic, and anti-coagulant bioactivities of high-molecular-weight fucoidan from *Fucus vesiculosus* using in vitro models [21]. Fucoidan can reduce cell proliferation, inhibit migration of cancer cells, and induce cell apoptosis. The anti-cancer effects and the bioavailability of fucodian are related to various fucoidan-mediated pathways including PI3K/AKT, the MAPK pathway, and the caspase pathway [22]. In addition, several case studies of fucoidan as an alternative medicine in animal and human clinical trials have proved that combining fucoidan with clinical therapeutic agents can alleviate side effects of anti-cancer chemotherapy [21,23]. Recently, Abdollah et al. [24] reported that fucoidan prolonged the circulation time of dextran-coated iron oxide nanoparticles (IONs) with a doubling in tumor uptake. Ikeguchi et al. [25] examined the synergistic effect of a high-molecular-weight fucoidan with colorectal cancer chemotherapy agents, oxaliplatin plus 5-fluorouracil/leucovorin (FOLFOX) or irinotecan plus 5-fluorouracil/leucovorin (FOLFIRI). In addition, it was reported that the degree of sulfation was one of the factors associated with the anticancer activity of fucoidan. Thus, highly sulfated fucoidans, mainly containing fucose residues, possess higher anticancer activities than heterofucans with a low degree of sulfation [26–28]. Fucoidan used in most anticancer studies is a commercially available and highly sulfated type extracted from *Fucus vesiculosus* [18]. The pharmacokinetic of fucoidan concentration was further evaluated using competitive ELISA or a more sensitive sandwich ELISA with fucoidan-specific antibodies (NCT03422055 and NCT0313082), which showed that the maximum concentration of fucodian was reached 4 hr after administration of a single dose in a rat model, and the relative bioavailability was very low [29]. Nagamine et al. demonstrated the uptake and distribution of 2% w/w dietary *Cladosiphon okamuranus* fucoidan in a rat mode [30]. The result showed that only 0.1% could be absorbed in Caco-2 cells. However, Kimura et al. [31] found that liposome NPs could improve the bioavailability of sulfated polysaccharide. Therefore, nanosystems or nanoparticles have been developed to promote the bioavailability of fucoidan. Thus, Fucoidan with PLGA was chosen in this study to develop nanoparticles as a drug delivery system.

Docetaxel (DTX), used as a model drug in this study, has shown highly cytotoxic activity in several types of cancer including breast, lung, prostate, and ovarian cancers [32,33], but its clinical application is restricted owing to its poor aqueous solubility, low bioavailability, and cumulative systemic toxicity after prolonged and high-dose therapy [34]. Therefore, DTX is usually dissolved in Tween80: ethanol (50:50, v/v) to enhance its solubility, but these solvent-based DTX formulations easily cause toxic effects, including neutropenia, hypersensitivity, fluid retention, nail toxicities, and neuropathy. To enhance the bioavailability and anticancer activity, research has focused on entrapping DTX in nanocarriers such as polymeric micelles poly(lactic-co-glycolic acid) (PLGA) nanoparticles, and liposomes. Badran et al. reported that DTX loaded in chitosan(CS)-decorated PLGA NPs can maintain a higher concentration in the plasma with a longer terminal half-life and showed more than 4-fold the area under the plasma drug concentration-time curve (AUC) in CS-decorated PLGA NP compared to DTX solution [35]. Bowerman et al. [36] showed that DTX loaded in PLGA-nanoparticles can increase docetaxel circulation time. An in vivo antitumor efficacy study further demonstrated that DTX-NPs are expected to increase the therapeutic efficacy of chemotherapy and reduce systemic toxicity. Therefore, the DTX-encapsulated fucoidan-PLGA (FPNs–DTX) nanoparticles were developed to improve the therapy because fucoidan served as not only the anticancer agent but also one of the main components for stabilizing the nanoparticle structure. In addition, FPNs–DTX nanoparticles exhibit highly uniform particle size and excellent colloidal stability. As an inherently therapeutic nanomedicine with long-term circulation and high colloidal stability, FPNs–DTX are demonstrated to be potential candidate for cancer treatments.

2. Materials and Methods

2.1. Materials

Fucoidan from *Fucus vesiculosus* ($\geq$95%, Mw 20–200 kDa [37], 27.0% sulfate content [29], monosaccharides [38], Sigma, St. Louis, MO, USA), Resomer® RG 502 H poly(D,L-lactide-co-glycolide) (PLGA, acid terminated, Mw = 7000–17,000), chloroform, acetonitrile (ACN, HPLC-grade), dialysis tubing cellulose membrane (molecular weight cut-off = 14,000), fetal bovine serum (FBS), bovine serum albumin (BSA), Trypsin-EDTA, paraformaldehyde, and dimethyl sulfoxide (DMSO) were purchased from the Sigma–Aldrich Chemical Co. (St. Louis, MO, USA). Triton-X 100 was purchased from J.T. Baker. PELCO® Center-Marked Grids (200 mesh, 3.0 mm O.D., Copper) were purchased from Ted Pella Inc. (Redding, CA, USA). CdSe/ZnS core/shell quantum dots (QDs, 620 nm and 520 nm) were purchased from Ocean Nanotech (Springdale, AR, USA). Docetaxel ($\geq$99.5%) (DTX) was purchased from Scinopharm (Shan-Hua, Taiwan). 3-(4,5-dimethylthiazol-2-yl)-5-(3-carboxymethoxyphenyl)-2-(4-sulfophenyl)-2H-tetrazolium (MTS reagent) was purchased from Promega Co. (Madison, WI, USA). Alexa Fluor® 488 Phalloidin and Gibco® DMEM (Dulbecco's Modified Eagle Medium) were purchased from Thermo Fisher Scientific (Waltham, MA, USA) and supplemented with 10% FBS (Gibco), 2 mM l-glutamine, penicillin/streptomycin (Sigma); Triple-negative breast cancer cells (MDA-MB-231) and Umbilical Cord-Derived Mesenchymal Stem Cells were kindly donated by Dr. Woei-Cherng Shyu (China Medical University, Taichung, Taiwan); 4′,6-diamidino-2-phenylindole (DAPI) was purchased from Invitrogen (Waltham, MA, USA). Fluorescence mounting medium was purchased from Dako (Agilent Technologies, Wood Dale, IL, USA). Phosphate-buffered saline (PBS) was purchased from AMRESCO (Solon, OH, USA). All chemicals and solvents were of analytical reagent grade.

2.2. Methods

2.2.1. Preparation of Docetaxel-Encapsulated Fucoidan–PLGA Nanoparticles (FPNs–DTX)

Docetaxel (DTX) was used as a model drug to form FPNs–DTX and was useful to evaluate the drug loading and release behavior of these nanoparticles. The FPNs–DTX were fabricated by the emulsion and solvent evaporation method. The organic solvent (50 μL chloroform) comprising PLGA (3 mg) and DTX (1 mg) was prepared and added to the fucoidan (0.5% *w/v*, 2 mL) aqueous solution. The mixed solution was subsequently emulsified by a probe ultrasonicator (UP200Ht-Handheld Ultrasonic Homogenizer, Hielscher Ultrasonics, Teltow, Germany) at the power of 10 W and the frequency of 26 kHz to form an oil-in-water emulsion. The organic solvent was then removed at 55 °C and the FPNs–DTX were washed with deionized water three times. Finally, the yield of FPN nanoparticles was about 18.4% (*w/w*). The abbreviations and compositions of all formulations used in this study are further summarized in Table 1.

Table 1. Abbreviations and composition of all formulations.

Formulations	Fucoidan	Poly(lactic-co-glycolic Acid)	Docetaxel	DTX-Loaded PLGA NPs	DTX-Loaded Fucoidan-PLGA NPs
Abbreviations	Fu	PLGA	DTX	PLGA 3-DTX	FPN 3-DTX
Composition	Fucoidan	PLGA	DTX	PLGA (3 mg) DTX (1 mg)	Fucoidan (10 mg) PLGA (3 mg) DTX (1 mg)

2.2.2. Characterization of FPNs–DTX

The average sizes and zeta potentials of the samples were measured by dynamic light scattering and electrophoretic light scattering, respectively (DLS and ELS, Delsa^{TM} Nano C analyzer, Beckman Coulter, Irving, TX, USA). The nanostructure of the FPNs–DTX was observed by field-emission scanning electron microscopy (FE-SEM, JSM-6700F, JEOL, Tokyo, Japan) and transmission electron microscopy (TEM, JSM-2100, JEOL, Tokyo, Japan). For SEM observation, the sample solution was dropped on silicon wafers, dried in a vacuum desiccator at room temperature, and coated with platinum sputtering. The element analysis was performed using energy-dispersive X-ray spectroscopy (EDS), with which SEM was equipped. For TEM observation, the sample solution was dropped on a carbon-coated copper grid. Residues of the droplet were removed, and then the copper grid was dried in a vacuum desiccator at room temperature. The chemical state of nanoparticles containing docetaxel was analyzed in the KBr mixture pellets using Fourier transform infrared spectroscopy (FTIR, spectrum 100, PerkinElmer, Waltham, MA, USA).

For the X-ray photoelectron spectroscopy (XPS) detection, the sample-coated silicon wafers were prepared using the method similar to the SEM sample process. The chemical states of the specific element on the sample surface were determined with XPS at beamline 24A at NSRRC (National Synchrotron Radiation Research Center, Hsinchu, Taiwan). The XPS data were processed and peak-fitted using Origin software (OriginLab Corporation, Northampton, MA, USA).

The amount of fucoidan adsorbed onto the FPNs–DTX was evaluated by referring to the method in the study of Garti et al. [39]. In brief, after centrifugation of the FPNs–DTX emulsion for separation of the supernatant aqueous phase and the nanoparticles, the amount of excess fucoidan in the supernatant phase was lyophilized and then weighed. The amount of adsorbed fucoidan was estimated by the difference in weight of the preparation and the excess fucoidan in the supernatant phase.

2.2.3. Drug-Loading Ability and In Vitro Drug Release

The loading capacity and encapsulation efficiency (EE) of FPNs–DTX were examined using high-performance liquid chromatography (HPLC). The sample solution was lyophilized and then weighed. The encapsulated DTX was extracted from nanoparticles with acetonitrile (ACN) and then quantified using HPLC (Agilent Technologies 1200 series, San Diego, CA, USA) with a 4.6 × 250 mm Inertsil® ODS-2 (5 μm) column. The detector was operated at a wavelength of 227 nm, and the sample was eluted using a mobile phase composed of 40% deionized water and 60% ACN at a flow rate of 1.0 mL/min and ambient temperature. Peak identity was confirmed by the retention time of DTX of about 6.5 min. The loading capacity and EE of the DTX in the FPNs–DTX were calculated as follows:

$$\text{Loading capacity\%} = A/B \times 100\% \tag{1}$$

where A is the mass of encapsulated drug and B is the mass of all FPNs–DTX, while

$$\text{EE\%} = C/D \times 100\% \tag{2}$$

where C is the amount of encapsulated drug and D is the total amount of drug in the feeding formula.

For the drug release test, a mixture of FPNs–DTX (0.2 mL) suspension containing equivalent DTX and phosphate-buffered saline (PBS, 3.8 mL) in a volume ratio of 1:19 was added to a dialysis tubing cellulose membrane with a cutoff molecular weight of 14,000. The dialysis bags were dialyzed in phosphate-buffered saline (PBS, 10 mL) with 0.5% Tween 80 at 37 °C with gentle shaking, and aliquots of the incubation medium were collected at predetermined time points. The amounts of released drug were quantified using the HPLC method. The solution was maintained at a constant volume by replacing the original solution with PBS. All experiments were performed in triplicate.

2.2.4. Comparison between PLGA Particles and FPNs

To study the role of fucoidan in forming a stable nanoemulsion, the appearance of the three emulsions containing fucoidan alone (Fucoidan), PLGA alone (PLGA), and fucoidan accompanying PLGA (Fu + PLGA) was observed and recorded over time after emulsification without removing the organic solvent. The amounts of ingredients and the preparation method were the same as those for FPNs–DTX prepared with PLGA (1 mg) but without DTX. For further comparing DTX-encapsulated PLGA particles (PLGA-DTX) with FPNs–DTX, characterizations were carried out by SEM, DLS, and ELS. The PLGA-DTX were fabricated according to the method of FPNs–DTX but in the absence of fucoidan.

To evaluate the stability of particles prepared with fucoidan (FPNs–DTX prepared with a fucoidan to PLGA ratio (10:3), abbreviated as FPN 3-DTX) or without fucoidan (PLGA-DTX corresponding to FPN 3-DTX, abbreviated as PLGA 3-DTX), particles were suspended in double distilled water (DDW) and fetal bovine serum (FBS). The particles were resuspended in DDW and FBS after centrifugation. Particle sizes were recorded over time by DLS measurement. The resuspended state of FPN 3-DTX versus PLGA 3-DTX in DDW after centrifugation and 10 min bath sonication was examined as well.

2.2.5. Stability of FPN 3-DTX

To estimate the stability of FPN 3-DTX in terms of drug encapsulation and particle size, DTX concentrations and particle sizes were measured and recorded over time for four weeks by HPLC and DLS methods, respectively. The samples were stored at 4 °C. All experiments were performed in triplicate.

2.2.6. Cell Culture

MDA-MB-231 breast cancer cells were seeded in 75T culture flasks containing Dulbecco's Modified Eagle Medium (DMEM) supplemented with 10% FBS and 1% penicillin–streptomycin–neomycin solution at 37 °C under 5% CO_2.

2.2.7. Cellular Uptake and Subcellular Localization of the FPNs

To facilitate the observation of cellular uptake, hydrophobic red fluorescent QDs were added to the oil phase and then incorporated into FPNs prepared with a fucoidan to PLGA ratio (10:3) (FPN 3-QD) and PLGA particles corresponding to FPN 3-QD (PLGA 3-QD) according to the FPN 3-DTX fabrication process. For cellular association studied by flow cytometry, 2×10^5 cells per well were seeded in 6-well plates and grown for 24 h. They were then incubated with nanoparticles for different periods of time. After incubation, cells were harvested by trypsin-EDTA and washed with PBS twice to remove free nanoparticles followed by analysis using a flow cytometer (NovoCyte®®, ACEA biosciences, San Diego, CA, USA). The data were acquired from 10,000 cumulatively gated events (cells) by measuring their fluorescence intensity along with the number of events.

In fluorescence microscopy and confocal laser scanning microscope (CLSM) experiments, cells were seeded on glass coverslips placed in well plates. After incubation, cells were incubated with nanoparticles for different periods of time. Subsequently, the cells were washed several times with PBS and fixed with 3.7% paraformaldehyde solution (in PBS) for 20 min. Cellular permeability was performed with 0.1% triton X-100 (in PBS) for 20 min. Blocking buffer (1% BSA in PBS) was added to the wells for 1 h. Finally, the cells were stained with blue-fluorescent 4′,6-diamidino-2-phenylindole (DAPI, 5 µg/mL) and green-fluorescent phalloidin (5 unit/mL) to visualize nuclei and F-actin, respectively, followed by mounting coverslips on glass slides using fluorescence mounting medium.

Subcellular localization of FPN 3-QD and PLGA 3-QD was investigated using a fluorescence microscope (Eclipse TE2000-U, Nikon) and confocal laser scanning microscope (CLSM, D-Eclipse C1, Nikon).

2.2.8. Cell Viability

In vitro cytotoxicity of free DTX, FPNs (FPN 3), PLGA particles (PLGA 3), DTX-loaded FPNs (FPN 3-DTX), and DTX-loaded PLGA particles (PLGA 3-DTX) was examined on MDA-MB-231 cells using the MTS method as follows: 2×10^3 cells per well were first seeded in 96-well plates and exposed to serial concentrations of DTX or equivalent particles at 37 °C for 72 h. Subsequently, a 200 µL mixture of MTS reagent and FBS-free medium (1:5) replaced the original medium, and the cells were incubated for an additional 4 h. The absorbance was detected at a wavelength of 450 nm using a Sunrise microplate absorbance reader (Tecan, Austria). Finally, the cell viability was determined by comparison of sample absorbance (A $_{sample}$) with the untreated control (A $_{control}$) and calculated using the following equation:

$$\text{Cell Viability (\%)} = (A_{sample}/A_{control}) \times 100\% \tag{3}$$

The half maximal inhibitory concentration (IC_{50}), which is defined as the dosage of a compound that inhibits 50% of cell growth, was calculated from the obtained viability curves using CompuSyn software (Version 1.0, 2004, ComboSyn Inc., Paramus, NJ, USA). All experiments were performed at least in triplicate.

2.2.9. Statistical Analysis

All means are presented with their standard deviation (SD). Analysis of variance (ANOVA) was conducted, and $p < 0.05$ was regarded as statistically significant.

3. Results

3.1. Preparation of Docetaxel-Encapsulated Fucoidan-PLGA Nanoparticles (FPNs–DTX)

Scheme 1 illustrates the fabrication process of the docetaxel-encapsulated fucoidan-PLGA nanoparticles (FPNs–DTX). To encapsulate the hydrophobic drug, various amounts of PLGA, which served as the main hydrophobic structure of the particle core, were adopted to form various FPNs–DTX. The formulas prepared with ratios 10:1, 10:3, and 10:5 (fucoidan to PLGA) were selected for further studies and are indicated as FPN 1-DTX, FPN 3-DTX, and FPN 5-DTX respectively. By mixing and emulsifying aqueous fucoidan solution in chloroform solution comprising PLGA and docetaxel, FPNs–DTX was obtained after removal of chloroform where the hydrophobic docetaxel drug was encapsulated within the PLGA matrix and fucoidan was exposed on the outside surface. To encapsulate the hydrophobic drug, various amounts of PLGA, which served as the main hydrophobic structure of particle core, were adopted to form various FPNs–DTX (Table 2). The encapsulation efficiencies (EE) and loading capacities for docetaxel showed that there was no positive correlation between encapsulation ability and PLGA content. As a result, the formulas prepared with ratios 10:1, 10:3, and 10:5 (fucoidan to PLGA) were selected for further studies and indicated as FPN 1-DTX, FPN 3-DTX and FPN 5-DTX, respectively.

Table 2. Formulations and their corresponding encapsulated amounts, encapsulation efficiency (EE), and loading capacity of FPNs–DTX for docetaxel. Data in the table are the mean ± SD, n = 3.

Fucoidan:PLGA (*w:w*)	Encapsulated DTX (µg)	Encapsulation Efficiency (%)	Loading Capacity (%X)
10:1	551 ± 37	55.1 ± 3.7%	34 ± 2.3%
10:3	687 ± 42	68.7 ± 4.2%	28.7 ± 1.8%
10:5	611 ± 23	61.1 ± 2.3%	18.8 ± 0.7%
10:10	446 ± 16	44.6 ± 1.6%	6.6 ± 0.2%

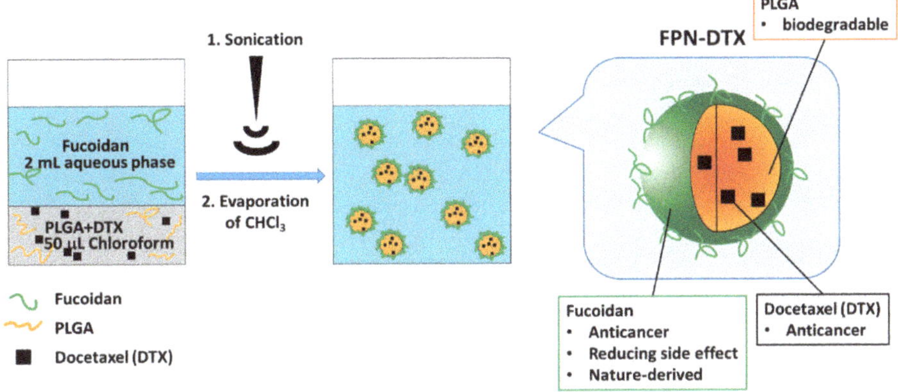

Scheme 1. Illustration of FPNs–DTX preparation using a one-step emulsification process (Fucoidan in 2 mL aqueous phase and PLGA+DTX in 50 µL Chloroform phase).

3.2. Effects of Fucoidan on Nanoemulsions and Nanoparticles

To study the effects of fucoidan on nanoemulsions, three emulsions containing fucoidan alone (Fucoidan), PLGA alone (PLGA), and a combination of fucoidan and PLGA (Fu + PLGA) were examined after emulsification without removing the organic solvent. By observing the variation of emulsions over time, it was found that white turbid layers in all three emulsions gradually sunk. The phase separation phenomenon called "creaming" or "sedimentation" is the migration of the dispersed phase in an emulsion as a reflection of instability. The sinking rate (sedimentation rate) can be an indicator of stability of the emulsions, while higher stability makes droplets sink more slowly [19,40]. In Figure 1, droplets with the highest to lowest sinking rate were ranked as fucoidan emulsion, PLGA emulsion, and Fu + PLGA emulsion. This indicated that the Fu + PLGA emulsion is the most stable among the three emulsions, implying the assistance of fucoidan in the stability of droplets or particles in the emulsion. However, it was also noted that the emulsion containing fucoidan alone was not sufficiently stabilized, so the hydrophobic material PLGA was necessary to form the rigid structure of the particles.

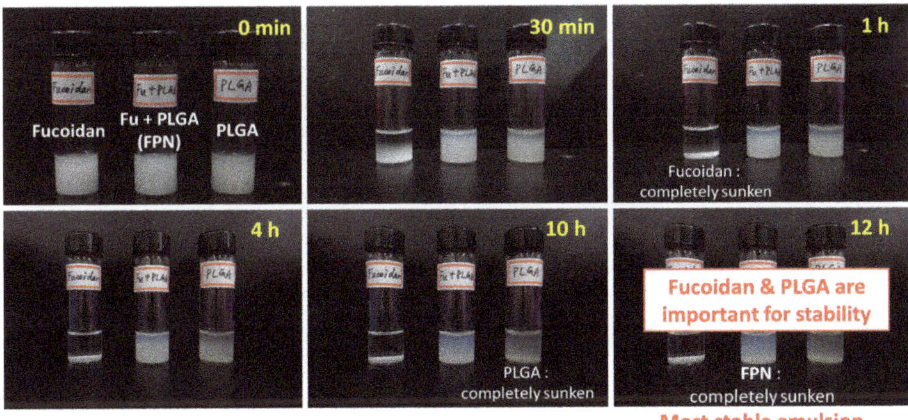

Figure 1. Observation of variation in emulsions containing fucoidan alone (Fucoidan), fucoidan accompanying PLGA (Fu + PLGA), and PLGA alone (PLGA) over time from the left to the right bottle in each image.

To further study the interaction between PLGA and Fucoidan in nanoparticles, SEM and TEM were used to observe the morphology of NPs. As illustrated in Figure 2a,b, the SEM image of particle morphology and distribution demonstrated the contribution of fucoidan to stabilize nanoparticles in suspension. Particles prepared without fucoidan (PLGA-DTX) did not exhibit uniform sizes and some of them aggregated together compared with those particles prepared with fucoidan (FPNs–DTX) as mentioned above. To investigate the surface hydrophobicity of particles prepared with or without fucoidan, the states of FPN 3-DTX (F) and PLGA 3-DTX (P) suspensions after centrifugation were observed in Figure 2d,e. Despite the similar appearance of both FPN 3-DTX and PLGA 3-DTX suspensions before and after centrifugation, FPN 3-DTX seemed to suspend and disperse well again in contrast to the sediment of PLGA 3-DTX for 10 min after resuspension. This indicates that the fucoidan of FPN 3-DTX can modify the surface hydrophobicity of PLGA and make particles disperse more easiily in an aqueous environment than PLGA 3-DTX, which is beneficial for delivery in the human circulatory system.

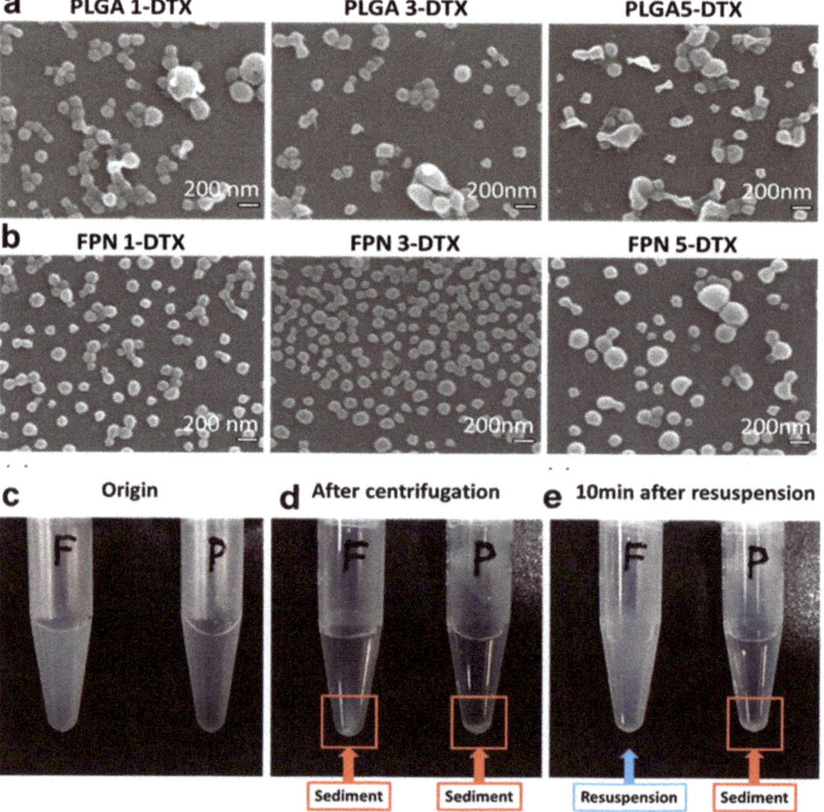

Figure 2. (**a**) SEM images of particles prepared without fucoidan (PLGA-DTX) compared with particles prepared with (**b**) fucoidan (FPNs–DTX). The concentration of PLGA in PLGA 1, PLGA 3, and PLGA 5 is equal to FPN 1, FPN 3, and FPN 5 respectively. Photograph of nanoparticles (FPN 3-DTX (F) and PLGA 3-DTX (P)) dispersive in DDW. In different states (**c**) the origins, (**d**) after centrifugation, and (**e**) 10 min after resuspension.

As observed from the DLS, particle size increased with an increasing amount of PLGA involved in preparation. FPN 3-DTX displayed the sharpest peak of size distribution and the smallest polydispersity index among the three formulas in Figure 3 based on DLS measurement, which is likely attributed to the negatively-charged fucoidan with Zeta potential (−60 mV) at the surface compared to those particles without fucoidan (PLGA-DTX = −25.2 mV), enhancing colloidal stability and the retention ability of FPNs–DTX in the human circulatory system.

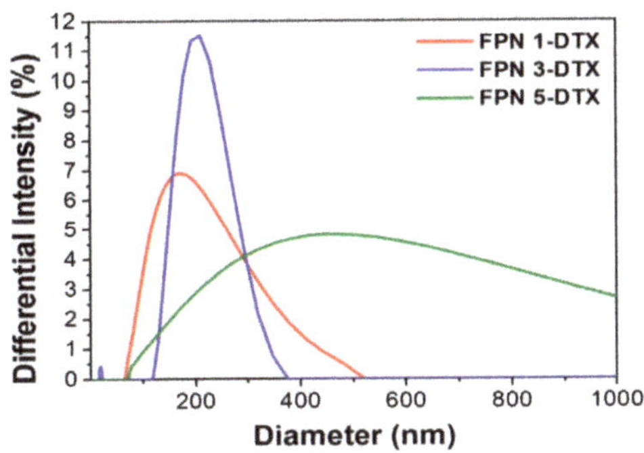

Figure 3. Size distributions of FPN 1-DTX, FPN 3-DTX, and FPN 5-DTX.

3.3. Drug Release Behavior and Stability of FPNs–DTX

To further study the release behavior of FPN nanoparticles, HPLC was used to detect the release of DTX from different ratios of FPN-DTX nanoparticles. Figure 4a shows in vitro natural drug release of FPNs–DTX in PBS. An initial burst release from FPN 1-DTX prepared with the least amount of PLGA was found, but the release exhibited a slower release rate as the amount of PLGA increased, which is possibly related to the change in the compactness of the particle structure. Therefore, FPN 3-DTX was selected for further evaluation because of three attractive features: (1) higher encapsulation ability for the drug, (2) excellent colloidal stability with uniform size, and (3) proper drug carrying ability rather than an initial burst release. To estimate the stability of FPN 3-DTX in suspension, particle sizes, morphologies, and the concentrations of drug encapsulation were measured and recorded over time for four weeks under 4 °C storage. The DTX concentration at various times was expressed as percentages of the initial concentration as shown in Figure 4b. Throughout this period, both indicators varied insignificantly ($p > 0.05$). The initial sample bore a great resemblance to the sample stored at 4 °C for about four weeks under SEM. In other words, FPN 3-DTX remained stable as a drug delivery system at least for about one month, which was probably attributed to the electrostatic interaction and steric hindrance from the nature of its anionic polysaccharide. Although fluctuations in FPN 3-DTX size were observed, the long-term stability of FPN 3-DTX was confirmed in Figure 4b.

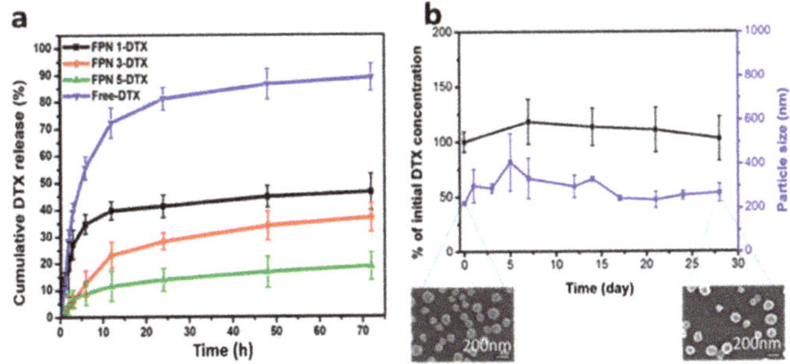

Figure 4. (a) Cumulative DTX release from FPN 1-DTX, FPN 3-DTX, FPN 5-DTX, and free DTX. (b) Variation in docetaxel concentration and particle size of FPN 3-DTX over time for four weeks.

3.4. Component Verification of FPN 3-DTX

The chemical bonds of PLGA, fucoidan, docetaxel, the Fu+PLGA+DTX mixed powder, and FPN 3-DTX were further measured (Figure 5a) using Fourier transform infrared spectra (FTIR). PLGA showed a major characteristic peak at 1763 cm^{-1} (COOH vibration). The spectrum of fucoidan presented a broad band at 3466 cm^{-1} (O-H groups stretching), a peak at 1644 cm^{-1} (COOH stretching vibration), a peak around 1247 cm^{-1} (S=O stretching), and a peak at 842 cm^{-1} (C-O-S bending). The major peak of docetaxel was around 1713 cm^{-1} (ester stretching). The spectrum of the powdered Fu+PLGA+DTX mixture showed major peaks of PLGA, fucoidan, and docetaxel even though some superpositions occurred. However, in the spectrum of docetaxel-encapsulated nanoparticles (FPN 3-DTX), the major peaks of docetaxel were significantly lowered, indicating the successful encapsulation of DTX drug in particles due to the chemical interaction between the carboxylic acid in PLGA and the ester group in docetaxel [41,42].

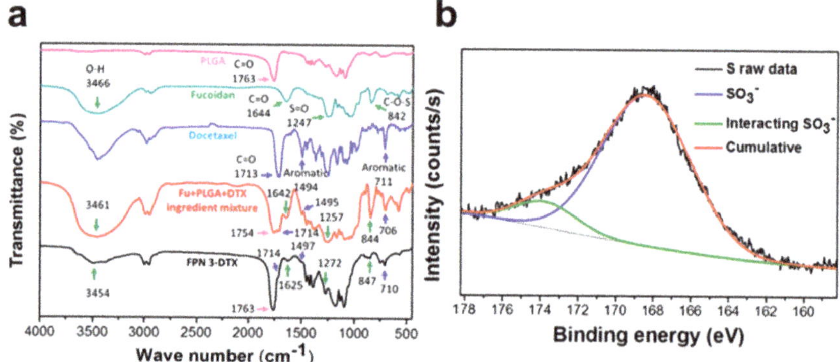

Figure 5. (**a**) Fourier transform infrared (FTIR) spectra of PLGA, fucoidan, docetaxel, powdered mixture of three main ingredients, and FPN 3-DTX. (**b**) Cumulative sulfur 2p (S 2p) X-ray photoelectron spectrum of FPN 3-DTX based on X-ray photoelectron spectroscopy (XPS).

X-ray photoelectron spectroscopy (XPS) is a technique for identifying the composition of material surface and can be used to confirm the surface adsorption of fucoidan on nanoparticles. The XPS spectrum of the sulfur 2p (S2p) from FPN 3-DTX samples can be detected and decomposed into two bands by a curve fitting technique (Figure 5b). The binding energy peak around 168 eV represents sulfur at a higher oxidation state assigned to sulfur atoms bonded to oxygen atoms such as sulfone, sulfonate or sulfate [43,44]. In addition to the main peak around 168 eV, there was another one in a different position. The alteration of binding energy peak could indicate the interactions of some sulfate groups on fucoidan with the other components in FPN 3-DTX [45].

3.5. Cellular Uptake Efficiency and Subcellular Localization of Fluorescence-Labeled FPNs

One of the important concerns of cancer therapy is to deliver anticancer agents to the cancer cells in efficient and appropriate ways. Fluorescence microscopy images shown in Figure 6 demonstrated similar uptake behavior for FPN 3-QD and PLGA 3-QD. The QDs with the red fluorescence implied the subcellular localization of particles. Increasing red fluorescence intensity from both particles associated with cells was observed within 24 h incubation, when the amount of cell uptake seemed to reach saturation.

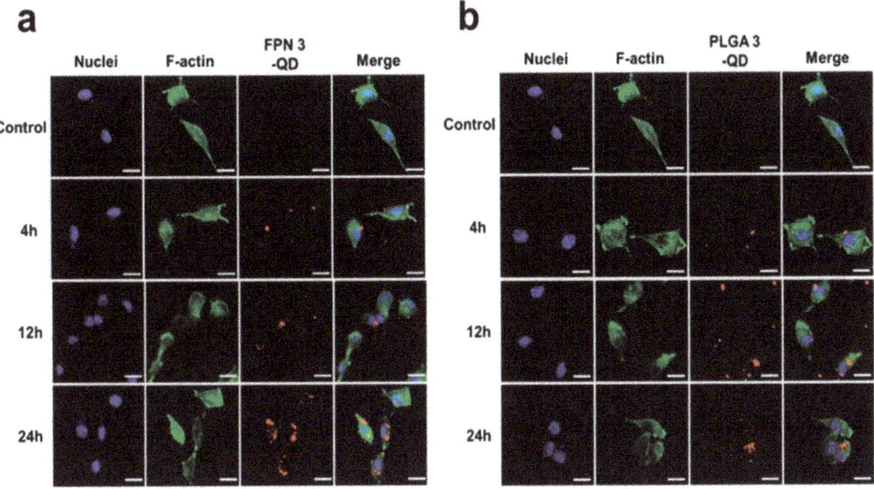

Figure 6. (**a**) Fluorescence images of cellular uptake of quantum dot-loaded FPN (FPN 3-QD) and (**b**) quantum dot-loaded PLGA particles (PLGA 3-QD) by MDA-MB-231 breast cancer cells at different times after feeding FPN 3-QD or PLGA 3-QD. Blue is DAPI for labeling the nuclei, green is phalloidin for labeling the cytoskeleton (F-actin), and red is the quantum dot from FPN 3-QD or PLGA 3-QD. (Scale bar: 20 μm).

Quantitative evaluation of cellular association was performed by flow cytometry as shown in Figure 7. The results were in line with the observations by fluorescence microscopy. The fluorescence intensity was observed to reach saturation at 24 h incubation for FPN 3-QD and at 12 h for PLGA 3-QD. The median fluorescence intensity (MFI) detected for PLGA 3-QD was higher than that for FPN 3-QD at all of the incubation times. On the other hand, higher fluorescence intensity compared with the untreated control indicates cell association with fluorescence-labeled particles.

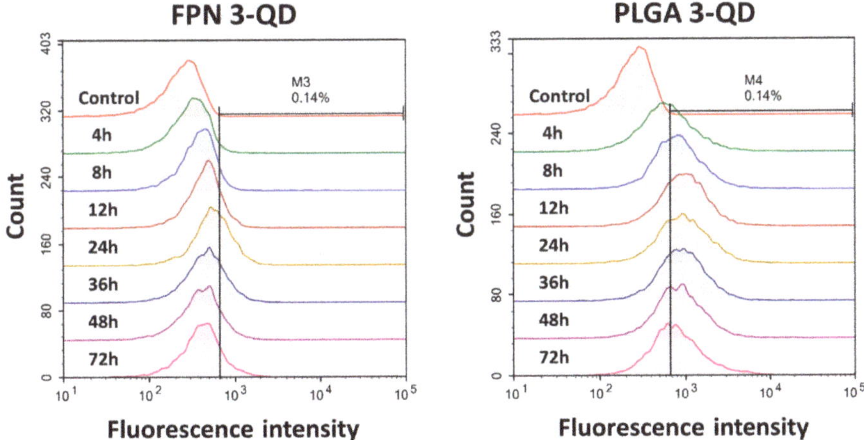

Figure 7. Flow cytometry for detecting cell associations with FPN 3-QD (F) and PLGA 3-QD (P) at different times after feeding particles. MFI in the tables stands for median fluorescence intensity. M3% (FPN 3-QD) and M4% (PLGA 3-QD) stand for the percent of the cell population emitting a fluorescence intensity in the range of the M3 gate and M4 gate, respectively.

The behavior of cellular uptake assessed by fluorescence microscopy and flow cytometry indicated that the uptake of FPN 3-QD was inferior to that of PLGA 3-QD, which is possibly correlated with the higher degree of the negative zeta potential and hydrophilicity of FPN 3-DTX.

3.6. Cytotoxicity of FPN 3-DTX

The cell viability of FPN 3-DTX in a non-tumor cell line is very important and was first evaluated using umbilical cord-derived mesenchymal stem cells in this study. Figure S1 shows that FPN 3-DTX had higher cell viability compared to free DTX and blank nanoparticles (FPN) in the non-tumor cell line (mesenchymal stem cells (MSCs)) under the higher DTX concentration due to the protection of DTX within FPN. Subsequently, cytotoxicity of FPN 3-DTX was assessed in MDA-MB-231 breast cancer cells and compared with the other formulations as shown in Figure 8. Without DTX drug loading, it was found that FPN-3 displayed high biocompatibility in the absence of DTX compared to PLGA-3 (Figure 8a). With DTX loading, Figure 8b shows that the cell viability and IC_{50} indicated that FPN 3-DTX had a slightly lower cytotoxicity compared with PLGA 3-DTX at a very low dose because FPN 3-DTX possesses higher biocompatibility. However, FPN 3-DTX exhibited better cytotoxicity compared with PLGA 3-DTX as the DTX concentration was higher than 10nM. This suggested that by carefully assessing the therapeutic index, a proper dosage of FPN 3-DTX could be applied to achieve enhanced therapeutic efficacy using a lower DTX amount to reduce the side effect in comparison with free DTX and PLGA 3-DTX.

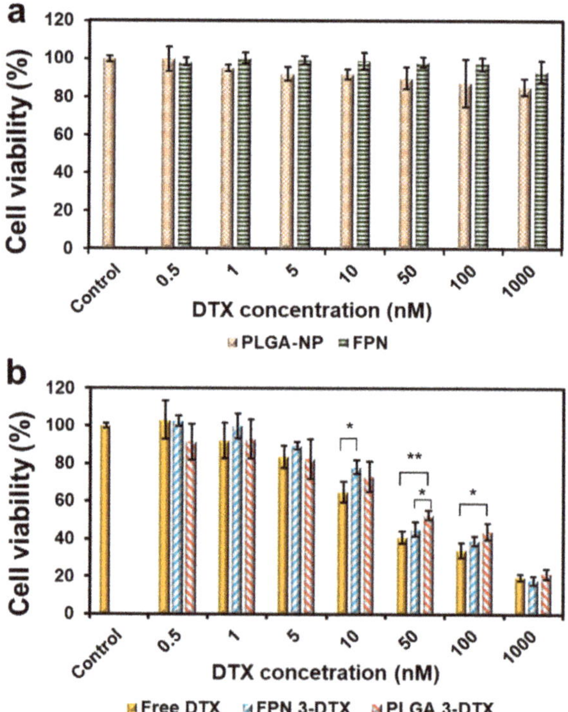

Figure 8. Cell viability of MDA-MB-231 breast cancer cells treated with (**a**) drug-loaded nanoparticles (FPN 3-DTX and PLGA 3-DTX) and free DTX. (**b**) blank nanoparticles (FPN and PLGA) in which the concentration of nanoparticles was equal to drug-loaded nanoparticles after 72 h treatment. Results are expressed as the mean ± SD of three replicates for each treatment (n = 3). Marked asterisks (*) with * $p < 0.05$ and ** $p < 0.01$ represent statistically significant differences.

4. Discussion

Fucoidan (Fu), a natural product from fucus vesiculosus, has recently drawn considerable attention because of its anti-coagulant, anti-virus, anti-thrombotic, and anticancer activities. Therefore, it was adopted as an excipient to produce FPN 3-DTX nanomedicine. The in vivo toxicity of the Fucodain-based drug delivery system was evaluated in our previous published paper [46] and demonstrated no obvious lesions in the liver, colon, lung or kidney based on histological analysis four weeks after the treatment of Fucoidan-base nanoparticles. Similar results were reported by Cai et al. when assessing the effect of PLGA with layer-by-layer (PLO/fucoidan) NPs on cell viability, acridine orange/ethidium bromide staining, hemolysis, and mouse systemic toxicity of cells and mice [47]. In addition, FPN 3-DTX with fucoidan modification on the surface of hydrophobic PLGA to encapsulate the hydrophobic drug DTX can produce more uniform and smaller particle sizes and also makes it easier for particles to disperse in an aqueous environment compared with PLGA 3-DTX, which is attributed to the higher hydrophilicity of FPN (Figure 2). The XPS result further confirmed the adsorption of fucoidan on the surface of FPN 3-DTX particles. In addition to interactions of the sulfate group, the formation of intermolecular hydrogen bonds could occur between the carbonyl groups of PLGA and the hydroxyl groups of fucoidan [48–50]. The advantage of using Fucoidan with PLGA addition can be further observed in Figure 4a which shows that FPN 3-DTX exhibited a slower release rate as the amount of PLGA increased, which is possibly related to the change in the compactness of the particle structure. Moreover, the FPN 3-DTX had high stability with hydrogen bonds which avoided the leakage of DTX and benefited long-time preservation as evidenced in Figure 4b. Barbosa et al. and Huang et al. further found that polysaccharide-base nanoparticles could improve the bioavailability [51,52]. Therefore, the FPN can be expected to improve the therapeutic effects with a high stability and low release rate.

To visualize and quantify the cellular uptake of FPN3 for studying the delivery to cells, the quantum dots (QD) were introduced and encapsulated in the FPN3 to form the FPN 3-QD for fluorescence microscopy and flow cytometry using MDA-MB-231 breast cancer cells. We also tested QDs-loaded PLGA particles (PLGA 3-QD) to identify how zeta potential affects cellular uptake of particles. The behavior of cellular uptake assessed by fluorescence microscopy and flow cytometry indicated that the uptake of FPN 3-QD was inferior to that of PLGA 3-QD which is possibly correlated with the higher degree of negative zeta potential and hydrophilicity of FPN 3-DTX [53,54]. Without DTX drug loading, it was found that FPN-3 displayed high biocompatibility in the absence of DTX compared to the PLGA-3. The toxicity of FPN 3-DTX to normal cells is presented in Figure S1. The results showed that FPN had high biocompatibility and well protected DTX to prevent release from the NPs in the circulation system [55]. This indicated that FPN 3-DTX could reduce the side effect of DTX when transported to the tumor. These results were attributed to the slow release rate and lower cell uptake under the shorter incubation time (24 h). Figure 8b further shows that FPN 3-DTX exhibited better cytotoxicity compared with PLGA 3-DTX as the DTX concentration was higher than 10 nM. This indicated that FPN 3-DTX effectively exerted anticancer ability against MDA-MB-231 cells. In terms of better retention ability and tumor killing effect, FPNs–DTX can be regarded as a good potential drug delivery system (DDS) compared with PLGA 3-DTX. Moreover, the immune-related benefits of fucoidan such as activation of nature killer cells, an increase of IL-2 secretion, as well as inhibition of angiogenesis and metastasis will further enhance immunotherapy at a lower dose of the chemotherapy drug, which could not be observed and shown in vitro. Therefore, further experiments in vivo are necessary to evaluate the therapeutic efficacy of FPN 3-DTX.

5. Conclusions

In this study, a novel type of anticancer fucoidan-PLGA nanoparticle (FPNs–DTX) was developed that is capable of encapsulating hydrophobic docetaxel (DTX) as a drug delivery system. With the addition of PLGA, FPNs–DTX formed smaller nanoparticles and dispersed well without obvious aggregations over long-term duration. Furthermore, FPNs–DTX nanoparticles displayed low cytotoxicity and biocompatibility in the absence of docetaxel (FPN3) compared to PLGA. The more negatively charged and more hydrophilic properties of FPN 3-DTX could lead to lower cellular uptake but could effectively exert a better anticancer ability. Therefore, FPN 3-DTX is regarded as a better potential drug delivery system than PLGA 3-DTX in terms of better suspension and anti-cancer ability. The further achievements in vivo and the clinical applications of FPNs–DTX are promising in the biomaterial and cancer therapy fields.

Supplementary Materials: The following are available online at http://www.mdpi.com/2218-273X/10/7/970/s1, Figure S1: Cell viability of umbilical cord-derived mesenchymal stem cells treated with free DTX, drug loaded nanoparticles (FPN 3-DTX) and blank nanoparticles (FPN).

Author Contributions: Conceptualization, Y.-H.L.; Data curation, Y.-H.L., C.-S.C. and C.-H.H.; Investigation, Y.-H.L. and H.-W.C.; Project administration, S.-Y.C.; Writing–original draft, Y.-H.L. All authors have read and agree to the published version of the manuscript.

Funding: This research was funded by Ministry of Science and Technology, Taiwan, grant number MOST 107-2221-E-214-012-MY3, MOST 108-2218-E-080-001, MOST 106-2221-E-009-065-MY3.

Conflicts of Interest: The authors declare no conflict of interest.

References

1. Sun, T.; Zhang, Y.S.; Pang, B.; Hyun, D.C.; Yang, M.; Xia, Y. Engineered Nanoparticles for Drug Delivery in Cancer Therapy. *Angew. Chem. Int. Ed.* **2014**, *53*, 12320–12364. [CrossRef] [PubMed]
2. Peer, D.; Karp, J.M.; Hong, S.; Farokhzad, O.C.; Margalit, R.; Langer, R. Nanocarriers as an emerging platform for cancer therapy. *Nat. Nanotechnol.* **2007**, *2*, 751–760. [CrossRef]
3. Chapman, M.; Pascu, S.I. Nanomedicines design: Approaches towards the imaging and therapy of brain tumours. *J. Nanomed. Nanotechnol.* **2012**, *4*, 6. [CrossRef]
4. Chiang, C.S.; Hu, S.H.; Liao, B.J.; Chang, Y.C.; Chen, S.Y. Enhancement of cancer therapy efficacy by trastuzumab-conjugated and pH-sensitive nanocapsules with the simultaneous encapsulation of hydrophilic and hydrophobic compounds. *Nanomedicine* **2014**, *10*, 99–107. [CrossRef] [PubMed]
5. Matsumura, Y.; Maeda, H. A new concept for macromolecular therapeutics in cancer chemotherapy: Mechanism of tumoritropic accumulation of proteins and the antitumor agent smancs. *Cancer Res.* **1986**, *46*, 6387–6392. [PubMed]
6. Torchilin, V.P. Micellar Nanocarriers: Pharmaceutical Perspectives. *Pharm. Res.* **2006**, *24*, 1. [CrossRef]
7. Maeda, H.; Bharate, G.Y.; Daruwalla, J. Polymeric drugs for efficient tumor-targeted drug delivery based on EPR-effect. *Eur. J. Pharm. Biopharm.* **2009**, *71*, 409–419. [CrossRef]
8. Abu Lila, A.S.; Ishida, T. Liposomal Delivery Systems: Design Optimization and Current Applications. *Biol. Pharm. Bull.* **2017**, *40*, 1–10. [CrossRef]
9. Barenholz, Y. Doxil®®–the first FDA-approved nano-drug: Lessons learned. *J. Control. Release* **2012**, *160*, 117–134. [CrossRef]
10. Chen, J.; Lu, W.L.; Gu, W.; Lu, S.S.; Chen, Z.P.; Cai, B.C.; Yang, X.X. Drug-in-cyclodextrin-in-liposomes: A promising delivery system for hydrophobic drugs. *Exp. Opin. Drug Deliv.* **2014**, *11*, 565–577. [CrossRef]
11. Gharib, R.; Greige-Gerges, H.; Fourmentin, S.; Charcosset, C.; Auezova, L. Liposomes incorporating cyclodextrin-drug inclusion complexes: Current state of knowledge. *Carbohydr. Polym.* **2015**, *129*, 175–186. [CrossRef] [PubMed]
12. Hong, S.S.; Kim, S.H.; Lim, S.J. Effects of triglycerides on the hydrophobic drug loading capacity of saturated phosphatidylcholine-based liposomes. *Int. J. Pharm.* **2015**, *483*, 142–150. [CrossRef] [PubMed]
13. Jain, A.K.; Das, M.; Swarnakar, N.K.; Jain, S. Engineered PLGA nanoparticles: An emerging delivery tool in cancer therapeutics. *Crit. Rev. Drug Carr. Syst.* **2011**, *28*, 1–45. [CrossRef] [PubMed]
14. Kalaria, D.R.; Sharma, G.; Beniwal, V.; Ravi Kumar, M.N. Design of biodegradable nanoparticles for oral delivery of doxorubicin: In vivo pharmacokinetics and toxicity studies in rats. *Pharm. Res.* **2009**, *26*, 492–501. [CrossRef]
15. Govender, T.; Stolnik, S.; Garnett, M.C.; Illum, L.; Davis, S.S. PLGA nanoparticles prepared by nanoprecipitation: Drug loading and release studies of a water soluble drug. *J. Control. Release* **1999**, *57*, 171–185. [CrossRef]

16. Avgoustakis, K.; Beletsi, A.; Panagi, Z.; Klepetsanis, P.; Karydas, A.G.; Ithakissios, D.S. PLGA-mPEG nanoparticles of cisplatin: In vitro nanoparticle degradation, in vitro drug release and in vivo drug residence in blood properties. *J. Control. Release* **2002**, *79*, 123–135. [CrossRef]

17. Song, C.X.; Labhasetwar, V.; Murphy, H.; Qu, X.; Humphrey, W.R.; Shebuski, R.J.; Levy, R.J. Formulation and characterization of biodegradable nanoparticles for intravascular local drug delivery. *J. Control. Release* **1997**, *43*, 197–212. [CrossRef]

18. Atashrazm, F.; Lowenthal, R.M.; Woods, G.M.; Holloway, A.F.; Dickinson, J.L. Fucoidan and cancer: A multifunctional molecule with anti-tumor potential. *Mar. Drugs* **2015**, *13*, 2327–2346. [CrossRef]

19. Kim, D.-Y.; Shin, W.-S. Roles of fucoidan, an anionic sulfated polysaccharide on BSA-stabilized oil-in-water emulsion. *Macromol. Res.* **2009**, *17*, 128–132. [CrossRef]

20. Cardoso, M.J.; Costa, R.R.; Mano, J.F. Marine Origin Polysaccharides in Drug Delivery Systems. *Mar. Drugs* **2016**. [CrossRef]

21. Pozharitskaya, O.N.; Obluchinskaya, E.D.; Shikov, A.N. Mechanisms of Bioactivities of Fucoidan from the Brown Seaweed Fucus vesiculosus L. of the Barents Sea. *Mar. Drugs* **2020**, *18*, 275. [CrossRef] [PubMed]

22. Van Weelden, G.; Bobiński, M.; Okła, K.; Van Weelden, W.J.; Romano, A.; Pijnenborg, J.M.A. Fucoidan Structure and Activity in Relation to Anti-Cancer Mechanisms. *Mar. Drugs* **2019**, *17*, 32. [CrossRef] [PubMed]

23. Hsu, H.Y.; Hwang, P.A. Clinical applications of fucoidan in translational medicine for adjuvant cancer therapy. *Clin. Transl. Med.* **2019**, *8*, 15. [CrossRef] [PubMed]

24. Abdollah, M.R.A.; Carter, T.J.; Jones, C.; Kalber, T.L.; Rajkumar, V.; Tolner, B.; Gruettner, C.; Zaw-Thin, M.; Baguña Torres, J.; Ellis, M.; et al. Fucoidan Prolongs the Circulation Time of Dextran-Coated Iron Oxide Nanoparticles. *ACS Nano* **2018**, *12*, 1156–1169. [CrossRef]

25. Ikeguchi, M.; Yamamoto, M.; Arai, Y.; Maeta, Y.; Ashida, K.; Katano, K.; Miki, Y.; Kimura, T. Fucoidan reduces the toxicities of chemotherapy for patients with unresectable advanced or recurrent colorectal cancer. *Oncol. Lett.* **2011**, *2*, 319–322. [CrossRef] [PubMed]

26. Haroun-Bouhedja, F.; Ellouali, M.; Sinquin, C.; Boisson-Vidal, C. Relationship between sulfate groups and biological activities of fucans. *Thromb. Res.* **2000**, *100*, 453–459. [CrossRef]

27. Koyanagi, S.; Tanigawa, N.; Nakagawa, H.; Soeda, S.; Shimeno, H. Oversulfation of fucoidan enhances its anti-angiogenic and antitumor activities. *Biochem. Pharmacol.* **2003**, *65*, 173–179. [CrossRef]

28. Anastyuk, S.D.; Shevchenko, N.M.; Ermakova, S.P.; Vishchuk, O.S.; Nazarenko, E.L.; Dmitrenok, P.S.; Zvyagintseva, T.N. Anticancer activity in vitro of a fucoidan from the brown alga Fucus evanescens and its low-molecular fragments, structurally characterized by tandem mass-spectrometry. *Carbohydr. Polym.* **2012**, *87*, 186–194. [CrossRef]

29. Pozharitskaya, O.N.; Shikov, A.N.; Faustova, N.M.; Obluchinskaya, E.D.; Kosman, V.M.; Vuorela, H.; Makarov, V.G. Pharmacokinetic and Tissue Distribution of Fucoidan from Fucus vesiculosus after Oral Administration to Rats. *Mar. Drugs* **2018**, *16*, 132. [CrossRef]

30. Nagamine, T.; Nakazato, K.; Tomioka, S.; Iha, M.; Nakajima, K. Intestinal absorption of fucoidan extracted from the brown seaweed, Cladosiphon okamuranus. *Mar. Drugs* **2014**, *13*, 48–64. [CrossRef]

31. Kimura, R.; Rokkaku, T.; Takeda, S.; Senba, M.; Mori, N. Cytotoxic effects of fucoidan nanoparticles against osteosarcoma. *Mar. Drugs* **2013**, *11*, 4267–4278. [CrossRef] [PubMed]

32. Kruh, G.D. Ins and outs of taxanes. *Cancer Biol.* **2005**, *4*, 1030–1032. [CrossRef] [PubMed]

33. Hernández-Vargas, H.; Palacios, J.; Moreno-Bueno, G. Molecular profiling of docetaxel cytotoxicity in breast cancer cells: Uncoupling of aberrant mitosis and apoptosis. *Oncogene* **2007**, *26*, 2902–2913. [CrossRef]

34. Kulhari, H.; Pooja, D.; Shrivastava, S.; Kuncha, M.; Naidu, V.G.M.; Bansal, V.; Sistla, R.; Adams, D.J. Trastuzumab-grafted PAMAM dendrimers for the selective delivery of anticancer drugs to HER2-positive breast cancer. *Sci. Rep.* **2016**, *6*, 23179. [CrossRef] [PubMed]

35. Badran, M.M.; Alomrani, A.H.; Harisa, G.I.; Ashour, A.E.; Kumar, A.; Yassin, A.E. Novel docetaxel chitosan-coated PLGA/PCL nanoparticles with magnified cytotoxicity and bioavailability. *Biomed. Pharmacother.* **2018**, *106*, 1461–1468. [CrossRef] [PubMed]

36. Bowerman, C.J.; Byrne, J.D.; Chu, K.S.; Schorzman, A.N.; Keeler, A.W.; Sherwood, C.A.; Perry, J.L.; Luft, J.C.; Darr, D.B.; Deal, A.M.; et al. Docetaxel-Loaded PLGA Nanoparticles Improve Efficacy in Taxane-Resistant Triple-Negative Breast Cancer. *Nano Lett.* **2017**, *17*, 242–248. [CrossRef]

37. Burchell, S.R.; Iniaghe, L.O.; Zhang, J.H.; Tang, J. Fucoidan from Fucus vesiculosus Fails to Improve Outcomes Following Intracerebral Hemorrhage in Mice. *Acta Neurochir. Suppl.* **2016**, *121*, 191–198. [CrossRef]

38. Fletcher, H.R.; Biller, P.; Ross, A.B.; Adams, J.M.M. The seasonal variation of fucoidan within three species of brown macroalgae. *Algal. Res.* **2017**, *22*, 79–86. [CrossRef]

39. Garti, N.; Madar, Z.; Aserin, A.; Sternheim, B. Fenugreek Galactomannans as Food Emulsifiers. *Lwt-Food Sci. Technol.* **1997**, *30*, 305–311. [CrossRef]

40. Robins, M.M.; Watson, A.D.; Wilde, P.J. Emulsions—creaming and rheology. *Curr. Opin. Colloid Interface Sci.* **2002**, *7*, 419–425. [CrossRef]

41. El-Houssiny, A.S.; Ward, A.A.; Mostafa, D.M.; Abd-El-Messieh, S.L.; Abdel-Nour, K.N.; Darwish, M.M.; Khalil, W.A. Drug–polymer interaction between glucosamine sulfate and alginate nanoparticles: FTIR, DSC and dielectric spectroscopy studies. *Adv. Nat. Sci. Nanosci. Nanotechnol.* **2016**, *7*, 025014. [CrossRef]

42. Keum, C.G.; Noh, Y.W.; Baek, J.S.; Lim, J.H.; Hwang, C.J.; Na, Y.G.; Shin, S.C.; Cho, C.W. Practical preparation procedures for docetaxel-loaded nanoparticles using polylactic acid-co-glycolic acid. *Int. J. Nanomed.* **2011**, *6*, 2225–2234. [CrossRef]

43. Elahi, M.F.; Guan, G.; Wang, L.; King, M.W. Influence of Layer-by-Layer Polyelectrolyte Deposition and EDC/NHS Activated Heparin Immobilization onto Silk Fibroin Fabric. *Materials* **2014**, *7*, 2956–2977. [CrossRef] [PubMed]

44. Gu, J.; Yang, X.; Zhu, H. Surface sulfonation of silk fibroin film by plasma treatment and in vitro antithrombogenicity study. *Mater. Sci. Eng. C* **2002**, *20*, 199–202. [CrossRef]

45. Ho, T.T.; Bremmell, K.E.; Krasowska, M.; Stringer, D.N.; Thierry, B.; Beattie, D.A. Tuning polyelectrolyte multilayer structure by exploiting natural variation in fucoidan chemistry. *Soft Matter* **2015**, *11*, 2110–2124. [CrossRef] [PubMed]

46. Chiang, C.-S.; Lin, Y.-J.; Lee, R.; Lai, Y.-H.; Cheng, H.-W.; Hsieh, C.-H.; Shyu, W.-C.; Chen, S.-Y. Combination of fucoidan-based magnetic nanoparticles and immunomodulators enhances tumour-localized immunotherapy. *Nat. Nanotechnol.* **2018**, *13*, 746–754. [CrossRef]

47. Cai, D.; Fan, J.; Wang, S.; Long, R.; Zhou, X.; Liu, Y. Primary biocompatibility tests of poly(lactide-co-glycolide)-(poly-L-orithine/fucoidan) core-shell nanocarriers. *R. Soc. Open Sci.* **2018**, *5*, 180320. [CrossRef]

48. Jin, G.; Kim, G. Rapid-prototyped PCL/fucoidan composite scaffolds for bone tissue regeneration: Design, fabrication, and physical/biological properties. *J. Mater. Chem.* **2011**, *21*, 17710–17718. [CrossRef]

49. Senda, T.; He, Y.; Inoue, Y. Biodegradable blends of poly (ε-caprolactone) with α-chitin and chitosan: Specific interactions, thermal properties and crystallization behavior. *Polym. Int.* **2002**, *51*, 33–39. [CrossRef]

50. Honma, T.; Senda, T.; Inoue, Y. Thermal properties and crystallization behaviour of blends of poly (ε-caprolactone) with chitin and chitosan. *Polym. Int.* **2003**, *52*, 1839–1846. [CrossRef]

51. Huang, Y.-C.; Kuo, T.-H. O-carboxymethyl chitosan/fucoidan nanoparticles increase cellular curcumin uptake. *Food Hydrocoll.* **2016**, *53*, 261–269. [CrossRef]

52. Barbosa, A.I.; Costa Lima, S.A.; Reis, S. Application of pH-Responsive Fucoidan/Chitosan Nanoparticles to Improve Oral Quercetin Delivery. *Molecules (Basel Switz.)* **2019**, *24*, 346. [CrossRef]

53. Kulkarni, S.A.; Feng, S.S. Effects of surface modification on delivery efficiency of biodegradable nanoparticles across the blood-brain barrier. *Nanomed. (Lond. Engl.)* **2011**, *6*, 377–394. [CrossRef]

54. Li, J.; Sabliov, C. PLA/PLGA nanoparticles for delivery of drugs across the blood-brain barrier. *Nanotechnol. Rev.* **2013**, *2*, 241. [CrossRef]

55. Yu, T.; Malugin, A.; Ghandehari, H. Impact of silica nanoparticle design on cellular toxicity and hemolytic activity. *ACS Nano* **2011**, *5*, 5717–5728. [CrossRef] [PubMed]

Article

Lipid Modulation in the Formation of β-Sheet Structures. Implications for De Novo Design of Human Islet Amyloid Polypeptide and the Impact on β-Cell Homeostasis

Israel Martínez-Navarro [1,†], Raúl Díaz-Molina [1], Angel Pulido-Capiz [1,2,†], Jaime Mas-Oliva [3], Ismael Luna-Reyes [3], Eustolia Rodríguez-Velázquez [4,5], Ignacio A. Rivero [6], Marco A. Ramos-Ibarra [7], Manuel Alatorre-Meda [8] and Victor García-González [1,*]

[1] Departamento de Bioquímica, Facultad de Medicina Mexicali, Universidad Autónoma de Baja California, Mexicali 21000, Baja California, Mexico; israel.martinez.navarro@uabc.edu.mx (I.M.-N.); rauldiaz@uabc.edu.mx (R.D.-M.); pulido.angel@uabc.edu.mx (A.P.-C.)

[2] Laboratorio de Biología Molecular, Facultad de Medicina Mexicali, Universidad Autónoma de Baja California, Mexicali 21000, Baja California, Mexico.

[3] Instituto de Fisiología Celular, Universidad Nacional Autónoma de México, Ciudad de Mexico 04510, Mexico; jmas@ifc.unam.mx (J.M.-O.); ismaelluna.biome@gmail.com (I.L.-R.)

[4] Facultad de Odontología, Universidad Autónoma de Baja California, Tijuana 22390, Mexico; eustolia.rodriguez@uabc.edu.mx

[5] Tecnológico Nacional de México/I.T. Tijuana, Centro de Graduados e Investigación en Química-Grupo de Biomateriales y Nanomedicina, Tijuana 22510, Mexico

[6] Tecnológico Nacional de México/Instituto Tecnológico de Tijuana, Centro de Graduados e Investigación en Química, Tijuana 22510, Baja California, Mexico; irivero@tectijuana.mx

[7] Facultad de Ciencias Químicas e Ingeniería, Universidad Autónoma de Baja California, Tijuana 22390, Baja California, Mexico; mramos@uabc.edu.mx

[8] Cátedras CONACyT- Tecnológico Nacional de México/I.T. Tijuana, Centro de Graduados e Investigación en Química-Grupo de Biomateriales y Nanomedicina, Tijuana 22510, Mexico; manuel_alatorre@yahoo.com.mx

[*] Correspondence: vgarcia62@uabc.edu.mx; Tel.: +52-68-6557-1622

[†] These authors contributed equally to this work.

Received: 17 July 2020; Accepted: 4 August 2020; Published: 19 August 2020

Abstract: Human islet amyloid polypeptide (hIAPP) corresponds to a 37-residue hormone present in insulin granules that maintains a high propensity to form β-sheet structures during co-secretion with insulin. Previously, employing a biomimetic approach, we proposed a panel of optimized IAPP sequences with only one residue substitution that shows the capability to reduce amyloidogenesis. Taking into account that specific membrane lipids have been considered as a key factor in the induction of cytotoxicity, in this study, following the same design strategy, we characterize the effect of a series of lipids upon several polypeptide domains that show the highest aggregation propensity. The characterization of the C-native segment of hIAPP (residues F_{23}-Y_{37}), together with novel variants $F_{23}R$ and $I_{26}A$ allowed us to demonstrate an effect upon the formation of β-sheet structures. Our results suggest that zwitterionic phospholipids promote adsorption of the C-native segments at the lipid-interface and β-sheet formation with the exception of the $F_{23}R$ variant. Moreover, the presence of cholesterol did not modify this behavior, and the β-sheet structural transitions were not registered when the N-terminal domain of hIAPP (K_1-S_{20}) was characterized. Considering that insulin granules are enriched in phosphatidylserine (PS), the property of lipid vesicles containing negatively charged lipids was also evaluated. We found that these types of lipids promote β-sheet conformational transitions in both the C-native segment and the new variants. Furthermore, these PS/peptides arrangements are internalized in Langerhans islet β-cells, localized in the endoplasmic reticulum, and trigger critical pathways such as unfolded protein response (UPR), affecting insulin secretion. Since this phenomenon was associated with the presence of cytotoxicity on Langerhans islet β-cells, it can be concluded that the anionic lipid environment and degree of solvation are critical conditions for the stability of segments with the propensity to form β-sheet structures, a situation that will eventually affect the structural characteristics and stability of IAPP within insulin granules, thus modifying the insulin secretion.

Keywords: hIAPP; amyloidogenesis; insulin granules; endoplasmic reticulum; anionic lipids; $F_{23}R$ variant; β-sheet transitions; β-cell cytotoxicity; unfolded protein response

---

1. Introduction

Human islet amyloid polypeptide (hIAPP) is a hormone that slows down gastric emptying and participates in the regulation of plasmatic glucose associated with functions such as glucagon-release inhibition and leptin sensitization [1,2]. hIAPP is a monomeric peptide processed in the Golgi complex and secreted in insulin granules in response to β-cell secretagogues [3]. However, amyloid fibril formation in hIAPP could trigger an amplified toxicity response that leads to failure of pancreatic β-cells, a hallmark of type 2 diabetes mellitus (T2DM). Several variants of hIAPP modify their stability accelerating amyloid formation; for instance, in Asian populations, the $S_{20}G$ mutation has been associated with early-stage cases of DM2 [4]. Likewise, variants have been described in the Maori populations of New Zealand [5]. By contrast, rat IAPP (rIAPP) containing structural differences with hIAPP in six residues situated in region 18–29 ($H_{18}R$, $F_{23}L$, $A_{25}P$, $I_{26}V$, $S_{28}P$, and $S_{29}P$) show a diminished trend to produce amyloid fibrils [6]. Although these changes allowed the development of pramlintide, an amylinomimetic peptide with three variants ($A_{25}P$, $S_{28}P$, and $S_{29}P$) used in DM2 therapy; the propensity to aggregation of this sequence is not completely avoided.

Several reports suggest that the C-native segment of IAPP (residues 23–37) is a critical domain in the structural transitions that trigger amyloid formation [6], hence, it could be a target for the development of strategies to reduce aggregation. Based on multiple sequence alignment among N- and C-domains on 240 sequences of different species, the N-domain (residues 1–20) has been described as a conserved sequence among a wide variety of organisms, while the C-domain (residues 22–37) has been restricted to phylogenetically close groups [7]. Therefore, we proposed a panel of optimized IAPP sequences, which could reduce aggregation with only one residue substitution [7].

Localized changes in the secondary structure of proteins and peptides are believed to work as a molecular switch regulating function or, in some cases, as a trigger for misfolding. In this context, we have described these conditions by studying the cholesteryl-ester transfer protein (CETP) and a series of apolipoproteins [8–13]. In addition, other studies have reported that phosphatidylserine (PS) vesicles could increase the peptide/aggregation ratio, suggesting an electrostatic factor as a triggering condition for the induction of conformational changes [14,15]. In this sense, within insulin secretory granules derived from the endoplasmic reticulum (ER) [16], the concentration of PS and phosphatidylinositol has been described to be fivefold higher compared to that of the cell membrane [17]. Therefore, lipids could be a factor that promotes structural changes in hIAPP, inducing misfolding phenomena.

However, highly sophisticated mechanisms that modulate protein structure and function have evolved to maintain cellular homeostasis and counteract misfolding [11]. Perturbations in these mechanisms can lead to protein dysfunction as well as deleterious cell processes. Specifically, imbalances in secretory protein synthesis pathways lead to a condition known as ER stress, which elicits the adaptive unfolded protein response (UPR) [11], a phenomenon that could be critical during insulin maturation. Importantly, proinsulin is folded in the ER by chaperones such as protein disulfide isomerases (PDI) and binding immunoglobulin protein (BiP) or GRP-78 [18]. Therefore, the transducers of the UPR pathway, IRE1, ATF6α, and PERK, promote the translation of target chaperones through XBPIs, ATF6α, and CHOP transcription factors, respectively, when unfolded proteins accumulate in the lumen [19]. The three branches of UPR are essential in cell homeostasis to reduce ER stress and to ensure adequate synthesis of peptides such as insulin [20]. Indeed, in several reports, we have described the effect of metabolic overload on the dysregulation of UPR arms [21,22].

Given this situation, the initial events of misfolding and amyloid aggregation promote a cascade of pathological processes considered the hallmark in the progression of several chronic degenerative diseases [8–13] that might be associated with conditions related to metabolic overload [21,22]. Having this in mind, we herein characterized the role of lipid systems on the conformational transitions of the most aggregative C-terminal domain of hIAPP, and through a biomimetic approach, evaluated this condition on variants $F_{23}R$ and $I_{26}A$ that potentially could reduce aggregation. In addition, we also characterized the association with cell responses involved in protein homeostasis such as UPR.

2. Materials and Methods

2.1. De Novo IAPP Sequences

Based on network analysis, different residues from 240 species reported in the NCBI database were replaced on the hIAPP sequence. The effect of the substitutions was characterized through physicochemical assays, as well as by the computational identification of regions with a high intrinsic propensity for aggregation [7]. Aggregation propensity range was obtained considering the aggregation value of hIAPP as a reference, based on the AGGRESCAN algorithm (Na4vSS) [23].

2.2. Materials

Cell culture reagents were purchased from Thermo Fisher (Carlsbad, CA, USA), while tissue culture plates and other plastic materials were obtained from Corning Inc. (Corning, NY, USA). Salts and buffers were obtained from Sigma-Aldrich (St. Luis, MI, USA), as well as Thioflavin T (ThT), black Sudan B, Congo red, sodium dodecyl sulfate (SDS), and 3-(4,5-dimethylthiazol-2-yl)-2,5-diphenyltetrazolium bromide (MTT). L-α-phosphatidylcholine (PC), L-α-phosphatidylserine (PS), L-α-phosphatidyl-ethanolamine (PE), 1-oleoyl-2-hydroxy-sn-glycero-3-phosphate (LPA), 1-palmitoyl-2-oleoyl-sn-glycerol (POPG), and cholesterol were obtained from Avanti Polar Lipids, Inc. (Alabaster, AL, USA). Antibodies anti-XBP1s and anti-BiP/GRP78 were purchased from Abcam (Cambridge, UK) and anti-β-actin was obtained from Santa Cruz Biotechnology (Dallas, TX, USA). Anti-PDI was donated by Dr. Marco A. Ramos Ibarra.

2.3. Peptide Synthesis and Preparation

Several peptides were synthesized considering the physicochemical properties of hIAPP such as: N-native segment (^{1}KCNTATCATQRLANFLVHSS^{20}); C-native segment (^{23}FGAILSSTNVGSNTY^{37}); $F_{23}R$ variant (^{23}RGAILSSTNVGSNTY^{37}); and I26A variant (^{23}FGAALSSTNVGSNTY^{37}) (Figure 1). Likewise, the aggregative core of amyloid beta (Aβ) peptide was used as a control ^{25}GSNKGAIIGLM$^{35}_{.}$. All solutions were filtered through 0.22 μm membrane filters (Millipore, Burlington, MA, USA) before the experiments. Peptide purity greater than 98% was confirmed by mass spectrometry and HPLC (GenScript, Piscataway, NY, USA). The best condition for peptide solubilization was the use of ultrapure H_2O (600 μg/mL), subsequently diluted in phosphate buffer pH 7.4 (60 μg/mL).

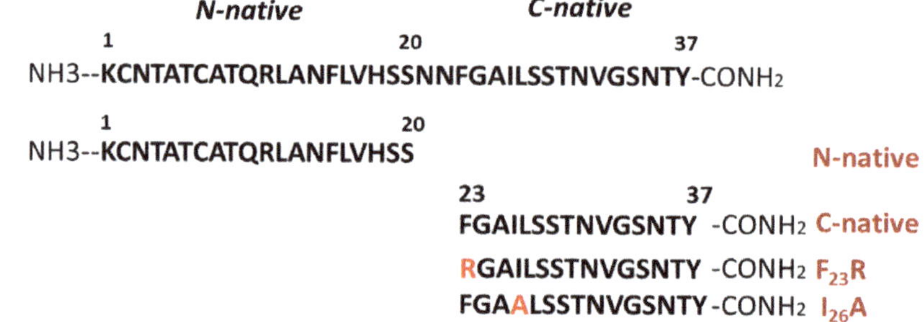

Figure 1. Representation of the primary structure of human islet amyloid polypeptide (hIAPP) and peptides corresponding to N-native and C-native domains, as well as new variants of hIAPP.

2.4. Preparation of Small Unilamellar Vesicles (SUVs)

SUVs were prepared from PC, PS, PE, cholesterol, and POPG (600 µg/mL) upon their mixing and conditioning at varying concentrations. PC, PS, PE, cholesterol, and POPG were dissolved in chloroform and dried for 90 min under a gentle stream of N_2 with an additional incubation of 5 h at 30 °C in an Eppendorf Vacufuge concentrator (Eppendorf, Hamburg, Germany), according to protocols established by our working group [9]. After drying, samples were hydrated in phosphate buffer to pH 7.4 and processed through 4 cycles of freezing in liquid N_2 and thawing at 37 °C, and finally subjected to a sonication process (for 10 min under 15 s on/30 s off cycles at 9.5–10 W) in a Cole-Parmer Ultrasonic Homogenizer (Vernon Hills, IL, USA). Samples were stabilized for 1 h at 25 °C and centrifuged at 13,000 rpm for 10 min.

LPA samples in chloroform were placed under a gentle flow of N_2 for 90 min and additional 12 h in a vacuum equipment. The samples were hydrated in phosphate buffer and afterwards processed through 4 cycles of freezing in liquid N_2 and thawing at 37 °C. Solutions were left to equilibrate for 2 h and centrifuged at 13,000 rpm for 10 min.

2.5. Preparation of Large Unilamellar Vesicles (LUVs)

LUVs were prepared from PC and PS by the reverse-phase evaporation methodology [24], with some adaptations. Specifically, PC LUVs were prepared by dissolving the lipid (600 µg/mL) in a 1:3 mixture of diethylether:phosphate buffer (pH 7.4). Then, the solution was sonicated for 5 min in an ultrasonic homogenizer. Finally, the solution was added to a rotary evaporator working at an initial pressure of 400 mmHg for 5 min, followed by a final pressure of 650 mmHg until complete removal of diethylether. Likewise, PS LUVs were prepared by dissolving the lipid (600 µg/mL) in chloroform and drying the resulting solution for 90 min under a gentle stream of N_2 with an additional incubation of 2 h at 30 °C in an Eppendorf Vacufuge concentrator, according to protocols established by our working group [9]. After drying, samples were hydrated in phosphate buffer, pH 7.4, and processed through 4 cycles of freezing in liquid N_2 and thawing at 37 °C. Finally, samples were subjected to a sonication process (for 5 min under 15 s on/30 s off cycles at 9.5–10 W) in an ultrasonic homogenizer.

2.6. Dynamic Light Scattering Experimentation and Optical Density Characterization

In the characterization of vesicle size, PC (300 µM) and PS (300 µM) vesicles were evaluated. The hydrodynamic diameter (Dh) and Z-Potential of the vesicles were assessed by dynamic light scattering (DLS), employing a Zetasizer Nanoseries spectrophotometer (Malvern Instruments, Malvern, UK). All measurements were carried out at 25 °C using polystyrene disposable cells and folded capillary cells for the size and ζ potential measurements, respectively. Before their characterization, the samples were vortexed for 10 s and left to rest for 30 min at 25 °C. The results are the average of five measurements. In another batch of vesicles (PS), we completed the characterization by using a

Microtrac equipment. In a complementary way, we performed the optical density characterization of vesicles employing a BioRad Smart spectrophotometer with diode array (Hercules, CA, USA).

2.7. Peptide Bond Conformational Changes

Experiments were performed through the characterization of optical density at 218 nm, which is associated with conformational changes along the formation of β-sheet structures [8,10]. The effect of lipid vesicles composed of several lipids on conformational changes of hIAPP-derived peptides was evaluated. Measurements were obtained using the above-described BioRad Smart spectrophotometer, employing a peptide concentration of 60 µg/mL and then evaluating the effect of the lipid vesicles.

2.8. Congo Red Birefringence Spectroscopy

Assays were performed based on a previous protocol [8], employing 10.6 µM Congo red and 60 µg/mL peptides solutions. The optical density was measured at 494 nm, employing the above-described BioRad Smart spectrophotometer, under varying solution conditions.

2.9. ThT-Fluorescence Assay

β-sheet structures of peptides were characterized through the ThT-fluorescence assay. Samples were incubated for 12 h at 37 °C and monitored with the ThT (20 µM) treatment. Fluorescence emission spectra were registered at 25 °C from 460 to 610 nm with an excitation wavelength of 450 nm in a Cary Eclipse Fluorescence spectrophotometer (Agilent Technologies, Inc., Santa Clara, CA, USA).

2.10. Circular Dichroism (CD)

CD experiments were performed at a peptide concentration of 120 µg/mL in a 1-mm path length quartz cuvette, using the CD neural network (CDNN) based software. Spectra were recorded with a 1-mm bandwidth, using 1 nm increments and 2.5 s accumulation time. CD spectra were recorded with an AVIV 62DS spectropolarimeter (AVIV Instruments, Lakewood, NJ, USA) at 25 °C employing far UV wavelengths (190–260 nm). CD results were reported as mean molar ellipticity (deg cm^2 dmol^{-1}).

2.11. Lipid–Peptide Interactions

Lipid/peptide samples were analyzed with a nondenaturing electrophoresis technique adapted by our group for lipid–peptide characterization. We established a new methodology through the use of 0.8–15% native gradient gel electrophoresis [8]. Later, gels were stained following the Sudan black and silver nitrate protocols [8].

2.12. Molecular Dynamics

Peptide-membrane systems were generated with the CHARMM-GUI input generator. For all systems, a MARTINI force field for polarizable amino acids and water was used. In our assays, we used the peptides in a simple lipid bilayer system, each lipid bilayer consisting of a homogenous array of 8×8 lipid molecules of PS (DIPS 18:2–18:2) or PC (3:1, DPOC 16:1–18:1 and POPC 16:0–18:0). Peptide models of N-native, C-native, and variants $F_{23}R$ and $I_{26}A$ were generated using the ITASSER online server; for all peptides, we used the predicted model with highest TM value. The systems were minimized using steepest descent and conjugate gradient methods. Then, five equilibration steps were performed for each system. Simulations were conducted in a 1:1 peptide-membrane system during 3000 ns at 303.15 K and 1 atm pressure. For each system, we evaluated the lipid bilayer lateral displacement from lipids as an indicator of bilayer fluidity. Our analysis was based on the GROMACS built-in function for MSD analysis. To this end, we used the built-in functions of GROMACS, which are calculated using the following equation:

$$D_A t = \frac{1}{6} \lim_{t \to \infty} < \| \mathbf{r}_i(t) - \mathbf{r}_i(0) \|^2 >_{i \in a}$$

where r_i (*t*) indicates molecule position a t time, r_i (0) indicates the position at molecules at time zero, and is calculated for molecules included in *A* set of molecules. Additionally, $6D_A{}^t$ represents diffusion coefficient over time (*t*) for *A* set of molecules, using Einstein correlation adjusted for long simulations.

In this sense, the analysis along the membrane was performed in one single plane. Before analysis of MSD, periodic boundary conditions were converted used built-in conversion functions to ensure the continuous trajectory of molecules.

In a complementary way, we performed molecular dynamics reruns, started from previously equilibrated systems that were used in original reported simulations. The systems were evaluated under the same conditions of temperature, pressure, and number of molecules. The simulation time was reduced to 30,000 ps; 100 consecutive simulations were performed for each system where the previous simulation was used as a starting point for the subsequent one. All trajectories obtained from short simulations were joined using built-in functions of GROMACS to reach the same simulation time reported previously; the resulting trajectory file was converted into a continuous trajectory file. MSD analysis for this system is reported in Figures S3 and S6. This strategy was based on previous reports [25,26].

2.13. Cell Culture

β-cell line RIN-m5F (American Type Culture Collection) was grown using RPMI-1640 culture medium supplemented with 10% fetal bovine serum, 10 U/mL penicillin, 10 μg/mL streptomycin, and 25 μg/mL amphotericin B. Cultures were maintained at 37 °C in a humidified atmosphere of 95% air and 5% CO_2.

2.14. Cell Viability Assay

Peptide cytotoxicity was assessed through MTT assays RIN-m5F cells, under different peptide and lipid–peptide treatments. Cells were seeded into 96-well plates at a density of 20,000 cells/well and allowed to grow to 90% of confluence. Next, the culture medium was replaced with Opti-MEM medium. After 1 h under this condition, cells were incubated under the different treatments and subsequently processed according to previous protocols [8].

2.15. Western Blotting Analysis

Under different peptide and PS-SUVs treatments, the expression of proteins associated with the UPR pathway and insulin folding were evaluated. After experimentation on RIN-m5F cells, proteins were extracted from cell cultures using ice-cold protein lysis buffer (150 mM NaCl, 10 mM Tris, pH 7.4, 1% Triton X-100, 0.5% NP40, 1 mM EDTA, 1 mM EGTA, 0.2 mM sodium orthovanadate, 10 mM benzamidine, 10 μg/mL leupeptin, 10 μg/mL aprotinin, and 250 μM PMSF). An average of 25 μg of protein lysates were separated on 8% SDS-PAGE electrophoresis and transferred to polyvinylidene difluoride (PVDF) membranes. The membranes were blocked with 5% nonfat milk in Tris-buffered saline 0.1% Tween-20 (TBS-T) for 1 h at 37 °C and incubated at 4 °C overnight with primary antibody (anti-XBP1s, anti-BiP/GRP78, anti-PDI, anti-SERCA2, and anti-β-actin). Following washing with TBS-T, the membranes were further incubated for 1.5 h at 37 °C with the corresponding horseradish peroxidase-conjugated secondary antibodies. Proteins were detected with the enhanced chemiluminescence reagent (Immobilon Western from Millipore, Burlington, MA, USA).

2.16. Endoplasmic Reticulum Isolation

ER fractions were obtained from RIN-m5F cells under several PS/peptides treatments. The methodology was based on the report of Prajapati et al. [27]. Cells were proliferated in 100 mm cell culture plates at a density of 2.3×10^5 cells/mL. Cells were maintained in proliferation for 72 h to reach 95% of confluence, and later, different treatments were performed on a volume of 5 mL. Later, culture cells were washed with PBS 1x, recovered, and treated with 1 mL of homogenizer buffer (30 mM Tris–HCl pH 7.4, 225 mM mannitol, 75 mM sucrose, 0.5 mM EGTA, protease inhibitor, and 0.5% BSA). Homogenates were briefly sonicated (two cycles of 15 s on/30 s off at 9.5–10 W). Then, the homogenate was centrifuged at 630× g for 5 min at 4 °C. Supernatant was collected and conserved, and the pellet was newly processed by sonication and centrifuged under the same conditions. The combined supernatant was centrifuged again at 630× g for 5 min at 4 °C (nuclei-free lysate), and 150 μL was conserved. Then, the nuclei-free lysate fraction was centrifuged at 6300× g for 10 min. The supernatant was transferred into a new tube, and then, it was centrifuged at 20,000× g for 30 min at 4 °C. The supernatant was recovered and centrifuged at 100,000× g for 60 min at 4 °C, using a S140-AT 2555 rotor. The supernatant corresponds to the cytoplasm fraction, whereas the pellet corresponds to the ER fraction. Protein markers (anti-BiP, anti-PDI, anti-SERCA2, and anti-β-actin) were used to evaluate the quality of the isolations (Figure S9).

2.17. Fabrication of Vesicles Composed of PS (30 μM) and the Fluorescent Probe BODIPY-Leu (6 μM)

BODIPY-Leu probe was synthesized according to protocols developed by our group [22]. Then, the required quantities of L-α-phosphatidylserine and BODIPY-Leu were dissolved in chloroform and mixed vigorously to obtain a clear solution. Then, the solution was dried for 90 min under a gentle stream of N_2 with an additional incubation of 5 h at 30 °C in an Eppendorf Vacufuge concentrator, according to protocols established by our working group [9]. After drying, samples were hydrated in phosphate buffer pH 7.4 and processed through 4 cycles of freezing in liquid N_2 and thawing at 37 °C, and finally subjected to a sonication process (for 10 min under 15 s on/30 s off cycles at 9.5–10 W) in an ultrasonic homogenizer. Samples were stabilized for 1 h at 25 °C and centrifuged at 13,000 rpm for 10 min.

2.18. Confocal Microscopy

A LEICA TCS-SP8 confocal scanning biological microscope (LEICA Microsystems Heidelberg GmbH, Nussloch, Germany) was employed in the characterization of the subcellular localization of PS/BODIPY-Leu vesicles (PS 30 μM/BODIPY-Leu 6 μM). RIN-m5F cells were proliferated to 90% of confluence and treated with PS/BODIPY-Leu vesicles and hIAPP-derived peptides for 20 h. Later, culture cells were washed with HBSS buffer, and then, the ER-tracker probe (1 μM) was added and incubated for 25 min at 37 °C. Cells were washed once with HBSS buffer and fixed with 4% formaldehyde for 2 min at 37 °C and mounted for observation. Macroscopically different zones were recorded, preferentially at the center of the specimens, to depict representative images. Images were recorded at excitation/emission wavelengths of 488/495-545 and 552/562-700 nm for detection of PS 30 μM/BODIPY-Leu (green) and ER-tracker (red), respectively.

2.19. Insulin ELISA Assays

Cells were proliferated in 100 mm cell culture plates at a density of 2.3×10^5 cells/mL. Cells were maintained in proliferation for 72 h, and later, different treatments were performed on a volume of 5 mL. Cell culture medium was recovered and centrifuged for 5 min at 5000 rpm. The supernatant medium was recovered and diluted (1/3) in PBS. Insulin concentrations were quantified with the Rat Ultrasensitive Insulin ELISA kit (80-INSRTU-E01, E10; ALPCO Diagnostics, Salem, NH, USA) through several adaptations according to manufacturer's recommendations. Absorbance readings were performed at 450 nm, and results were reported as ng/mL.

2.20. Statistical Analysis

Data were expressed as mean ± SD. The statistical analyses were conducted with one-way ANOVA. In MTT assays data were expressed as mean ± SD.

3. Results

3.1. Secondary Structure Characterization of hIAPP Peptides

In a previous report, we proposed a panel of 113 hIAPP variants, 30 of which could reduce aggregation propensity with only one substitution on the hIAPP sequence [7], compared to the pramlintide drug (3 substitutions) and several hIAPP analogs generated by substituting up to four arginine residues at positions F_{23}-I_{26} [28]. Our results suggested that the 23 and 26 positions maintain secondary structure stability [7]. To assess changes in these positions, $F_{23}R$ and $I_{26}A$ variants were evaluated and compared to the highly aggregative hIAPP C-native residues (^{23}FGAILSSTNVGSNTY^{37}). Through ThT-fluorescence assays, the $F_{23}R$ variant showed the lowest fluorescence values corresponding to a reduced β-sheet structure formation compared with the C-native and the $I_{26}A$ variant (Figure 2A). Most probably, the result found with $F_{23}R$ is associated with the electrostatic charge of the arginine-side chain (position 23) that exerts a repulsive effect among peptide monomers.

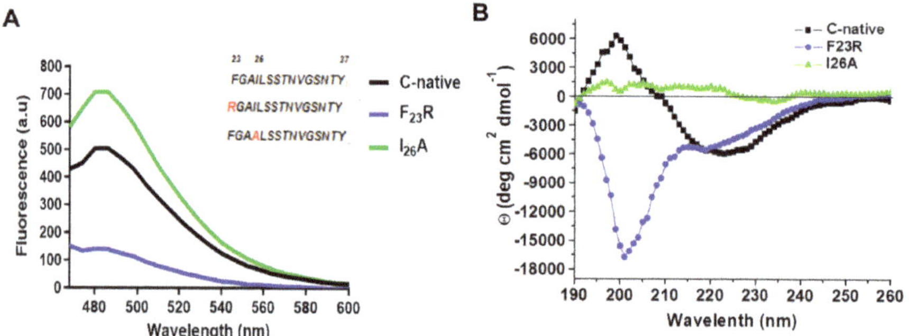

Figure 2. Structural characterization of C-terminal IAPP-derived peptides shows the inhibition of β-structures by the new $F_{23}R$ variant. (**A**) Fluorescence emission spectra through ThT assays on IAPP variants (60 μg/mL). The characteristic emission peak was registered at 482 nm. (**B**) Circular dichroism (CD) spectra of hIAPP variants (190–260 nm). In this representative experiment, the peptide concentration was 120 μg/mL.

The substitution of only $F_{23}R$ could reduce β-sheet structure formation. In the $I_{26}A$ variant, as opposed to the expected, a slightly higher signal than in the C-native segment was detected (Figure 2A). Although aggregation propensity values (AGGRESCAN Index) for $I_{26}A$ (−8.2) suggest a low tendency to generate β-structures, parameters such as hydrophobicity and isoelectric point are similar in both fragments (Table 1). Characterization of peptides was completed by CD analysis (Figure 2B), in which the C-native spectrum indicated a β-sheet structure formation. For the $F_{23}R$ variant, a peak at 190 nm and two minima at 202 and 219 nm were registered, typical of a mixture of α-helix, disordered structures, and minimal β-sheet structures. $I_{26}A$ showed an atypical behavior since the spectrum did not represent a defined secondary structure, possibly associated with amorphous aggregation.

Table 1. Physicochemical parameters of evaluated peptides.

ID	Sequence	AGGRESCAN Value [1]	Hydropathy (kcal/mol)	Charge pH 7.4	Hydrophobicity (kcal/mol)	pI	μH kcal/mol
C-native	^{23}FGAILSSTNGSNTY^{37}	6.6	0.28	0	0.29	5.92	0.18
$F_{23}R$ variant	^{23}RGAILSSTNGSNTY^{37}	−0.3	−0.21	1	0.21	9.84	0.04
$I_{26}A$ variant	^{23}FGAALSSTNGSNTY^{37}	−8.2	0.1	0	0.24	5.92	0.13
N-native	^{1}KCNTATCATQRLANFLVHSS^{20}	−6.3	−0.04	3	−0.02	9.14	0.27

[1] Negative values are associated with a low propensity to form amyloid fibrils, whereas positive ones are associated with higher propensity. Red identifies the residue variants.

In amyloidogenic peptides, the stability of antiparallel β-sheet structures is determined by hydrogen bonds, salt bridges, and weak polar interactions among side chains of residues [29]. Under our working conditions, the net charge of +1 for the $F_{23}R$ sequence at pH 7.4 could be a determining factor in the low propensity to aggregation, due to the likely appearance of electrostatic repulsions. In turn, $I_{26}A$ and the C-native segment showed a neutral charge at a pH range of 3.0 to 8.5 with an intrinsic tendency to aggregation, a situation that has been discussed in several reports as a problem with the therapeutics of hIAPP [30]. Therefore, a balance among the dynamic secondary structure of the hIAPP C-native segment, the net charge, and the physicochemical properties of the lipid microenvironment could define the type of the adopted secondary structure [10].

3.2. Effect of Lipids on the Secondary Structure of hIAPP Peptides

The effect of lipids on the secondary structure of hIAPP peptides was characterized through the use of SUVs. In a first approach, PC-SUVs were employed as a model system and their effect on the C-native segment as well as the $F_{23}R$ and $I_{26}A$ variants was assessed through spectroscopy at 218 nm and birefringence with the Congo red assay (Figure 3). PC was chosen in the first instance since it is the most abundant zwitterionic phospholipid of the outer plasma membrane. Under incubation with increasing concentrations of PC-SUVs, the $F_{23}R$ variant did not show significant changes in the β-sheet structure content where minimal absorbance values were recorded for peptide-bond characterization. In addition, an evident difference in birefringence values compared to the C-native segment was registered (Figure 3A,B). The behavior of the $I_{26}A$ variant was comparable to the C-native segment suggesting β-sheet formation, a phenomenon associated with physicochemical parameters such as isoelectric point, hydrophobicity, and electrostatic charge (Table 1). Our data agree with the results of Cho et al., 2008, employing PC liposomes, wherein the aggregation effect of IAPP was evident [31]. By contrast, when assessing the effect of these PC SUVs on the N-native segment (^{1}KCNTATCATQRLANFLVHSS^{20}), β-structure formation was not found (Figure S1A–C), therefore, suggesting that structural transitions leading to β-sheet formation in hIAPP fundamentally occur on the C-native domain. Remarkably, this β-sheet structural transition exclusive to the C-native segment was likewise observed when working with PC-based large unilamellar vesicles (PC LUVs), also studied to extend the characterization of the effect of this lipid (Figure S2). While stored in β-cell granules at a pH near 5.5, the aggregation of hIAPP is inhibited; however, aggregation propensity increases when hIAPP is released into the extracellular compartment at pH 7.4 [32]. Taking into account that the effect of lipids was studied at pH 7.4, the induction of β-structures was evident on the C-native segment when incubated with PC vesicles, both SUVs and LUVs (Figure 3C,D). Interestingly, this phenomenon did not occur when the F23R variant was used (Figure 3E,F). Since several studies have reported that cholesterol (Chol) functions as a trigger factor of amyloid formation [33,34], unilamellar PC/Chol vesicles (prepared at a 3/2 molar ratio) were evaluated. Despite the increase of PC/Chol SUVs concentration (0–180 μM), our data suggest that PC/Chol vesicles do not promote β-sheet formation on the $F_{23}R$ variant (Figure 3G,H), whereas the $I_{26}A$ variant shows the formation of β-sheet structures (Figure 3G,H).

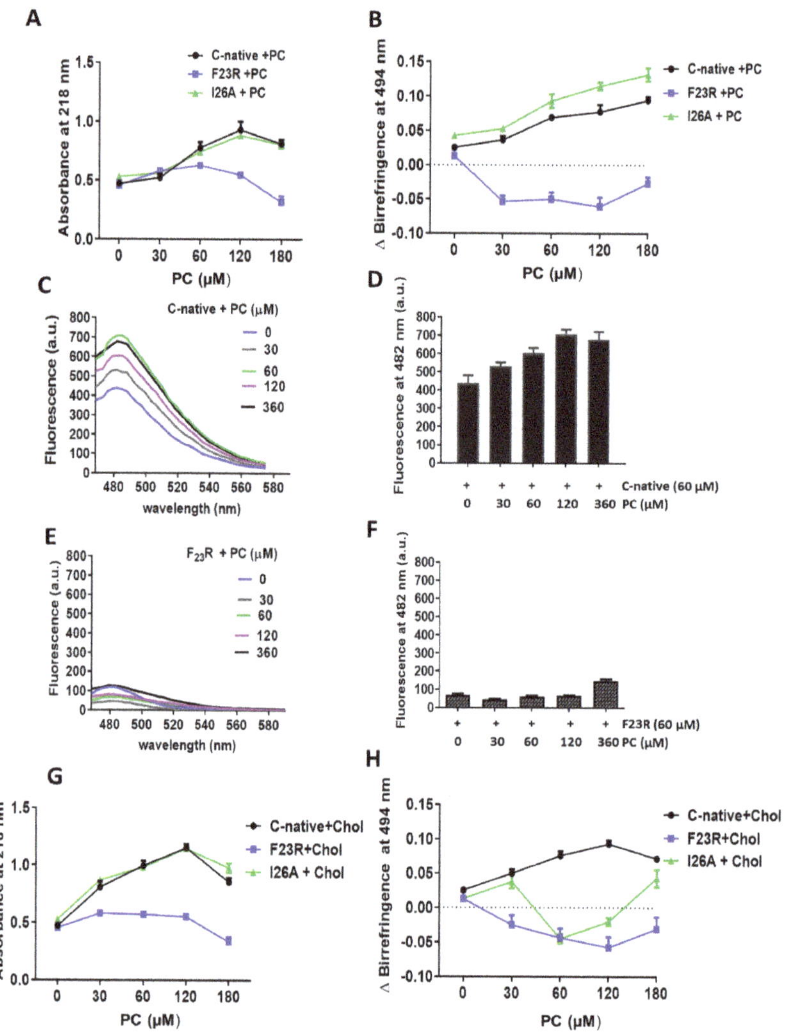

Figure 3. Effect of zwitterionic L-α-phosphatidylcholine-small unilamellar vesicles (PC-SUVs) on the secondary structure of IAPP-peptides. (**A**) Peptide bond absorbance of the C-native segment, $F_{23}R$, and $I_{26}A$ under incubation with PC-SUVs. (**B**) Peptide characterization by birefringence with Congo red assay. (**C**) Interaction of the C-native segment under increasing concentrations of PC-vesicles evaluated by ThT assay. (**D**) Values of ThT-fluorescence emission at 482 nm. (**E**) Interaction between $F_{23}R$ under increasing concentrations of PC. (**F**) Values of ThT-fluorescence emission at 482 nm. Characterization of the C-native segment, $F_{23}R$, and $I_{26}A$ under the incubation of SUVs composed of PC and cholesterol (Chol) through peptide bond absorbance at 218 nm (**G**) and Congo red assay (**H**).

In a complementary way, we performed molecular dynamic simulations according to our experimental conditions using PC molecules. Simulations were conducted in the presence of modeled peptides (C-native segment, variants $F_{23}R$, $I_{26}A$, and N-native segment) and lipid bilayers with no peptides used as control. We ran simulations during 3000 ns in all systems, and the peptide molecule was situated at 30 Å from the top of the membrane. Although the three peptides derived from the hIAPP C-terminal domain (C-native, $F_{23}R$, and $I_{26}A$) adsorb at the membrane surface at less than

500 ns (Figure 4A–F), we found a slightly lower displacement for the C-native segment concerning the $F_{23}R$ and $I_{26}A$ during the evaluation of the displacement of phospholipids and peptide MSD (mean square deviation) on the z-axis, suggesting that interactions between C-native/lipids generate a movement restriction (Figure 4I). This phenomenon could be related to peptide interactions on the hydrophilic/hydrophobic interface of membranes. Moreover, to robust our experimental system and in order to discard false-positive bias, complementary MSD analysis (Figure S3) was performed through 100 consecutive short simulations (30 ns) for each system. According to the obtained results, the displacement of phospholipids and peptides derived from IAPP on the z-axis, follows the same behavior described in Figure 4I (Figure S3).

Several reports indicate that the region that initially interacts with cell membranes is the N-terminal of hIAPP [35]; our simulations suggest that the adsorption phenomenon of the N-native segment at the lipid interface (Figure 4G,H) presents an even greater restriction of movement with respect to peptides derived from the C-domain segment (Figure 4I).

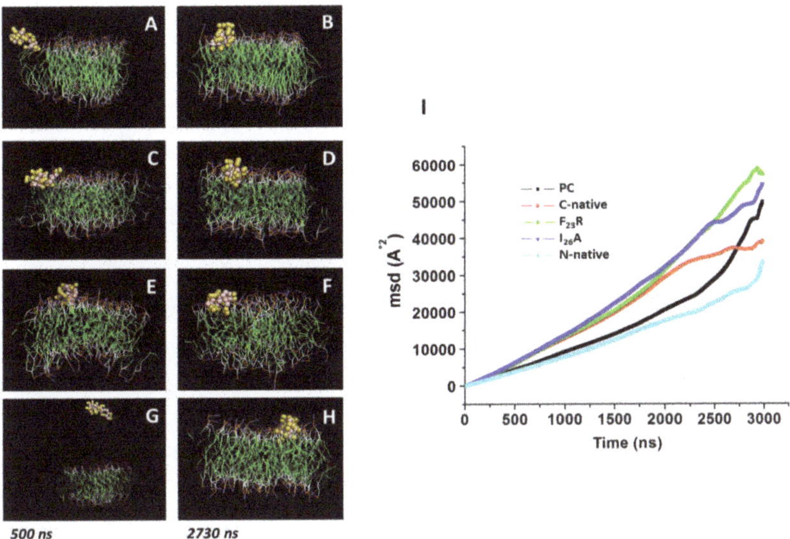

Figure 4. Displacement of peptides derived from IAPP on the z-axis of PC-bilayers (mean square deviation, MSD). Representative snapshots of PC bilayers associated with the C-native fragment of IAPP at 500 ns (**A**) and 2730 ns (**B**), $F_{23}R$ (**C,D**), $I_{26}A$ (**E,F**), and the N-native fragment (**G,H**). (**I**) Behavior of mean square deviation ($Å^2$) through a 3000 ns simulation employing coarse grain methodology.

3.3. Structure Modulation Dependent on Negative Electrostatic Surface in hIAPP Peptides

Insulin secretory granules are derived from the ER membrane, contain high levels of PS and phosphatidylinositol in comparison to the plasmatic membrane [16], and show an important dynamic behavior [17]. Therefore, in order to evaluate the impact of anionic electrostatic charge, PS and POPG vesicles were evaluated (Figure 5). In a first assay, our results suggest a modulation by PS-SUVs on the C-native segment, wherein a significant increase in ThT-fluorescence values dependent on PS concentration was identified (Figure 5A,B). The $F_{23}R$ variant showed the same β-sheet formation from the lowest PS concentration (30 µM) (Figure 5C,D), a result that contrasts with the effect found when zwitterionic PC and PC/Chol vesicles are used. In complementary experimentation, the same tendency was registered through peptide bond spectrophotometry and Congo red assays (Figure 5E,F). Results confirm the propensity of the $F_{23}R$ variant to show β-sheet transitions induced by PS-SUVs, supporting the fact that the fibrillation process begins when monomers change to oligomers and that, in general, the formation of mature fibrils follow a sigmoidal behavior [36].

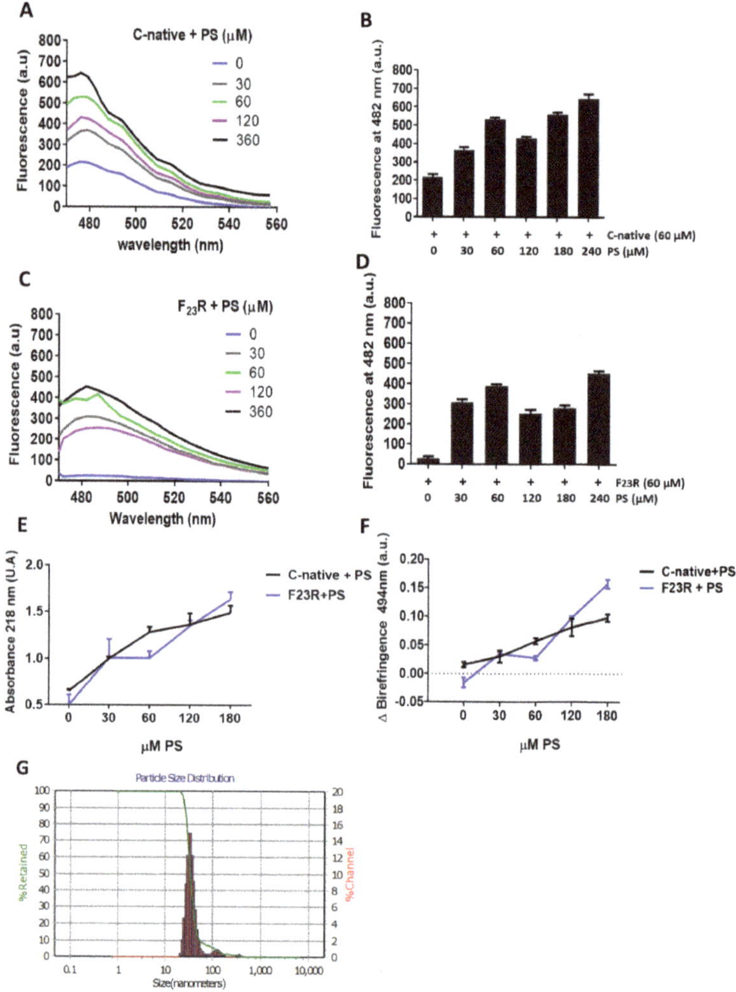

Figure 5. PS-SUVs promote the formation of β-sheet structures on the C-terminal-derived peptides of IAPP. (**A**) Interaction of the C-native segment at increasing concentrations of PS evaluated by ThT-fluorescence assays. (**B**) Same experimentation as in A, ThT-fluorescence at 482 nm. (**C**) Fluorescence emission spectra obtained through ThT assays on the $F_{23}R$ variant under increasing concentrations of PS. (**D**) Same experimentation as in (**C**), ThT-fluorescence at 482 nm. (**E**) Evaluation of the PS-effect on the secondary structure of peptides through peptide-bond absorbance at 218 nm, and by Congo red birefringence at 494 nm (**F**). (**G**) Dispersion of the size of PS-SUVs used in this experimentation by DLS.

Moreover, although we registered the effect of PS-SUVs on β-sheet conformational transitions in hIAPP-derived peptides, we performed several efforts to obtain a homogenate LUV-preparation, however, considering the intrinsic PS-property to form SUVs, we designed a protocol based in freezing/thawing and a slightly sonication process (Materials and Methods). Then, we obtained a mixture of SUVs and LUVs. Results suggest the presence of the same β-sheet transitions registered in the C-native and $F_{23}R$ variant under treatment with SUVs/LUVs mixtures (Figure S4).

To perform an approximation to the nucleation phenomenon of hIAPP-derived peptides at a 1/2 peptide/PS ratio, peptide bond conformation was monitored along with time (Figure 6A).

Data suggest that the incubation with PS-SUVs triggers a conformational transition to β-sheet formation, shortening the lag phase [37,38]. Variant $F_{23}R$ shows the same tendency under the interaction with anionic PS-vesicles (Figure 6B). Complementary results obtained through CD confirmed the same phenomenon with the C-native segment and the $F_{23}R$ variant (Figure 6D,E). Once the lag phase is surpassed, the oligomeric structures establish interchain electrostatic interactions to evolve into more ordered structures and amyloid fibrils [36]. In this case, this phenomenon was critical for the $F_{23}R$ variant, showing an eightfold increase in the CD signal at 222 nm, corresponding to the formation of β-sheet structures. In this sense, arginine and cationic residues have been described to interact with PS membranes due to high affinity of the side chains for anionic lipids [39]. These β-sheet transitions were not evident in the N-domain segment despite of positive charge content (Figure 6C) at neutral pH (Table 1); however, according to CD spectra, this interaction facilitates α-helix formation (Figure 6F). Furthermore, the structural transitions promoted by PS (Figure 6A,B) were not evidenced when vesicles composed of phosphatidyl-ethanolamine in both the C-native and N-native segments were evaluated (Figure S5). Then, the cationic lipid surface is not a critical factor for β-sheet aggregation on hIAPP segments.

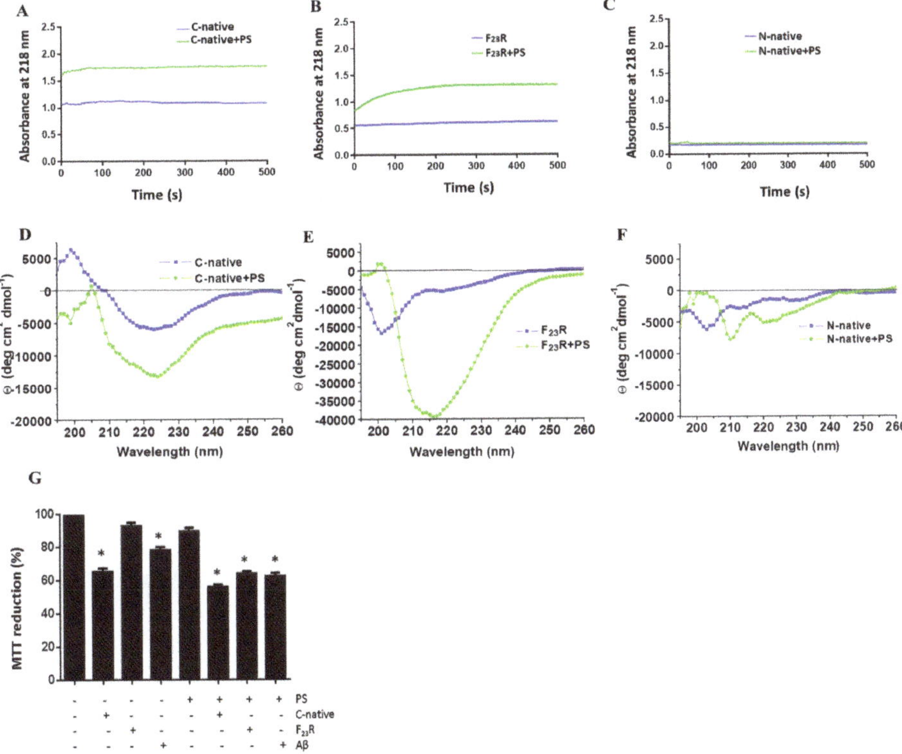

Figure 6. The anionic lipid surface as the critical factor for β-sheet aggregation on C-terminal peptides. Effect of PS on nucleation of the C-native segment (**A**), $F_{23}R$ variant (**B**), and the N-native segment (**C**). Same experimentation as in A, B, and C, the effect of PS on the secondary structure of the C-native segment (**D**), $F_{23}R$ variant (**E**), and N-native segment (**F**), evaluated by CD. (**G**) Cell viability evaluation on RIN-m5F cells treated under different stimuli of peptides/PS vesicles (7.5/15 μM). Data are expressed as mean ± SD ($n = 6$) * $p < 0.005$.

Strikingly, on the other hand, variants $F_{23}R$ and $I_{26}A$ showed aggregation upon their interaction with the anionic lipid system (PS), as evidenced by CD spectra (Figure 6E). These results appear to be contradictory considering the AGGRESCAN values (Na4vSS), −5.3 for $F_{23}R$ and −8.2 for $I_{26}A$ (Table 1), which predict that the formation of β-structures should be inhibited, in association with reports suggesting that membranes of β-cells are composed of anionic lipids (2.5–13%) [40]. Taking into account this rather unexpected outcome, cell viability assays were performed in a model of Langerhans islet β-cells (RIN-m5F cells) to assess the effect of peptide/PS vesicles mixtures at the cellular level. Results pointed out that $F_{23}R$/PS mixtures decreased cell viability compared to controls. Likewise, when the anionic vesicles are present, the C-native segment and the $F_{23}R$ fragment showed cytotoxicity properties (Figure 6G). In this experimental design, the Aβ peptide was used as a control.

As a complementary test, we performed dynamic simulations of IAPP-derived peptides in the presence of a preassembled PS bilayer during 3000 ns, wherein the peptide molecule was situated at 30 Å above the top of the membrane in all systems (Figure 7). In the evaluation of the displacement of the interacting species (PS and peptides), we found a considerable diminution in MSD for the $F_{23}R$ variant with respect to the C-terminal segment and $I_{26}A$ (Figure 7A), which is possibly associated with an electrostatic factor. Indeed, in the analysis of the trend line, the $F_{23}R$ slope showed the lowest value compared to results obtained with other peptides (data not showed). In the same way that of the PC system, a complementary MSD analysis through 100 consecutive short simulations (30 ns) for each PS-system was performed, and results obtained from these short simulations follow the same tendency (Figure S6). Moreover, Aβ-peptide was incorporated as a control. Importantly, through the simulation time, the collapse of the simulations was not registered. Thus, interactions between the $F_{23}R$ and lipids generate a restriction in movement modifying the lipid packing and show the $F_{23}R$ peptide situated deeper within the lipid bilayer (Figure 7D,E). According to our CD experimentation, this is possibly a critical condition for the propensity modulation towards the formation of β-sheet structures.

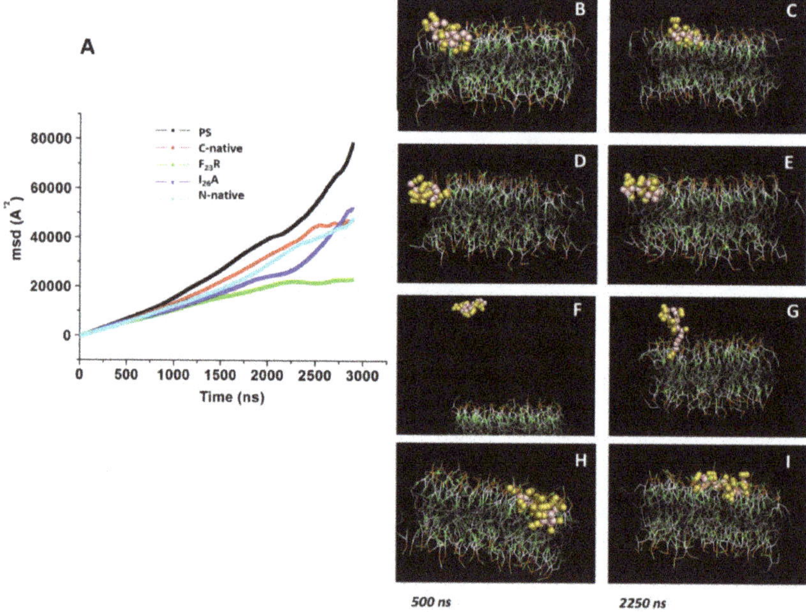

Figure 7. Displacement of peptides derived from the IAPP C-terminal on the z-axis of PS-bilayer. (**A**) MDS values on PS bilayer under the incubation of peptides-derived of IAPP through 3000 ns of simulation. Representative snapshots of the PS model with the C-native segment at 500 ns (**B**) and 2250 ns (**C**). Snapshots of $F_{23}R$ (**D,E**), $I_{26}A$ (**F,G**), and the N-native segment (**H,I**) under the same conditions.

To determine whether the PS anionic environment is the factor that leads to the formation of β-structures, other anionic lipid vesicles were evaluated. To this end, vesicles of PC and anionic-POPG were prepared at a 3 to 1 molar ratio, for which, the characterization of the peptide bond at 218 nm showed for the C-native segment, $F_{23}R$, and $I_{26}A$ variants, an increase in peptide bond absorbance, demonstrating a phosphatidyl-glycerol (POPG)-modulation dependent response (Figure S7A). In the same way, the three fragments, C-native segment, $F_{23}R$, and $I_{26}A$, incubated with POPG enhanced cytotoxicity on β-cells (Figure S7B). Likewise, a control with Aβ-peptide was included. To extend this characterization using peptide-bond spectroscopy, anionic SUVs were evaluated by increasing the molar concentration of POPG (0–100%). A strong modulation of POPG on the β-structure of the C-native fragment and both $F_{23}R$ and $I_{26}A$ was observed (Figure S7C). Therefore, it seems that anionic electrostatic charges prove to play an important role in amyloidogenesis, as further demonstrated by the evaluation of a representative Aβ-peptide (Figure S7C).

Although the structure of IAPP could change according to the physicochemical properties of lipids [7], in an attempt to characterize the role of lipid surfaces on the conformational transitions of β-sheets, we evaluated the influence of solvation determined by the presence of a free hydroxyl group in position 2 of phospholipid heads. Our results indicate that treatment with lysophosphatidic acid (LPA) micelles under a neutral pH promotes conformational transitions towards β-sheet structures only in the C-native segment and $F_{23}R$ variant (Figure S8). For the case of the N-native segment, a spectrum corresponding to an α-helix signal was registered, reflecting the avoidance of β-sheet structure formation. These results appear to confirm that β-structure modulation is localized at the C-native domain of IAPP (Figure S8). Results that are in concordance with the proposal that amyloidogenicity and cytotoxicity are induced by two different regions of the hIAPP sequence [41]. Likewise, these findings are in agreement with previous results from our laboratory where we described that incubation with LPA promotes conformational transitions in the C-terminal domain of cholesteryl ester transfer protein (CETP) [10].

4. Discussion

During the course of the present investigation, employing a biomimetic approach, we have been able to develop and evaluate new variants of IAPP with the property to show fewer propensities to form β-sheet structures. New sequences such as $F_{23}R$ have shown to increase the stability of the C-native segment of IAPP, with an important involvement of the anionic nature lipids. This behavior can be associated at pH 7.4 with charged lipids that contributed to structural changes towards the β-sheet formation, whereas the $I_{26}A$ variant shows a neutral charge and an isoelectric point similar to the C-native segment, promoting a strong interaction with anionic phospholipids. Molecular dynamics showing an absorption phenomenon of $F_{23}R$ at the surface of lipid bilayer indicate the possibility for the formation of peptide-PS aggregates that, by means of lipotoxicity, could have contributed to the observed cytotoxicity (Figure 6G). Further experiments are nowadays in progress in our laboratories to confirm this assumption (data to be published).

The content of anionic phospholipids in β-cell membranes is reported to range from 2.5% to 13.2%, a proportion mainly situated in the inner leaflet of the membrane [38], proportion much similar as found in insulin secretory granules where a higher content of PS and phosphatidylinositol has been described [17,42], which are derived from ER membranes during insulin maturation. This contributes to the possibility that hIAPP aggregation could be increased at the intracellular space, especially during the process of maturation of insulin secretory granules. The phenomenon of membrane asymmetry that could enhance hIAPP amyloid formation and membrane damage in vivo [43] could modify the PS-biodisponibility and, therefore, triggers the β-sheets formation.

Therefore, considering the critical role of PS in the promotion of β-sheet conformational transitions on hIAPP-derived peptides, the impact of phospholipid negative-electrostatic charge on UPR regulation was characterized. In this sense, PS-SUVs with the hIAPP-derived peptides under a 1 to 2 peptide/PS ratio were incubated on β-cell cultures for 20 h. In the first case, critical UPR-targets were evaluated under the peptide/PS treatments in complete cellular lysates (Figure 8A) results suggest the slight activation of XBPIs, a transcription factor of the activation of IRE1-arm of UPR. Likewise, an increase in the expression of chaperone BiP under peptides/PS treatments was registered, and this phenomenon was more evident under the C-native treatment. Possibly, these modifications could be partly related to a cellular compensatory response aimed to maintain protein homeostasis in ER (Figure 8A), as well as critical functions in the physiology of β-cells. Importantly, when insulin concentrations were evaluated in extracellular media, we found a diminution of the insulin levels upon treatment with the peptide/PS mixtures (Figure 8B).

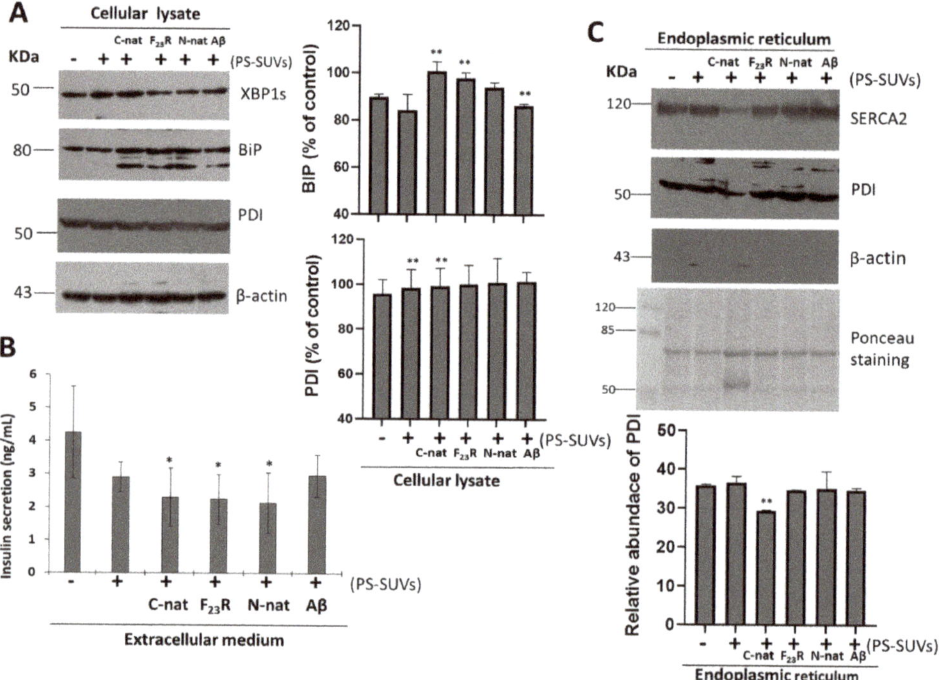

Figure 8. Unfolded protein response is activated under the treatment of hIAPP-derived peptides/PS vesicles mixtures. (**A**) Effect of hIAPP-derived peptides/PS on the expression of XBP1s and binding immunoglobulin protein (BiP) targets of unfolded protein response (UPR), as well as protein disulfide isomerases (PDI) expression on complete cell lysates. Densitometry analysis of (BiP) and PDI immunoblots, results are reported as means ± SD and expressed as % of control, ** $p < 0.05$. Aβ-peptide was used as a control. β-actin was used as a loading control. (**B**) Insulin concentrations (ng/mL) in extracellular media, * $p < 0.1$. (**C**) Under the same experimental conditions, the ER was purified and the expression of sarco/endoplasmic reticulum Ca^{2+}-ATPase-2 (SERCA2), PDI, and β-actin were characterized. Polyvinylidene difluoride (PVDF) membranes were stained with Ponceau. Densitometry analysis of PDI immunoblots; results are reported as means ± SD and expressed as relative abundance; ** $p < 0.05$.

PDI is a chaperone that regulates folding of proinsulin, participates in disulfide bond formation, and maintains ER redox homeostasis [44]. In our conditions, when total β-cell lysates were evaluated under peptide/PS treatments, we did not find changes in PDI expression (Figure 8A). Constituting a chaperone-protein with critical functions, high expression levels of PDI have been found in the ER lumen, to a lower extend in the cytosol, and also in different cellular membranes. [45]. In addition, we have also detected PDI in the extracellular medium of β-cells treated with peptide/PS mixtures (data not shown). However, more experimental evidence is required to establish a mechanistic proposal.

To dissect the role of PDI, we performed the isolation of ER of β-cells cultures, evaluating sarco/endoplasmic reticulum Ca^{2+}-ATPase-2 (SERCA2) and β-actin as controls of ER isolations. Results indicated the isolation of pure ER-fractions (Figure S9). Then, under the same treatments of hIAPP-peptides/PS-SUVs, we characterized the expression of PDI. We found that the PDI levels diminished upon treatment with C-native/PS, as well as the levels of SERCA2 resident of ER (Figure 8C). Results suggest that the affectation of SERCA2 and PDI under C-native/PS treatment might be related to both their ER localization and the activation of UPR, affecting insulin secretion.

ER-lumen and the function of chaperones BiP and PDI are critical during proinsulin folding. Considering our results, and in an attempt to characterize the effect of the PS vesicles and possibly trace their cellular localization, confocal microscopy experiments were carried out. To this end, we prepared PS and peptide/PS vesicles (30 μM) tagged with the green fluorescent probe BODIPY-Leu [46] (6 μM) (referred to as BODIPY-Leu/PS and BODIPY-Leu/peptides/PS vesicles, respectively), with which RIN-m5F cells were treated. In the first instance, our results demonstrated that BODIPY-Leu/PS vesicles are internalized in RIN-m5F cells (Figure 9A–C, stained in green). Then, we used ER-tracker (red staining) for characterize the colocalizing in ER sites (Figure 9D–I). Results suggest that the system is located in the ER. Importantly, when we evaluated the localization of the BODIPY-Leu/PS vesicles incubated with hIAPP-derived peptides (C-native and $F_{23}R$ variant) under a molar relationship (1/2; peptide/vesicle), the BODIPY-Leu/PS signal was localized under C-native treatment in ER sites (Figure 9J–L), however, the signal diminished slightly under $F_{23}R$ treatment (Figure 9M–O). In an important way, C-native/PS treatment promoted the higher levels of cytotoxicity in the studied cells (Figure 6G), affecting insulin secretion (Figure 8B). Moreover, these phenomena coincide with UPR activation and affectation of localization of PDI and SERCA2 in the ER (Figure 8C). Therefore, it appears that the localization of peptide/PS is a critical condition to induce the alterations in homeostasis of the ER. This phenomenon was evidenced after treatment with the C-native/PS system, whereupon the higher fluorescence signal of the BODIPY-Leu/PS vesicles along the ER very likely corresponds with the alterations in localization of PDI, SERCA2, and the insulin secretion. This phenomenon was not evident upon treatment with the $F_{23}R$ variant.

Therefore, our results suggest that the lipid-anionic electrostatic charge is a critical condition that could modulate the UPR pathway and the conformational transition of IAPP-derived peptides. Thus, a negative electrostatic-charge environment could be critical in insulin secretion as well. Interestingly, the content of anionic lipids of β-cells and insulin secretory granules has been related to an altered glucose-stimulated insulin exocytosis [17,43]. Therefore, a condition of metabolic overload also has been associated with the biodisponibility of fatty acids [22], very likely contributing to the deleterious phenomenon correlated with the concentration of anionic phospholipids, promoting misfolded transitions on hIAPP. Having in mind this phenomenon, our group has generated new materials of polymeric films of polyvinyl dimethylazlactone (PVDMA) and polyethylene imine (PEI) to evaluate the effect of fatty acids on β-cell membranes [47] and diverse critical physiological functions.

In a complementary way, current results of our laboratory suggest that the induction of oligomers at the C-native domain of IAPP accelerates β-sheet formation when treated with oleic acid/PC vesicles. By contrast, when palmitic acid/PC vesicles are used, this result is not found (data not shown). Then, considering that unsaturation and shorter fatty acids of phospholipids facilitate the curvature and fluidity of membranes favoring their fusion [17], although increasing the risk of aggregation, there is a subtle regulation in the conservation of the structure of the hIAPP. Moreover, a report reveals synergic implications of free fatty acids and hIAPP in ER stress and apoptosis of islet β-cells [48]. In this context, we have documented the role of metabolic overload by saturated fatty acids on proteostasis and its impact on insulin secretion, specifically the dysregulation of targets that control intracellular calcium homeostasis [22], as also documented in this report for SERCA2.

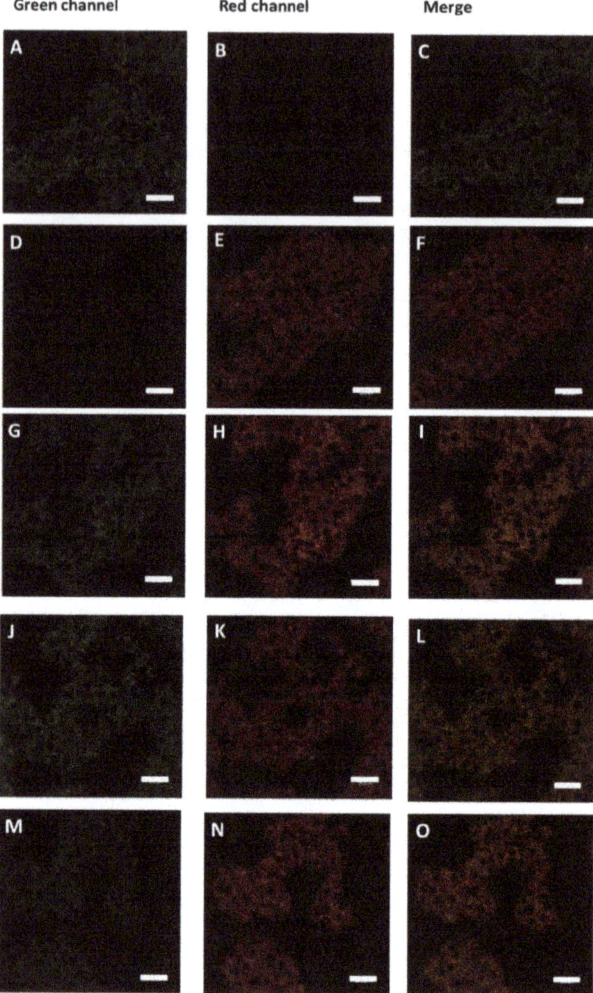

Figure 9. BODIPY-Leu/PS vesicles (green) colocalize with ER-tracker (red). (**A–C**) Langerhans islet β-cells (RIN-m5F cells) treated with BODIPY-Leu/PS vesicles. (**D–F**) Cells treated with ER-tracker. (**G–I**) Cells treated jointly with BODIPY-Leu/PS vesicles and ER-tracker. (**J–L**) Effect of C-native treatment on localization of BODIPY-Leu/PS vesicles and ER-tracker. (**M–O**) Effect of $F_{23}R$ peptide treatment on localization of BODIPY-Leu/PS vesicles and ER-tracker. Scale bars correspond to 25 μm.

Recently, following peptidomimetic design strategies, research has been developed to find a way to inhibit the formation of β-sheet structures in segments of an important series of polypeptides and proteins as a therapeutic way to fight amyloid disease. There is still a field of action in amyloidogenesis design. Although prolines residues in rIAPP promote disordered structures, results of coincubation of rIAPP and hIAPP suggest that rat amylin does not block β-sheet and also forms its own β-sheet, most probably on the outside of the human fibrils [49], revealing the complex behavior in the development of an amyloid fibril inhibitor. We and other authors have documented that it is critical to consider the impact of lipid environment. In the light of our results, the $F_{23}R$ variant of IAPP showed a low propensity to form β-sheet structures even under the effect of zwitterionic lipids. However, anionic charge of lipid vesicles and degree of solvation were factors for the modulation of β-sheet formation of the $F_{23}R$ and $I_{26}A$ variants, as well as in the C-native segment of IAPP, all associated to the cytotoxicity phenomena of β-cells. In conclusion, our results show the potential implications of modulating the structure and stability of IAPP for the design of analog therapeutics based on peptides and proteins.

Supplementary Materials: The following are available online at http://www.mdpi.com/2218-273X/10/9/1201/s1, Figure S1. Incubation with SUVs composed of phosphatidylcholine does not induce conformational transitions on the N-native segment (^{1}KCNTATCATQRLANFLVHSS^{20}) of hIAPP. Figure S2. Effect of PC-LUVs on the secondary structure of peptides derived from hIAPP. Figure S3. Displacement of peptides on the z-axis of PC-bilayers (MSD) obtained by short simulations. Figure S4. Mixtures of LUVs/SUVs composed of PS facilitate the formation of β-sheet structures. Figure S5. The cationic lipid surface is not a critical factor for β-sheet aggregation on hIAPP segments. Figure S6. Displacement of peptides on the z-axis of PS-bilayers (MSD) obtained by short simulations. Figure S7. Effect of SUVs composed of 1-palmitoyl-2-oleoyl-sn-glycerol (POPG) on the structure of IAPP variants. Figure S8. Effect of lysophosphatidic acid incubation on the secondary structure of C-native (**A**), $F_{23}R$ variant (**B**), and N-native (**C**). Figure S9. Characterization of several fractions obtained during endoplasmic reticulum (ER) isolation.

Author Contributions: I.M.-N., A.P.-C., and V.G.-G., conceived and designed the experiments; I.M.-N., A.P.-C., I.L.-R., E.R.-V., M.A.-M., and J.M.-O., performed experiments; I.M.-N., A.P.-C., J.M.-O., I.L.-R., R.D.-M., and V.G.-G., analyzed data; J.M.-O., M.A.R.-I., I.A.R., and V.G-G, contributed reagents/materials/analysis tools; A.P.C. and V.G.-G., wrote the paper. All authors have read and agreed to the published version of the manuscript.

Funding: This research was funded by the Coordinación de Posgrado e Investigación-UABC grant 106/2/N/57/1 (1983), Fondo Sectorial de Investigación para la Educación CB 2017-2018 (A1-S-28653/SEP/CONACYT), and 21a. Convocatoria Interna de Apoyo a Proyectos de Investigación (Coordinación General de Investigación y Posgrado/UABC). We thank DGTIC-UNAM for granting access to the supercomputing cluster MIZTLI (project: LANCAD-UNAM-DGTIC-352 granted to JM-O).

Acknowledgments: Authors thank Blanca Delgado-Coello for expert technical assistance and the administrative support of Josue Villegas-Sandoval and Mónica López-Valladares. Authors thank Ernesto Beltran-Partida and Benjamin Valdes-Salas for participation in DLS experimentation. Authors thank Santa Cruz Biotechnology for the donation of anti-SERCA2. M.A.-M. thanks funding from CONACyT (Mexico) through Research Projects INFR-2015-251863 and PDCPN-2015-89.

Conflicts of Interest: The authors declare no conflict of interest.

References

1. Koda, J.E.; Fineman, M.; Rink, T.J.; Dailey, G.E.; Muchmore, D.B.; Linarelli, L.G. Amylin concentrations and glucose control. *Lancet* **1992**, *339*, 1179–1180. [CrossRef]

2. Percy, A.J.; Trainor, D.A.; Rittenhouse, J.; Phelps, J.; Koda, J.E. Development of sensitive immunoassays to detect amylin and amylin-like peptides in unextracted plasma. *Clin. Chem.* **1996**, *42*, 576–585. [CrossRef] [PubMed]

3. Novials, A.; Sarri, Y.; Casamitjana, R.; Rivera, F.; Gomis, R. Regulation of islet amyloid polypeptide in human pancreatic islets. *Diabetes* **1993**, *42*, 1514–1519. [CrossRef] [PubMed]

4. Xu, W.; Jiang, P.; Mu, Y. Conformation preorganization: Effects of S20G mutation on the structure of human islet amyloid polypeptide segment. *J. Phys. Chem. B* **2009**, *113*, 7308–7314. [CrossRef] [PubMed]

5. Poa, N.R.; Cooper, G.J.; Edgar, P.F. Amylin gene promoter mutations predispose to Type 2 diabetes in New Zealand Maori. *Diabetologia* **2003**, *46*, 574–578. [CrossRef] [PubMed]

6. Berhanu, W.M.; Hansmann, U.H. Inter-species cross-seeding: Stability and assembly of rat-human amylin aggregates. *PLoS ONE* **2014**, *9*, e97051. [CrossRef]

7. Pulido-Capiz, A.; Diaz-Molina, R.; Martinez-Navarro, I.; Guevara-Olaya, L.A.; Casanueva-Perez, E.; Mas-Oliva, J.; Rivero, I.A.; Garcia-Gonzalez, V. Modulation of Amyloidogenesis Controlled by the C-Terminal Domain of Islet Amyloid Polypeptide Shows New Functions on Hepatocyte Cholesterol Metabolism. *Front. Endocrinol.* **2018**, *9*, 331. [CrossRef]

8. Garcia-Gonzalez, V.; Gutierrez-Quintanar, N.; Mas-Oliva, J. The C-terminal Domain Supports a Novel Function for CETPI as a New Plasma Lipopolysaccharide-Binding Protein. *Sci. Rep.* **2015**, *5*, 16091. [CrossRef]

9. Garcia-Gonzalez, V.; Gutierrez-Quintanar, N.; Mendoza-Espinosa, P.; Brocos, P.; Pineiro, A.; Mas-Oliva, J. Key structural arrangements at the C-terminus domain of CETP suggest a potential mechanism for lipid-transfer activity. *J. Struct. Biol.* **2014**, *186*, 19–27. [CrossRef]

10. Garcia-Gonzalez, V.; Mas-Oliva, J. Amyloid fibril formation of peptides derived from the C-terminus of CETP modulated by lipids. *Biochem. Biophys. Res. Commun.* **2013**, *434*, 54–59. [CrossRef]

11. Diaz-Villanueva, J.F.; Diaz-Molina, R.; Garcia-Gonzalez, V. Protein Folding and Mechanisms of Proteostasis. *Int. J. Mol. Sci.* **2015**, *16*, 17193–17230. [CrossRef] [PubMed]

12. Garcia-Gonzalez, V.; Mas-Oliva, J. Amyloidogenic properties of a D/N mutated 12 amino acid fragment of the C-terminal domain of the Cholesteryl-Ester Transfer Protein (CETP). *Int. J. Mol. Sci.* **2011**, *12*, 2019–2035. [CrossRef] [PubMed]

13. Mendoza-Espinosa, P.; Garcia-Gonzalez, V.; Moreno, A.; Castillo, R.; Mas-Oliva, J. Disorder-to-order conformational transitions in protein structure and its relationship to disease. *Mol. Cell. Biochem.* **2009**, *330*, 105–120. [CrossRef] [PubMed]

14. Knight, J.D.; Hebda, J.A.; Miranker, A.D. Conserved and cooperative assembly of membrane-bound alpha-helical states of islet amyloid polypeptide. *Biochemistry* **2006**, *45*, 9496–9508. [CrossRef]

15. Jayasinghe, S.A.; Langen, R. Lipid membranes modulate the structure of islet amyloid polypeptide. *Biochemistry* **2005**, *44*, 12113–12119. [CrossRef]

16. Suckale, J.; Solimena, M. The insulin secretory granule as a signaling hub. *Trends Endocrinol. Metab.* **2010**, *21*, 599–609. [CrossRef]

17. MacDonald, M.J.; Ade, L.; Ntambi, J.M.; Ansari, I.U.; Stoker, S.W. Characterization of phospholipids in insulin secretory granules and mitochondria in pancreatic beta cells and their changes with glucose stimulation. *J. Biol. Chem.* **2015**, *290*, 11075–11092. [CrossRef]

18. Saito Michiko, S.Y. ER Stress, Secretory Granule Biogenesis, and Insulin. In *Ultimate Guide to Insulin*; IntechOpen: London, UK, 2019. [CrossRef]

19. Schuck, S.; Prinz, W.A.; Thorn, K.S.; Voss, C.; Walter, P. Membrane expansion alleviates endoplasmic reticulum stress independently of the unfolded protein response. *J. Cell. Biol.* **2009**, *187*, 525–536. [CrossRef]

20. Meyerovich, K.; Ortis, F.; Allagnat, F.; Cardozo, A.K. Endoplasmic reticulum stress and the unfolded protein response in pancreatic islet inflammation. *J. Mol. Endocrinol.* **2016**, *57*, R1–R17. [CrossRef]

21. Galindo-Hernandez, O.; Cordova-Guerrero, I.; Diaz-Rubio, L.J.; Pulido-Capiz, A.; Diaz-Villanueva, J.F.; Castaneda-Sanchez, C.Y.; Serafin-Higuera, N.; Garcia-Gonzalez, V. Protein translation associated to PERK arm is a new target for regulation of metainflammation: A connection with hepatocyte cholesterol. *J. Cell. Biochem.* **2019**, *120*, 4158–4171. [CrossRef]

22. Acosta-Montano, P.; Rodriguez-Velazquez, E.; Ibarra-Lopez, E.; Frayde-Gomez, H.; Mas-Oliva, J.; Delgado-Coello, B.; Rivero, I.A.; Alatorre-Meda, M.; Aguilera, J.; Guevara-Olaya, L.; et al. Fatty Acid and Lipopolysaccharide Effect on Beta Cells Proteostasis and its Impact on Insulin Secretion. *Cells* **2019**, *8*, 884. [CrossRef] [PubMed]

23. Conchillo-Sole, O.; de Groot, N.S.; Aviles, F.X.; Vendrell, J.; Daura, X.; Ventura, S. AGGRESCAN: A server for the prediction and evaluation of "hot spots" of aggregation in polypeptides. *BMC Bioinform.* **2007**, *8*, 65. [CrossRef] [PubMed]

24. Szoka, F., Jr.; Papahadjopoulos, D. Comparative properties and methods of preparation of lipid vesicles (liposomes). *Annu. Rev. Biophys. Bioeng.* **1980**, *9*, 467–508. [CrossRef] [PubMed]

25. Knapp, B.; Ospina, L.; Deane, C.M. Avoiding False Positive Conclusions in Molecular Simulation: The Importance of Replicas. *J. Chem. Theory Comput.* **2018**, *14*, 6127–6138. [CrossRef]

26. Losasso, V.; Pietropaolo, A.; Zannoni, C.; Gustincich, S.; Carloni, P. Structural role of compensatory amino acid replacements in the alpha-synuclein protein. *Biochemistry* **2011**, *50*, 6994–7001. [CrossRef]

27. Prajapati, P.; Wang, W.X.; Nelson, P.T.; Springer, J.E. Methodology for Subcellular Fractionation and MicroRNA Examination of Mitochondria, Mitochondria Associated ER Membrane (MAM), ER, and Cytosol from Human Brain. *Methods Mol. Biol.* **2020**, *2063*, 139–154. [CrossRef]

28. Patil, S.M.; Alexandrescu, A.T. Charge-Based Inhibitors of Amylin Fibrillization and Toxicity. *J. Diabetes Res.* **2015**, *2015*, 946037. [CrossRef]

29. Balbach, J.J.; Ishii, Y.; Antzutkin, O.N.; Leapman, R.D.; Rizzo, N.W.; Dyda, F.; Reed, J.; Tycko, R. Amyloid fibril formation by A beta 16-22, a seven-residue fragment of the Alzheimer's beta-amyloid peptide, and structural characterization by solid state NMR. *Biochemistry* **2000**, *39*, 13748–13759. [CrossRef]

30. Smaoui, M.R.; Waldispuhl, J. Complete characterization of the mutation landscape reveals the effect on amylin stability and amyloidogenicity. *Proteins* **2015**, *83*, 1014–1026. [CrossRef]

31. Cho, W.J.; Jena, B.P.; Jeremic, A.M. Nano-scale imaging and dynamics of amylin-membrane interactions and its implication in type II diabetes mellitus. *Methods Cell. Biol.* **2008**, *90*, 267–286. [CrossRef]

32. Khemtemourian, L.; Domenech, E.; Doux, J.P.; Koorengevel, M.C.; Killian, J.A. Low pH acts as inhibitor of membrane damage induced by human islet amyloid polypeptide. *J. Am. Chem. Soc.* **2011**, *133*, 15598–15604. [CrossRef] [PubMed]

33. Cho, W.J.; Trikha, S.; Jeremic, A.M. Cholesterol regulates assembly of human islet amyloid polypeptide on model membranes. *J Mol Biol* **2009**, *393*, 765–775. [CrossRef] [PubMed]

34. Simons, K.; Toomre, D. Lipid rafts and signal transduction. *Nat. Rev. Mol. Cell. Biol.* **2000**, *1*, 31–39. [CrossRef] [PubMed]

35. Skeby, K.K.; Andersen, O.J.; Pogorelov, T.V.; Tajkhorshid, E.; Schiott, B. Conformational Dynamics of the Human Islet Amyloid Polypeptide in a Membrane Environment: Toward the Aggregation Prone Form. *Biochemistry* **2016**, *55*, 2031–2042. [CrossRef] [PubMed]

36. DeToma, A.S.; Salamekh, S.; Ramamoorthy, A.; Lim, M.H. Misfolded proteins in Alzheimer's disease and type II diabetes. *Chem. Soc. Rev.* **2012**, *41*, 608–621. [CrossRef]

37. Kelly, J.W. Mechanisms of amyloidogenesis. *Nat. Struct. Biol.* **2000**, *7*, 824–826. [CrossRef]

38. Xu, W.; Wei, G.; Su, H.; Nordenskiold, L.; Mu, Y. Effects of cholesterol on pore formation in lipid bilayers induced by human islet amyloid polypeptide fragments: A coarse-grained molecular dynamics study. *Phys. Rev. E* **2011**, *84*, 051922. [CrossRef]

39. Vorobyov, I.; Allen, T.W. On the role of anionic lipids in charged protein interactions with membranes. *Biochim. Biophys. Acta* **2011**, *1808*, 1673–1683. [CrossRef]

40. Platre, M.P.; Jaillais, Y. Anionic lipids and the maintenance of membrane electrostatics in eukaryotes. *Plant Signal. Behav.* **2017**, *12*, e1282022. [CrossRef]

41. Martel, A.; Antony, L.; Gerelli, Y.; Porcar, L.; Fluitt, A.; Hoffmann, K.; Kiesel, I.; Vivaudou, M.; Fragneto, G.; de Pablo, J.J. Membrane Permeation versus Amyloidogenicity: A Multitechnique Study of Islet Amyloid Polypeptide Interaction with Model Membranes. *J. Am. Chem. Soc.* **2017**, *139*, 137–148. [CrossRef]

42. Saitta, F.; Signorelli, M.; Fessas, D. Dissecting the effects of free fatty acids on the thermodynamic stability of complex model membranes mimicking insulin secretory granules. *Colloids Surf. B Biointerfaces* **2019**, *176*, 167–175. [CrossRef] [PubMed]

43. Zhang, X.; St Clair, J.R.; London, E.; Raleigh, D.P. Islet Amyloid Polypeptide Membrane Interactions: Effects of Membrane Composition. *Biochemistry* **2017**, *56*, 376–390. [CrossRef] [PubMed]

44. Jang, I.; Pottekat, A.; Poothong, J.; Yong, J.; Lagunas-Acosta, J.; Charbono, A.; Chen, Z.; Scheuner, D.L.; Liu, M.; Itkin-Ansari, P.; et al. PDIA1/P4HB is required for efficient proinsulin maturation and ss cell health in response to diet induced obesity. *eLife* **2019**, *8*. [CrossRef] [PubMed]

45. Ali Khan, H.; Mutus, B. Protein disulfide isomerase a multifunctional protein with multiple physiological roles. *Front Chem.* **2014**, *2*, 70. [CrossRef]

46. Avelino Jorge, G.-G.V.; Alatorre-Meda, M.; Rodríguez-Velázquez, E.; Rivero, I.A. Synthesis of BODIPY-amino acids and the potential applications as specific dyes for the cytoplasm of Langerhans β-cells. 2020, (in review).

47. Avila-Cossio, M.E.; Rivero, I.A.; Garcia-Gonzalez, V.; Alatorre-Meda, M.; Rodriguez-Velazquez, E.; Calva-Yanez, J.C.; Espinoza, K.A.; Pulido-Capiz, A. Preparation of Polymeric Films of PVDMA-PEI Functionalized with Fatty Acids for Studying the Adherence and Proliferation of Langerhans beta-Cells. *ACS Omega* **2020**, *5*, 5249–5257. [CrossRef]

48. Gao, L.P.; Chen, H.C.; Ma, Z.L.; Chen, A.D.; Du, H.L.; Yin, J.; Jing, Y.H. Fibrillation of human islet amyloid polypeptide and its toxicity to pancreatic beta-cells under lipid environment. *Biochim. Biophys. Acta Gen. Subj.* **2020**, *1864*, 129422. [CrossRef]

49. Middleton, C.T.; Marek, P.; Cao, P.; Chiu, C.C.; Singh, S.; Woys, A.M.; de Pablo, J.J.; Raleigh, D.P.; Zanni, M.T. Two-dimensional infrared spectroscopy reveals the complex behaviour of an amyloid fibril inhibitor. *Nat. Chem.* **2012**, *4*, 355–360. [CrossRef]

 *biomolecules*

Article

Evaluation of the In Vitro Oral Wound Healing Effects of Pomegranate (*Punica granatum*) Rind Extract and Punicalagin, in Combination with Zn (II)

**Vildan Celiksoy [1], Rachael L. Moses [2], Alastair J. Sloan [3], Ryan Moseley [2,* and Charles M. Heard [1,*

[1] School of Pharmacy and Pharmaceutical Sciences, Cardiff University, Cardiff CF10 3NB, UK;
   CeliksoyV@cardiff.ac.uk
[2] Oral and Biomedical Sciences, School of Dentistry, Cardiff University, Cardiff CF14 4XY, UK;
   MosesR@cardiff.ac.uk
[3] Melbourne Dental School, Faculty of Medicine, Dentistry and Health Sciences, Melbourne Dental School,
   University of Melbourne, Victoria 3010, Australia; alastair.sloan@unimelb.edu.au
* Correspondence: MoseleyR@cardiff.ac.uk (R.M.); Heard@cardiff.ac.uk (C.M.H.);
   Tel.: +44-(0)2022-510649 (R.M.); +44-(0)2920-875819 (C.M.H.)

Received: 28 July 2020; Accepted: 20 August 2020; Published: 25 August 2020

Abstract: Pomegranate (*Punica granatum*) is a well-established folklore medicine, demonstrating benefits in treating numerous conditions partly due to its antimicrobial and anti-inflammatory properties. Such desirable medicinal capabilities are attributed to a high hydrolysable tannin content, especially punicalagin. However, few studies have evaluated the abilities of pomegranate to promote oral healing, during situations such as periodontal disease or trauma. Therefore, this study evaluated the antioxidant and in vitro gingival wound healing effects of pomegranate rind extract (PRE) and punicalagin, alone and in combination with Zn (II). In vitro antioxidant activities were studied using DPPH and ABTS assays, with total PRE phenolic content measured by Folin–Ciocalteu assay. PRE, punicalagin and Zn (II) combination effects on human gingival fibroblast viability/proliferation and migration were investigated by MTT assay and scratch wounds, respectively. Punicalagin demonstrated superior antioxidant capacities to PRE, although Zn (II) exerted no additional influences. PRE, punicalagin and Zn (II) reduced gingival fibroblast viability and migration at high concentrations, but retained viability at lower concentrations without Zn (II). Fibroblast speed and distance travelled during migration were also enhanced by punicalagin with Zn (II) at low concentrations. Therefore, punicalagin in combination with Zn (II) may promote certain anti-inflammatory and fibroblast responses to aid oral healing.

Keywords: pomegranate; punicalagin; tannins; gingiva; fibroblasts; antioxidant; wound healing

---

1. Introduction

Wound healing is a complex process, involving a chain of well-orchestrated biochemical and cellular events that effect the repair of diseased or damaged tissues. Healing is mainly achieved through four precise and programmed phases: homeostasis, inflammation, proliferation and remodeling. These phases must occur in an orderly and suitable timeframe which is essential to normal healing, although disruption to these mechanisms by various factors may cause delayed or non-healing to occur [1].

Wounds within the oral cavity can be caused by many factors, including trauma, periodontal disease and surgery. Although the oral cavity harbors a wide variety of commensal microbial species [2], its distinct environment with continuous salivary flow helps prevent contamination and infection [3,4]. Furthermore, although oral and dermal wounds proceed through similar stages of healing, oral wounds are characterized by rapid healing with minimal scar formation, mediated in part via enhanced fibroblast and keratinocyte repair responses [5,6]. However, despite such superior healing properties, oral wounds are common and yet difficult to protect using conventional wound dressing approaches; and therefore are susceptible to microbial contamination and further trauma, such as during mastication [7]. For the treatment of the oral wounds, antibiotics, corticosteroids, non-steroidal anti-inflammatory drugs (NSAIDs) and disinfectants, such as chlorhexidine, have all been used to accelerate the healing process and prevent patient disconformity [8]. However, these drugs are commonly associated with various side effects, such as gastrointestinal damage, discoloration, dysgeusia and excessive sensitivity in the oral mucosa [9]. Therefore, alternative pharmaceutical therapies are needed for the promotion of oral wound healing, which overcome these issues. Natural compounds and formulations may offer such promise, as many have had strong historical roles in the treatment of many different diseases and conditions worldwide. As a result, natural products are favored by modern societies and consumer acceptance is high [10,11]. Indeed, it has been suggested that medicinal plants have more efficacious healing properties and less adverse effects than the other more synthetic pharmaceutical chemicals [12,13].

Pomegranate (*Punica granatum*), a part of the *punicaceae* family native to the Middle East, is a well-established folklore medicine, and mainly cultivated in Iran, India, USA and most of the near and far eastern countries. It has been used in the treatment of dysentery, diarrhoea and stomatitis in traditional medicine in many cultures which is documented in Egyptian Papyrus of Ebers [14,15]. Recent studies have shown that pomegranate demonstrates benefits in treating numerous conditions, due to its anticancer, antimicrobial, anti-inflammatory and antioxidant properties [16]. The different parts of the pomegranate have rich sources of secondary metabolites with potential biological activities [17]. The fruit exocarp (rind) is particularly abundant in hydrolysable tannins, in particular punicalagin, which is a large (mw 1,084.71 g/mol) molecule comprised of gallagic acid and ellagic acid linked via a glucose moiety (Figure 1) [18,19]. These compounds have been attributed as being the primary sources of bioactivity responsible for the desirable medicinal properties of pomegranate, including their dermal wound healing efficacies [19–24]. Indeed, from our previous work, pomegranate rind extract (PRE) and punicalagin itself have been shown to exhibit potent anti-inflammatory, antimicrobial and antiviral activities, which can be further potentiated by combination with Zn (II) ions [25–28]. Zn (II) itself also has a prominent role in all stages of wound repair, regulating immuno-inflammatory cell, endothelial cell, keratinocyte and fibroblast responses [29,30]. Indeed, the importance of Zn (II) to successful wound repair outcomes is supported by studies correlating delayed healing with deficient Zn (II) levels and enhanced repair following the topical application of Zn-containing compounds. Thus, it may be hypothesized that PRE and punicalagin supplementation with Zn (II) can promote additional beneficial wound healing effects. However, whereas beneficial PRE and punicalagin effects on dermal wound healing are supported in the literature, no studies have to date examined whether PRE and punicalagin could offer similar therapeutic wound healing benefits within the oral cavity. Therefore, the purpose of the present study was to evaluate the potential of PRE and punicalagin, with and without Zn (II), used to promote the healing of oral wounds caused by periodontal disease or trauma. Specifically, PRE, punicalagin, Zn (II) alone and Zn (II) in combination with PRE and punicalagin were assessed for their in vitro antioxidant activities, in addition to their effects on the viability, proliferation and migration of human primary gingival fibroblasts.

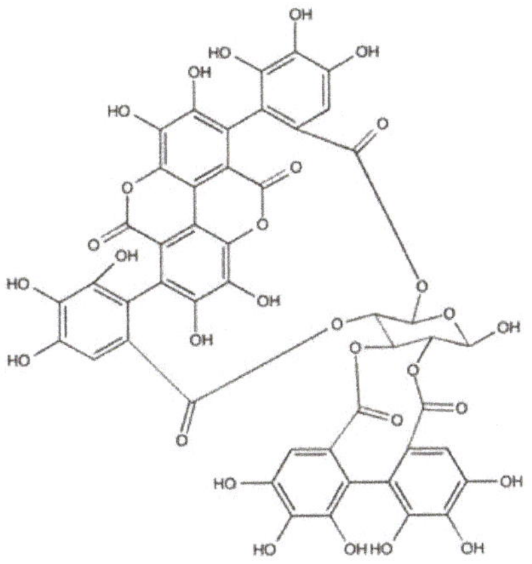

Figure 1. The chemical structure of punicalagin.

2. Materials and Methods

2.1. Materials

Pomegranates were obtained from a local supermarket and were of Spanish origin. Punicalagin (≥98%), [3-(4,5-dimethyl-2-thiazolyl)-2,5-diphenyltetrazolium bromide] (MTT), 2,2′-azino-bis(3-ethylbenzothiazoline-6-sulfonic acid diammonium salt (ABTS), 2,2-diphenyl-1-picrylhydrazyl (DPPH), Folin–Ciocalteu (F-C) reagent, potassium persulphate, (±)-6-hydroxy-2,5,7,8-tetramethylchromane-2-carboxylic acid (Trolox); dimethylsulfoxide (DMSO), ascorbic acid and sodium carbonate (Na_2CO_3) were all obtained from Sigma-Aldrich (Gillingham, UK). Zinc sulfate heptahydrate ($ZnSO_4 \cdot 7H_2O$), potassium hydrogen phthalate, Dulbecco's Modified Eagle Medium (DMEM), fetal calf serum (FCS), L-glutamine and antibiotics/antimycotics were obtained from ThermoFisher Scientific (Loughborough, UK).

2.2. Preparation of Pomegranate Rind Extract (PRE)

The rind of the pomegranates was peeled with a scalpel and cut to approximately 2 cm^2 pieces. The net weight of the rind was 300 g. This was blended (25%) *w/v* in deionized water in a standard blender until visibly homogenous. The blended rind in deionized water was boiled for 10 min and centrifuged (×4) using a Heraeus Multifuge 3S/3S-R centrifuge (5980× *g* at 4 °C for 30 min), before filtration through a Whatman 0.45 μm nylon membrane filter. The collected solution was freeze dried, protected from light and stored at −20 °C until required. The desired concentration of PRE was prepared in pH 4.5 phthalate buffer and sterilized by using a 0.45 μm Millex-FG syringe-driven filter [26]. Punicalagin concentrations were determined by HPLC (Thermo LCQ classic LCMS with ESI source) according to method of Seeram et al. [31]; and found that 1 mg/mL PRE contains approximately 17 μg punicalagin.

2.3. Determination of Total Phenolic Content

The Folin–Ciocalteu (F-C) colorimetric assay was used to quantify the total phenolic content in PRE, according to method by Ainsworth and Gillespie [32]. Briefly, 0.5 mg/mL PRE samples were prepared and 200 μL 10% (*v/v*) F-C reagent was added to 100 μL of the prepared PRE samples, followed by the addition of 800 μL 700 mM Na_2CO_3. Samples were incubated at room temperature for 2 h. After the incubation period, 200 μL of each sample was added to 96-well plates and the absorbance values read at 760 nm on a plate reader (Fluostar Optima, BMG Labtech, Aylesbury, UK).

The concentration of phenolic compounds in the PRE was shown as tannic acid equivalents (TAE) per gram of freeze-dried sample.

2.4. The 2,2-diphenyl-1-picrylhydrazyl (DPPH) Radical Scavenging Assay

The DPPH assay was used to evaluate the scavenging of stable radicals by PRE, punicalagin, Zn (II) and PRE and punicalagin in combination with Zn (II), as previously described [33]. Briefly, samples were initially prepared in 0.2 mM DPPH solution and two-fold serial dilutions made in 96-well plates for each sample. Plates were wrapped in foil and incubated for 30 min at room temperature. After 30 min, the absorbance values of each sample were read at 515 nm as above, versus samples containing only DPPH (negative control), with ascorbic acid used as a positive control. The % of the radical scavenging activities of each sample was calculated as follows:

$$\% \text{ DPPH scavenging} = 100 \times [1 - (OD_{sample}/OD_{control})]$$

The concentration of each sample which scavenged 50% of the initial DPPH radicals generated was calculated by interpolating the [(Abs of the sample) − (Abs sample blank)] into a calibration curve generated by the DPPH absorbance values at different sample concentrations. All assays were performed on 3 separate occasions, with each experiment including 3 replicates.

2.5. The 2,2'-azino-bis(3-ethylbenzothiazoline-6-sulfonic-acid (ABTS) Radical Scavenging Assay/Trolox Equivalent Antioxidant Activity (TEAC)

The antioxidant potential of PRE, punicalagin, Zn (II) and PRE and punicalagin in combination with Zn (II), was also assessed using the ABTS/TEAC assay, based on the study by Re et al. [34]. This assay is based on the ability of compounds to scavenge the ABTS radical, produced by the reaction between 7 mM ABTS and 2.45 mM potassium persulphate. The ABTS solution was prepared and diluted to a final absorbance of 0.7 ± 0.2 at 734 nm obtained using a plate reader, as described above. The antioxidant capacities of PRE and punicalagin (both 0.5 mg/mL) and 0.1 mM Zn (II) were determined. Trolox (0–400 µg/mL) was used as a positive control and to express the data as Trolox equivalent antioxidant capacity (TEAC). All assays were performed on 3 separate occasions, with each experiment including 3 replicates.

2.6. Cell Culture

Human primary gingival fibroblasts were obtained from the American Type Cell Culture Collection (ATCC, Manassas, VA, USA). Gingival fibroblasts were cultured in DMEM supplemented with 10% heat-inactivated FCS, 1% L-glutamine (2 mM) and 1% antibiotic/antimycotic solution. Cells were incubated at 37 °C in a humidified atmosphere of 5 % CO_2. The passage number of cells used in all experiments was between 2 and 7.

Samples were prepared fresh on the day of treatment. Different concentrations of PRE (0.1–100 µg/mL), punicalagin (0.1–10 µg/mL), $ZnSO_4.7H_2O$ (0.1 mM) and PRE and $ZnSO_4·7H_2O$ (0.1 mM), punicalagin and $ZnSO_4·7H_2O$ (0.1 mM) were prepared. Freeze-dried PRE, Zn (II) and punicalagin were firstly dissolved in phthalate buffer pH 4.5 to make the stock solutions, and then filtered using 0.2 µm Minisart syringe filters made of acrylic resin, methacrylate butadiene styrene (Sartorius Stedim Biotech GmbH, Göttingen, Germany), under sterile conditions. Compound concentrations were further prepared in DMEM containing 1% FCS, 1% L-glutamine and 1% antibiotics/antimycotics. Control culture medium was also supplemented with 1% phthalate buffer pH 4.5 to negate any influences on cellular activities by the buffer itself.

2.7. Cell Viability and Proliferation

The effects of PRE, punicalagin, Zn (II) and PRE and punicalagin in combination with Zn (II) on gingival fibroblast viability and proliferation were determined MTT assay [35]. Gingival fibroblasts were seeded into 96-well plates at a density of 2.5×10^3 cells/well and cultured at 37 °C/5% CO_2 for 24 h. After 24 h, the media was changed to serum-free DMEM and the cells cultured for a further 24 h. Cells were subsequently treated with various concentrations of the samples for 24, 48 and 72 h, with media changes every 24 h. At each time point, 25 µL MTT (5 mg/mL in phosphate buffered saline, PBS) was added to each well and cultured at 37 °C/5% CO_2 for 4 h. After 4 h incubation, the MTT was discarded and each well treated with 100 µL pure DMSO, followed by further incubation at 37 °C/5% CO_2 for 30 min, with light protection. The absorbance values of each well were then read at 570 nm. Sample effects on cell viability and proliferation were expressed as % viable cells versus untreated controls, which were arbitrarily assigned a viability of 100%. All assays were performed on 3 separate occasions, with each experiment including 6 replicates.

2.8. Cell Migration and Wound Repopulation

The effects of PRE, punicalagin, Zn (II) and PRE and punicalagin in combination with Zn (II), on fibroblast migration were assessed for the ability to stimulate in vitro scratch wound repopulation, as previously described [36]. Gingival fibroblasts were seeded into 24-well plates at a density of 2.5×10^4 cells/well and cultured at 37 °C/5% CO_2 for 48 h. After 48 h, the media was changed to serum-free DMEM and the cells cultured for a further 24 h. Serum-free DMEM was removed and scratch wounds made using sterile pipettes. Fibroblasts were subsequently treated with different concentrations of test sample, with untreated cells in serum-free media serving as negative controls. Cell migration and wound repopulation were monitored by automated time-lapse microscopy, using a Cell-IQ® Automated Cell Culture and Analysis System (Chip-Man Technologies Ltd., Tampere, Finland), at 37 °C/5% CO_2. Digital images taken every 20 min over a 48 h period, using Cell-IQ Analyser™ Software, whilst ImageJ® Software (Version 1.49, https://imagej.nih.gov/ij/), were used to quantify cell migration parameters, including: cell displacement (Td), overall velocity (Td/t), distance travelled (Tt) and migratory speed (Tt/t). Each experiment was performed on 3 separate occasions, with each experiment including 3 replicates.

2.9. Statistical Analysis

Data values were expressed as the average ± standard error of the mean (SEM). Statistical analysis of antioxidant data was performed using the Duncan's multiple range test. Statistical analysis of gingival fibroblast viability, proliferation and migration was performed by one-way ANOVA with Tukey's multiple comparison post-test. Statistical analyses were performed using the GraphPad Prism, Version 8.00 (GraphPad Software, San Diego, CA, USA). Significance was considered at $p < 0.05$.

3. Results

3.1. Total Phenolic Content

The quantitative determination of the total phenolic content of PRE was expressed in µg TAE per g of freeze-dried PRE. The results showed that, on average, freeze-dried PRE contained 496 mg TAE/g.

3.2. Antioxidant Activities Using DPPH and ABTS Assays

The antioxidant capacities of PRE, punicalagin and their combination with 0.1 mM Zn (II) were assessed using both the DPPH and ABTS assays (Figure 2). The results of the DPPH assay are expressed as % of DPPH inhibition and IC_{50} values (the sample concentration needed to inhibit 50% of the initial DPPH free radical flux). All samples studied showed a dose-dependent response in the % of free DPPH inhibition. PRE (10.69 ± 0.44%) and PRE in combination with 0.1 mM Zn (II) (8.1 ± 0.27%) were required at higher concentrations than the ascorbic acid positive controls (8.31 ± 0.64%), to induce 50% inhibition. However, punicalagin (6.04 ± 0.29%) and its combination with 0.1 mM Zn (II) (6.99 ± 0.20%) required lower concentrations than the ascorbic acid controls. While there was a slight difference between the compounds and their Zn (II) combinations, no statistically significant differences were observed ($p > 0.05$). The ABTS assay showed similar patterns of antioxidant capabilities to the DPPH assay. Punicalagin and punicalagin with 0.1 mM Zn (II) exhibited significantly higher TEAC values than PRE and PRE with 0.1 mM Zn (II) ($p < 0.001$). Similarly, 0.1 mM Zn (II) addition did not cause any significant change in the antioxidant scavenging activities of PRE or punicalagin ($p > 0.05$).

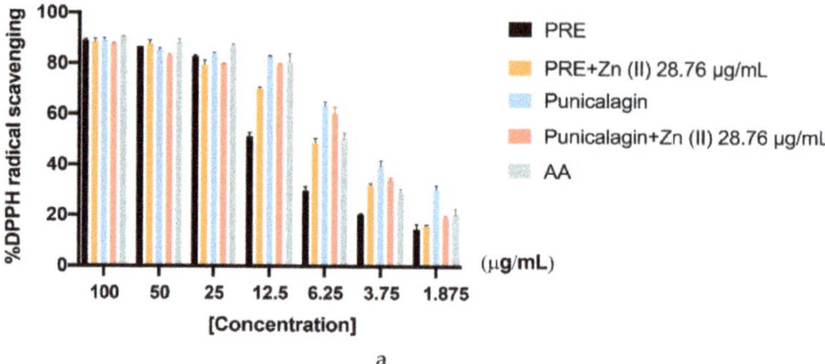

a

Compounds	DPPH IC50 (µg/mL)	ABTS/TEAC (mM/mg)
PRE	10.69±0.44[A]	3.27 ± 0.08[A]
PRE + Zn (II) 0.1mM	8.1±0.27[B]	3.47 ±0.20[A]
Punicalagin	6.04±0.29[C]	6.3±0.02[B]
Punicalagin + Zn (II) 0.1mM	6.99±0.20[C]	6.15±0.01[B]
Ascorbic acid (control)	8.31±0.64[B]	ND*

b

Figure 2. Antioxidant capabilities of PRE and punicalagin alone and in combination with 28.76 µg/mL (0.1 mM) Zn (II). (**a**) % DPPH antioxidant scavenging capacities at different sample concentrations. (**b**) TEAC values obtained for each sample, based on the finding of the ABTS assay. Values are presented as the mean ± SEM ($n = 3$). TEAC, Trolox equivalent antioxidant capacity. Values followed by the same capital letter within the same column are not significantly different ($p > 0.05$) between the compounds analyzed by Duncan's multiple range test. * ND; not determined.

3.3. Effects on Gingival Fibroblast Viability and Proliferation

The effects of PRE, punicalagin, Zn (II) and PRE and punicalagin in combination with Zn (II) on fibroblast viability and proliferation, were determined by MTT assay (Figure 3). According to the data obtained, 1 mM Zn (II) significantly reduced fibroblast viability and proliferation at all time-points ($p < 0.001$), while lower Zn (II) concentrations did not exhibit such decreases at 24 or 48 h ($p > 0.05$). When fibroblasts were treated with PRE or punicalagin alone, both showed dose-dependent decreases in cell viability from 24 h onwards, at the highest concentrations of PRE (100 µg/mL) and punicalagin (10 µg/mL) examined ($p < 0.001$). In contrast, lower concentrations of neither PRE nor punicalagin influenced cell viability, versus untreated negative controls (NC, $p > 0.05$). However, PRE and punicalagin combined with 0.1 mM Zn (II) significantly reduced fibroblast viability ($p < 0.001$), although treatment with 0.1 mM Zn (II) alone did not decrease viability ($p > 0.05$).

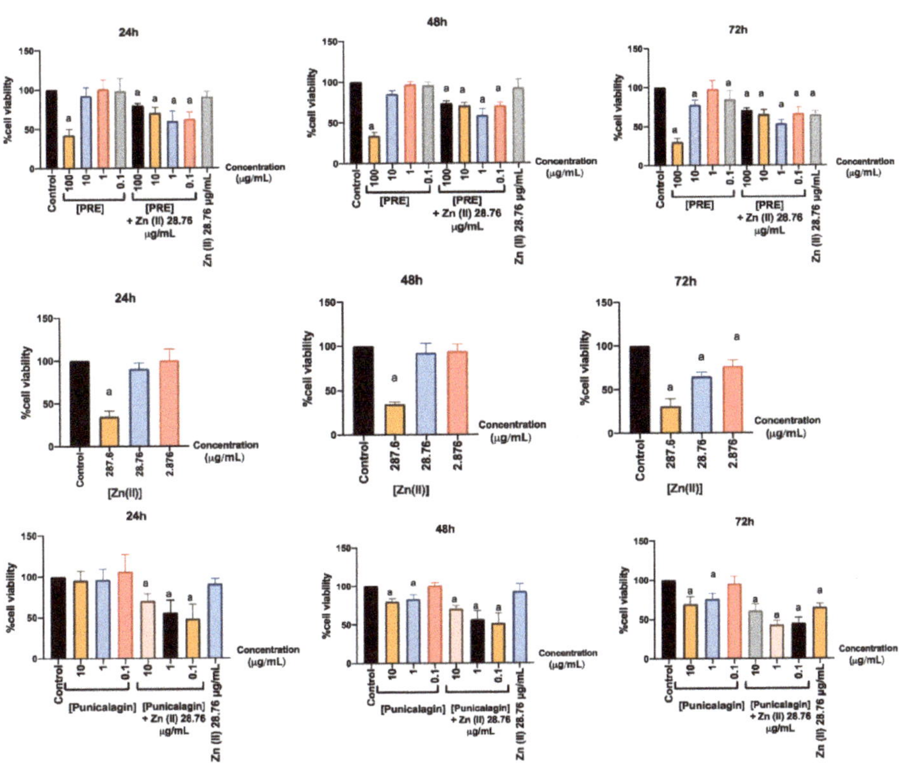

Figure 3. Effects of PRE (0.1–100 µg/mL), punicalagin (0.1–10 µg/mL) and 28.76 µg/mL (0.1 mM) Zn (II) alone and in combination with Zn (II), on human gingival fibroblast viability and proliferation at 24, 48 and 72 h, as determined by MTT assay. Values are presented as the mean % ± SEM ($n = 3$). Mean values with an "a" letter was significantly different than the untreated negative controls ($p < 0.001$).

3.4. Effects on Gingival Fibroblast Migration and Wound Repopulation

The effects of PRE, punicalagin, Zn (II) and PRE and punicalagin in combination with Zn (II) on fibroblast migration and wound repopulation were evaluated using the in vitro scratch wound assays, with cell displacement (Td), overall velocity (Td/t), distance travelled (Tt) and the migratory speed (Tt/t) of the gingival fibroblasts monitored. PRE (100 µg/mL) and punicalagin (10 µg/mL) reduced gingival fibroblast migration and wound repopulation, significantly decreasing fibroblast speed compared to untreated controls ($p < 0.001$, Figures 4 and 5). In contrast, lower PRE and punicalagin concentrations increased the speed, cell displacement, overall velocity and distance travelled. However, no significant differences in these cellular parameters were determined versus untreated controls (all $p > 0.05$).

Fibroblasts treated with 0.1 mM Zn (II) did not show any significant differences compared to untreated controls ($p > 0.05$, Figures 4 and 6). However, the combination of 0.1 µg/mL punicalagin and 0.1 mM Zn (II) induced a significant increase in cell speed and distance travelled, versus untreated controls ($p < 0.001$). Likewise, although there was a significant decrease in fibroblasts treated only with punicalagin at the highest concentration (10 µg/mL), when combined with 0.1 mM Zn (II), this inhibitory effect was not observed ($p > 0.05$).

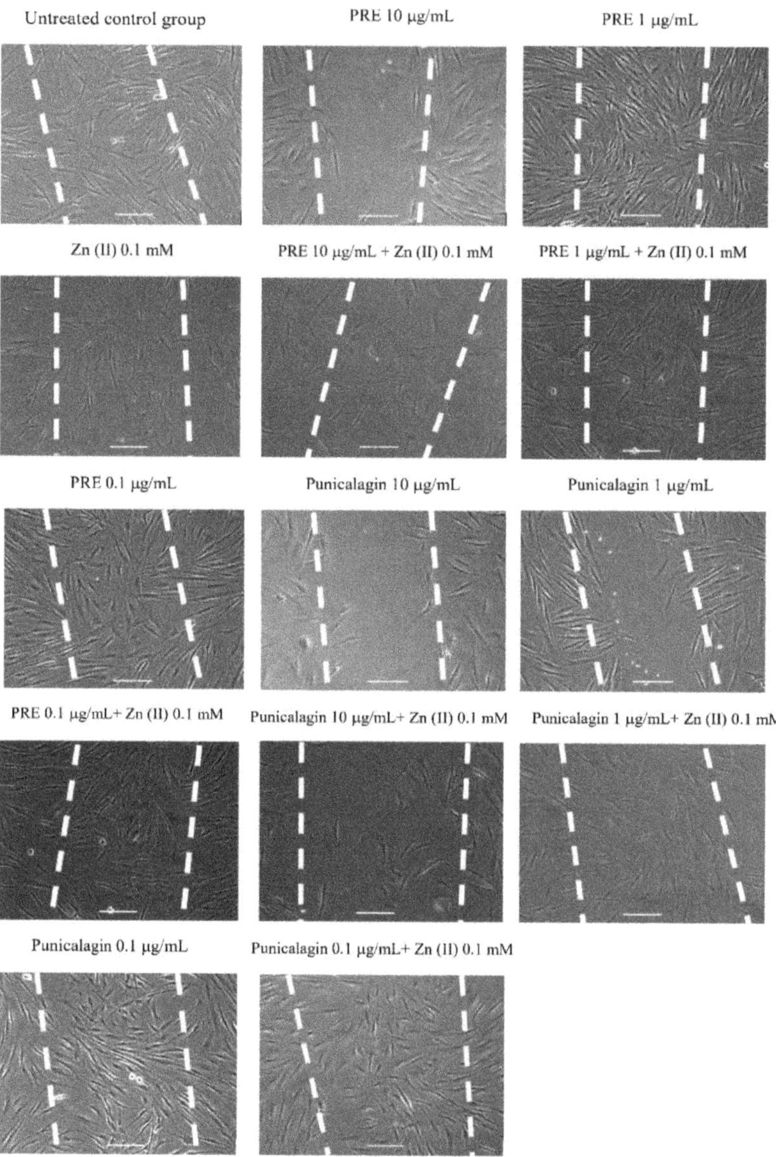

Figure 4. Representative time-lapse microscopy images of gingival fibroblast migration and wound repopulation at 48 h, following treatment with PRE and punicalagin (0.1–10 µg/mL) alone and with 28.76 µg/mL (0.1 mM) Zn (II). White dashed lines show original scratch wounds at 0 h. Scale bar = 100 µm.

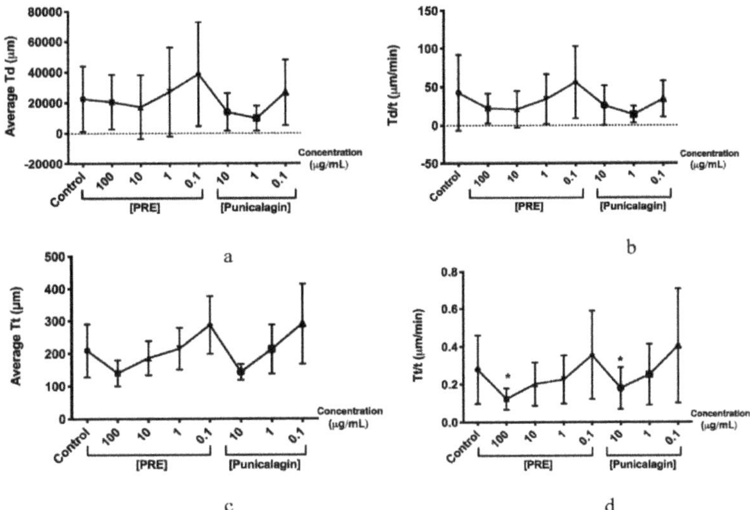

Figure 5. Effects of PRE (0.1–100 µg/mL) and punicalagin (0.1–10 µg/mL) on human gingival fibroblast scratch wound migration parameters, over 48 h. (**a**) Cell displacement (Td), (**b**) overall velocity (Td/t), (**c**) distance travelled (Tt), and (**d**) migration speed (Tt/t). Values are presented as the mean ± SEM ($n = 3$, * $p < 0.05$).

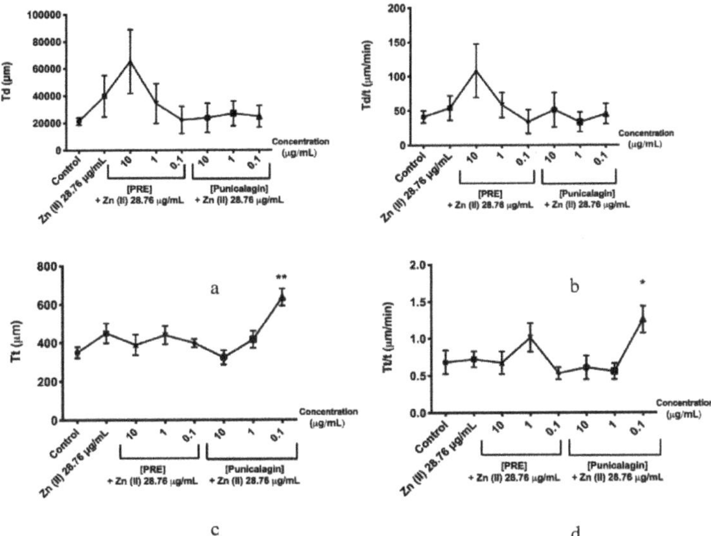

Figure 6. Effects of PRE (0.1–10 µg/mL) and punicalagin (0.1–10 µg/mL) in combination with 28.76 µg/mL (0.1 mM) Zn (II), on human gingival fibroblast scratch wound migration parameters, over 48 h. (**a**) Cell displacement (Td), (**b**) overall velocity (Td/t), (**c**) distance travelled (Tt), and (**d**) migration speed (Tt/t). Values are presented as the mean ± SEM ($n = 3$, * $p < 0.05$, ** $p < 0.01$).

4. Discussion

In light of its folklore medicinal status, crude pomegranate extracts and its constituent compounds, such as punicalagin, have received much biomedical attention given the considerable evidence supporting their efficacy against a wide range of diseases and conditions, ascribed to its various anticancer, antimicrobial, anti-inflammatory and antioxidant bioactivities [16]. Although numerous studies have previously endorsed the beneficial effects of PRE and punicalagin and advocated their application in the treatment of impaired wound healing responses in skin [19–24], a clinical area which has largely been overlook from a wound healing viewpoint are the potential abilities of PRE and punicalagin within the oral cavity, when tissue damage is commonly caused by periodontal disease and trauma. Indeed, periodontal diseases, comprising gingivitis and periodontitis, are regarded as being the most common disease of mankind, leading to huge economic burdens for healthcare providers worldwide [37]. As prevalence is also associated with risk factors such as age and diabetes, projections estimate further escalations in incidence with ever-increasing age demographics and diabetic rates worldwide. Although a wide array of therapeutic entities are available, these predominantly possess antibiotic, antimicrobial or anti-inflammatory properties, thereby indirectly promoting periodontal healing through the eradication of dental plaque/bacterial biofilm accumulation and/or the exacerbation of chronic inflammatory responses [8,37]. Furthermore, despite the development of a plethora of antibiotic and non-antibiotic-based drug delivery approaches to counteract microbial accumulation, biofilm formation or the inflammation associated with periodontal disease, few agents have fully progressed to routine clinical use [38,39]. Thus, in addition to addressing the side effects commonly associated with such therapeutics, the development of efficacious pharmaceutical options with established potent antimicrobial, anti-inflammatory and pro-healing properties, such as pomegranate, could meet a significant clinical and public health need in reducing the prevalence and severity of such conditions on a global scale.

The bioactivities of pomegranate extracts are generally attributed to its phenolic contents. Although the whole fruit comprises a large number of phenolic compounds, including anthocyanins, gallotannins, hydroxycinnamic acids, hydroxybenzoic acids and hydrolysable tannins. Compared to other parts of the fruit, pomegranate rind is known to contain the highest levels of bioactive polyphenolics, especially hydrolysable tannins such as punicalagin, which are responsible for the antioxidant activities of the PRE [18,40–43]. Indeed, potent antioxidant activity could play an important role in periodontal wound healing, as chronic inflammation, excessive reactive oxygen species (ROS) production and oxidative stress are key contributors to the host connective tissue damage associated with periodontal disease pathology [44,45]. In this study, it was shown that the total polyphenol content of the aqueous extract of PRE was 496 mg TAE/g freeze dried pomegranate rind. This result is similar to study by Malviya and Jha [46], who quantified the total polyphenol content of pomegranate rind using different solvents and found that water extract had the highest value, 435 mg TAE/g pomegranate rind. Furthermore, in this study, the antioxidant activity was evaluated using DPPH and ABTS assays. In both assays, punicalagin showed significantly higher antioxidant activity than PRE when at the same concentration as punicalagin, although addition of Zn (II) did not cause any significant changes in the ROS scavenging capacities of punicalagin and PRE, probably due to the stability of the Zn (II) ion in respect of redox reactions. Seeram et al. [47] found that pomegranate juice had higher antioxidant activity than punicalagin when they used the same concentrations of pomegranate juice and punicalagin and suggested a synergistic/additive activity of polyphenols than only one compound for this result. However, it is very difficult to assess the antioxidant activity, using a single method, since it can provide only basic information about antioxidant activity but using different methods can give more detail. There could be differences between the results because of extract and sample preparation, selection of endpoints and expression of results [48]. That said, it has been suggested that there is a correlation between the phenolic contents and antioxidant properties of PRE, with the most abundant polyphenol being punicalagin [46,49,50]. Thus, as the antioxidant assay data in this study provided a dose-dependent response, this may further imply that polyphenols are responsible for the antioxidant activity in PRE and punicalagin. Indeed, punicalagin showed higher antioxidant activity than PRE at the same concentrations in both DPPH and ABTS assays, as PRE contains a range of non-phenolic compounds. Therefore, it may be suggested that the antioxidant activity of PRE could be attributed to its punicalagin content, in line with previous findings [18,40–43,50].

Assessment of PRE and punicalagin effects on human gingival fibroblasts alone and in combination with Zn (II) showed no stimulation of fibroblast proliferation over the 72 h culture period. In contrast, PRE and punicalagin significantly reduced fibroblast viability at high concentrations (100 μg/mL and 10 μg/mL respectively) and when applied with Zn (II). Similar findings have been reported with other natural compounds and extracts, such as propolis, where despite its antimicrobial and antioxidant properties mediated through its polyphenol constituents, it can promote significant fibroblast cytotoxicity when co-administered with Zn (II) [51]. Furthermore, numerous studies have demonstrated the anti-proliferative or cytotoxic activities of PRE and punicalagin against a wide range of cancer cell types [22,52–55] and fibroblasts [56]. Similarly, although many studies have shown the stimulatory effects of Zn (II) on keratinocyte proliferation [29,31], negligible or inhibitory effects on fibroblast proliferative responses have been identified for Zn (II) [57,58]. However, such responses are likely to be concentration dependent, as fibroblasts are reported as being resistant to Zn (II) cytotoxicity <500 mM [59], as evident here.

Further studies evaluated PRE and punicalagin effects on gingival fibroblast migration and wound repopulation alone and in combination with Zn (II), via the analysis of relevant parameters including cell migration speed, cell displacement, overall velocity and distance travelled, over 48 h in culture. High concentrations of PRE (100 μg/mL) and punicalagin (10 μg/mL) significantly inhibited fibroblast migration and wound repopulation, presumably as a consequence of the cytotoxic effects identified above. However, lower PRE and punicalagin concentrations maintained or enhanced fibroblast migration and wound repopulation, equivalent to untreated controls. Furthermore, despite Zn (II) alone exerting no effects on fibroblast wound repopulation, 0.1 μg/mL punicalagin combined with 0.1 mM Zn (II) induced significant increases in cell speed and distance travelled, versus untreated controls; whilst 0.1 mM Zn (II) supplementation also attenuated the inhibitory effects of punicalagin (10 μg/mL) on cell speed. Such stimulatory effects on cell migration are significant, as previous studies have mostly described the inhibitory effects of PRE and punicalagin on cell motility/invasion, for instance, in cancer cells [22,52,53]. However, as Zn (II) and Zn-containing compounds can significantly enhance fibroblast migration and wound closure in vitro and in vivo [29,31,60], Zn (II) may actually be the key mediator of the increased fibroblast migratory responses identified herein.

Fibroblasts play a pivotal role in mediating wound healing responses, from initial cellular migration, proliferation and cytokine/growth factor production to subsequent extracellular matrix (ECM) synthesis/remodeling, wound contraction and closure. Thus, as wound repopulation is acknowledged to be dependent on the induction of both migratory and proliferative responses [61], the data presented herein would suggest that punicalagin and Zn (II) primarily promote oral fibroblast migration, rather than proliferation, in light of the absence of stimulated fibroblast proliferative responses induced by these concentrations alone or in combination. However, although oral and dermal wounds proceed through similar stages of healing, oral wounds are well- characterized by minimal inflammatory and angiogenic responses, rapid healing and minimal scar formation; unlike dermal wounds [5,6], with such superior healing responses attributed to the specialized genotypic and phenotypic properties of fibroblasts residing within oral tissues. In contrast, in non-healing skin wounds, such cellular responses are impaired, leading to failed wound closure [62]. Thus, fibroblast viability and induced proliferative and migratory responses are key events in normal repair processes, although differences in the responses of oral and dermal fibroblasts to specific PRE or punicalagin concentrations may be a consequence of the well-established differences in proliferative and migratory capabilities which exist between these distinct fibroblast populations [6].

Although the positive dermal wound healing activities of PRE and punicalagin have been recognized for some time [19–24], our findings are the first to report on the potential wound healing benefits of punicalagin in combination with Zn (II) to oral wounds caused by periodontal disease or trauma, in terms of alleviating ROS levels and oxidative stress and by stimulating gingival fibroblast migration. Whereas such particular antioxidant and pro-migratory responses could benefit gingival repair processes, it remains to be determined whether PRE or punicalagin alone or with Zn (II) supplementation possess any bactericidal, bacteriostatic or anti-biofilm properties versus the pathogenic Gram-negative bacterial species commonly associated with the initiation and progression of periodontal disease, such as *Porphyromonas gingivalis* [63], as established with microflora from other clinical situations [16,25,40,46]. However, such antimicrobial properties are currently under investigation. As uncontrolled biofilms initiate and sustain the inflammatory and resident connective tissue cell destruction in periodontal tissues, the therapeutic limitation or eradication of microbial biofilm accumulation by PRE or punicalagin would undoubtedly help inhibit the development and progression of periodontal disease evoking further tissue reparative responses.

5. Conclusions

Although pomegranate (*Punica granatum*) extracts and its bioactive constituents, such as punicalagin, have been used since ancient times to treat a broad range of diseases and conditions, only now have studies begun to assess its therapeutic potential for the treatment of wounds within the oral cavity, such as those manifested during periodontal disease or trauma. Both PRE and punicalagin were shown to possess potent antioxidant capabilities, whilst punicalagin combined with Zn (II) further induced human gingival fibroblast migration and wound repopulation responses, but exerted no stimulatory effects on fibroblast proliferation. Therefore, purified punicalagin in combination with Zn (II) may offer potential benefits as a natural compound-based therapy, aiding wound healing mechanisms within the oral cavity.

Author Contributions: Conceptualization, C.M.H.; methodology, C.M.H., A.J.S., R.M., and R.L.M.; software, V.C.; validation, V.C.; formal analysis, V.C.; investigation, V.C.; resources, C.M.H.; data curation, V.C.; writing—original draft preparation, V.C.; writing—review and editing, C.M.H., A.J.S., R.M., and R.L.M.; visualization, V.C.; supervision, C.M.H., A.J.S., R.M., and R.L.M.; project administration, C.M.H.; funding acquisition, V.C. and C.M.H. All authors have read and agreed to the published version of the manuscript.

Funding: This research has externally funded by Turkish Ministry of Education.

Acknowledgments: We would like to thank to Turkish Ministry of Education for supporting Vildan Celiksoy's PhD project.

Conflicts of Interest: The authors declare no conflict of interest.

References

1. Guo, S.A.; DiPietro, L.A. Factors affecting wound healing. *J. Dent. Res.* **2010**, *89*, 219–229. [CrossRef] [PubMed]
2. Socransky, S.S.; Haffajee, A.D. Periodontal microbial ecology. *Periodontology* **2005**, *38*, 135–187. [CrossRef] [PubMed]
3. Marcotte, H.; Lavoie, M.C. Oral microbial ecology and the role of salivary immunoglobulin A. *Microbiol. Mol. Biol. Rev.* **1998**, *62*, 71–109. [CrossRef] [PubMed]
4. Ebersole, J.L. Humoral immune responses in gingival crevice fluid: Local and systemic implications. *Periodontology* **2003**, *31*, 135–166. [CrossRef] [PubMed]
5. Szpaderska, A.M.; Zuckerman, J.D.; DiPietro, L.A. Differential injury responses in oral mucosal and cutaneous wounds. *J. Dent. Res.* **2003**, *82*, 621–626. [CrossRef]
6. Glim, J.E.; van Egmond, M.; Niessen, F.B.; Everts, V.; Beelen, R.H. Detrimental dermal wound healing: What can we learn from the oral mucosa? *Wound Repair Regen.* **2013**, *21*, 648–660. [CrossRef]
7. Politis, C.; Schoenaers, J.; Jacobs, R.; Agbaje, J.O. Wound healing problems in the mouth. *Front. Physiol.* **2016**, *7*, 507. [CrossRef]
8. Cleland, W.P., Jr. Opportunities and obstacles in veterinary dental drug delivery. *Adv. Drug Deliv. Rev.* **2001**, *50*, 261–275. [CrossRef]
9. Gjermo, P. Chlorhexidine and related compounds. *J. Dent. Res.* **1989**, *68*, 1602–1608.
10. Saraf, S. Formulating moisturizers using natural raw materials. In *Treatment of Dry Skin Syndrome*, 1st ed.; Lodén, M., Maibach, H.I., Eds.; Springer: Berlin, Germany, 2012; pp. 379–397.

11. Ghosh, P.K.; Gaba, A. Phyto-extracts in wound healing. *J. Pharm. Pharm. Sci.* **2013**, *16*, 760–820. [CrossRef]

12. Davis, R.H.; Maro, N.P. Aloe vera and gibberellin. Anti-inflammatory activity in diabetes. *J. Am. Podiatr. Med. Assoc.* **1989**, *79*, 24–26. [CrossRef] [PubMed]

13. Biswas, T.K.; Mukherjee, B. Plant medicines of Indian origin for wound healing activity: A review. *Int. J. Low. Extrem. Wounds* **2003**, *2*, 25–39. [CrossRef] [PubMed]

14. Wren, R.C. *Potter's New Cyclopedia of Botanical Drugs and Preparations*, 7th ed.; Wren, R.W., Ed.; CW Daniel Company Ltd.: Saffron Walden, UK, 1988; p. 112.

15. Seeram, N.P.; Zhang, Y.; Reed, J.; Krueger, C.; Vaya, J. Commercialization of pomegranates. In *Pomegranates: Ancient Roots to Modern Medicine*; Seeram, N.P., Schulman, R.N., Heber, D., Eds.; CRC Press: Boca Raton, FL, USA, 2006; Volume 43, pp. 193–195.

16. Ismail, T.; Sestili, P.; Akhtar, S. Pomegranate peel and fruit extracts: A review of potential anti-inflammatory and anti-infective effects. *J. Ethnopharmacol.* **2012**, *143*, 397–405. [CrossRef] [PubMed]

17. Bekir, J.; Mars, M.; Souchard, J.P.; Bouajila, J. Assessment of antioxidant, anti-inflammatory, anti-cholinesterase and cytotoxic activities of pomegranate (*Punica granatum*) leaves. *Food Chem. Toxicol.* **2013**, *55*, 470–475. [CrossRef]

18. Singh, K.; Jaggi, A.S.; Singh, N. Exploring the ameliorative potential of *Punica granatum* in dextran sulfate sodium induced ulcerative colitis in mice. *Phyther. Res.* **2009**, *23*, 1565–1574. [CrossRef]

19. Chidambara, M.K.N.; Reddy, V.K.; Veigas, J.M.; Murthy, U.D. Study on wound healing activity of *Punica granatum* peel. *J. Med. Food* **2004**, *7*, 256–259. [CrossRef]

20. Mo, J.; Panichayupakaranant, P.; Kaewnopparat, N.; Nitiruangjaras, A.; Reanmongkol, W. Wound healing activities of standardized pomegranate rind extract and its major antioxidant ellagic acid in rat dermal wounds. *J. Nat. Med.* **2014**, *68*, 377–386. [CrossRef]

21. Fleck, A.; Cabral, P.F.; Vieira, F.F.; Pinheiro, D.A.; Pereira, C.R.; Santos, W.C.; Machado, T.B. *Punica granatum* L. hydrogel for wound care treatment: From case study to phytomedicine standardization. *Molecules* **2016**, *21*, 1059. [CrossRef]

22. Tang, J.; Li, B.; Hong, S.; Liu, C.; Min, J.; Hu, M.; Li, Y.; Liu, Y.; Hong, L. Punicalagin suppresses the proliferation and invasion of cervical cancer cells through inhibition of the β-catenin pathway. *Mol. Med. Rep.* **2017**, *16*, 1439–1444. [CrossRef]

23. Nirwana, I.; Rachmadi, P.; Rianti, D. Potential of pomegranate fruit extract (*Punica granatum* Linn.) to increase vascular endothelial growth factor and platelet-derived growth factor expressions on the post-tooth extraction wound of *Cavia cobaya*. *Vet. World.* **2017**, *10*, 999. [CrossRef]

24. Lukiswanto, B.S.; Miranti, A.; Sudjarwo, S.A.; Primarizky, H.; Yuniarti, W.M. Evaluation of wound healing potential of pomegranate (*Punica granatum*) whole fruit extract on skin burn wound in rats (*Rattus norvegicus*). *J. Adv. Vet. Anim. Res.* **2019**, *6*, 202. [CrossRef] [PubMed]

25. McCarrell, E.M.; Gould, S.W.; Fielder, M.D.; Kelly, A.F.; El Sankary, W.; Naughton, D.P. Antimicrobial activities of pomegranate rind extracts: Enhancement by addition of metal salts and vitamin C. *BMC Complement. Altern. Med.* **2008**, *8*, 1–7. [CrossRef] [PubMed]

26. Houston, D.M.; Robins, B.; Bugert, J.J.; Denyer, S.P.; Heard, C.M. In vitro permeation and biological activity of punicalagin and zinc (II) across skin and mucous membranes prone to Herpes simplex virus infection. *Eur. J. Pharm. Sci.* **2017**, *96*, 99–106. [CrossRef]

27. Houston, D.M.; Bugert, J.; Denyer, S.P.; Heard, C.M. Anti-inflammatory activity of *Punica granatum* L. (Pomegranate) rind extracts applied topically to ex vivo skin. *Eur. J. Pharm. Biopharm.* **2017**, *112*, 30–37. [CrossRef] [PubMed]

28. Houston, D.M.; Bugert, J.J.; Denyer, S.P.; Heard, C.M. Correction: Potentiated virucidal activity of pomegranate rind extract (PRE) and punicalagin against Herpes simplex virus (HSV) when co-administered with zinc (II) ions, and antiviral activity of PRE against HSV and aciclovir-resistant HSV. *PLoS ONE.* **2017**, *12*, e0188609. [CrossRef] [PubMed]

29. Lansdown, A.B.; Mirastschijski, U.; Stubbs, N.; Scanlon, E.; Ågren, M.S. Zinc in wound healing: Theoretical, experimental, and clinical aspects. *Wound Repair Regen.* **2007**, *15*, 2–16. [CrossRef]

30. Seeram, N.; Lee, R.; Hardy, M.; Heber, D. Rapid large-scale purification of ellagitannins from pomegranate husk, a by-product of the commercial juice industry. *Sep. Purif. Technol.* **2005**, *41*, 49–55. [CrossRef]

31. Lin, P.H.; Sermersheim, M.; Li, H.; Lee, P.H.; Steinberg, S.M.; Ma, J. Zinc in wound healing modulation. *Nutrients* **2018**, *10*, 16. [CrossRef]

32. Ainsworth, E.A.; Gillespie, K.M. Estimation of total phenolic content and other oxidation substrates in plant tissues using Folin–Ciocalteu reagent. *Nat. Protoc.* **2007**, *2*, 875–877. [CrossRef]

33. Okonogi, S.; Duangrat, C.; Anuchpreeda, S.; Tachakittirungrod, S.; Chowwanapoonpohn, S. Comparison of antioxidant capacities and cytotoxicities of certain fruit peels. *Food Chem.* **2007**, *103*, 839–846. [CrossRef]

34. Re, R.; Pellegrini, N.; Proteggente, A.; Pannala, A.; Yang, M.; Rice-Evans, C. Antioxidant activity applying an improved ABTS radical cation decolorization assay. *Free Radic. Biol. Med.* **1999**, *26*, 1231–1237. [CrossRef]

35. Mosmann, T. Rapid colorimetric assay for cellular growth and survival: Application to proliferation and cytotoxicity assays. *J. Immunol. Methods* **1983**, *65*, 55–63. [CrossRef]

36. Hardwicke, J.; Moseley, R.; Stephens, P.; Harding, K.; Duncan, R.; Thomas, D.W. Bioresponsive dextrin–rhEGF conjugates: In vitro evaluation in models relevant to its proposed use as a treatment for chronic wounds. *Mol. Pharm.* **2010**, *7*, 699–707. [CrossRef] [PubMed]

37. Tonetti, M.S.; Jepsen, S.; Jin, L.; Otomo-Corgel, J. Impact of the global burden of periodontal diseases on health, nutrition and wellbeing of mankind: A call for global action. *J. Clin. Periodontol.* **2017**, *44*, 456–462. [CrossRef]

38. Rovai, E.S.; Souto, M.L.; Ganhito, J.A.; Holzhausen, M.; Chambrone, L.; Pannuti, C.M. Efficacy of local antimicrobials in the non-surgical treatment of patients with periodontitis and diabetes: A systematic review. *J. Periodontol.* **2016**, *87*, 1406–1417. [CrossRef]

39. Graziani, F.; Karapetsa, D.; Alonso, B.; Herrera, D. Nonsurgical and surgical treatment of periodontitis: How many options for one disease? *Periodontology* **2017**, *75*, 152–188. [CrossRef]

40. Negi, P.S.; Jayaprakasha, G.K. Antioxidant and antibacterial activities of *Punica granatum* peel extracts. *J. Food Sci.* **2003**, *68*, 1473–1477. [CrossRef]

41. Li, Y.; Guo, C.; Yang, J.; Wei, J.; Xu, J.; Cheng, S. Evaluation of antioxidant properties of pomegranate peel extract in comparison with pomegranate pulp extract. *Food Chem.* **2006**, *96*, 254–260. [CrossRef]

42. Sestili, P.; Martinelli, C.; Ricci, D.; Fraternale, D.; Bucchini, A.; Giamperi, L.; Curcio, R.; Piccoli, G.; Stocchi, V. Cytoprotective effect of preparations from various parts of *Punica granatum* L. fruits in oxidatively injured mammalian cells in comparison with their antioxidant capacity in cell free systems. *Pharm. Res.* **2007**, *56*, 18–26. [CrossRef]

43. Akhtar, S.; Ismail, T.; Fraternale, D.; Sestili, P. Pomegranate peel and peel extracts: Chemistry and food features. *Food Chem.* **2015**, *174*, 417–425. [CrossRef]

44. Waddington, R.J.; Moseley, R.; Embery, G. Periodontal Disease Mechanisms: Reactive oxygen species: A potential role in the pathogenesis of periodontal diseases. *Oral Dis.* **2000**, *6*, 138–151. [CrossRef] [PubMed]

45. Chapple, I.L.; Matthews, J.B. The role of reactive oxygen and antioxidant species in periodontal tissue destruction. *Periodontology* **2007**, *43*, 160–232. [CrossRef] [PubMed]

46. Malviya, S.; Jha, A.; Hettiarachchy, N. Antioxidant and antibacterial potential of pomegranate peel extracts. *J. Food Sci. Technol.* **2014**, *51*, 4132–4137. [CrossRef] [PubMed]

47. Seeram, N.P.; Adams, L.S.; Henning, S.M.; Niu, Y.; Zhang, Y.; Nair, M.G.; Heber, D. In vitro antiproliferative, apoptotic and antioxidant activities of punicalagin, ellagic acid and a total pomegranate tannin extract are enhanced in combination with other polyphenols as found in pomegranate juice. *J. Nutr. Biochem.* **2005**, *16*, 360–367. [CrossRef]

48. Číž, M.; Čížová, H.; Denev, P.; Kratchanova, M.; Slavov, A.; Lojek, A. Different methods for control and comparison of the antioxidant properties of vegetables. *Food Control* **2010**, *21*, 518–523. [CrossRef]

49. Viuda-Martos, M.; Fernández-López, J.; Pérez-Álvarez, J.A. Pomegranate and its many functional components as related to human health: A review. *Compr. Rev. Food Sci. Food Saf.* **2010**, *9*, 635–654. [CrossRef]

50. Gil, M.I.; Tomás-Barberán, F.A.; Hess-Pierce, B.; Holcroft, D.M.; Kader, A.A. Antioxidant activity of pomegranate juice and its relationship with phenolic composition and processing. *J. Agric. Food Chem.* **2000**, *48*, 4581–4589. [CrossRef]

51. Tyszka-Czochara, M.; Paśko, P.; Reczyński, W.; Szlósarczyk, M.; Bystrowska, B.; Opoka, W. Zinc and propolis reduces cytotoxicity and proliferation in skin fibroblast cell culture: Total polyphenol content and antioxidant capacity of propolis. *Biol. Trace Elem. Res.* **2014**, *160*, 123–131. [CrossRef]

52. Khan, G.N.; Gorin, M.A.; Rosenthal, D.; Pan, Q.; Bao, L.W.; Wu, Z.F.; Newman, R.A.; Pawlus, A.D.; Yang, P.; Lansky, E.P.; et al. Pomegranate fruit extract impairs invasion and motility in human breast cancer. *Integr. Cancer Ther.* **2009**, *8*, 242–253. [CrossRef]

53. Shirode, A.B.; Kovvuru, P.; Chittur, S.V.; Henning, S.M.; Heber, D.; Reliene, R. Antiproliferative effects of pomegranate extract in MCF-7 breast cancer cells are associated with reduced DNA repair gene expression and induction of double strand breaks. *Mol. Carcinog.* **2014**, *53*, 458–470. [CrossRef]

54. Adaramoye, O.; Erguen, B.; Nitzsche, B.; Höpfner, M.; Jung, K.; Rabien, A. Punicalagin, a polyphenol from pomegranate fruit, induces growth inhibition and apoptosis in human PC-3 and LNCaP cells. *Chem. Biol. Interact.* **2017**, *274*, 100–106. [CrossRef] [PubMed]

55. Khwairakpam, A.D.; Bordoloi, D.; Thakur, K.K.; Monisha, J.; Arfuso, F.; Sethi, G.; Mishra, S.; Kumar, A.P.; Kunnumakkara, A.B. Possible use of *Punica granatum* (Pomegranate) in cancer therapy. *Pharmacol. Res.* **2018**, *133*, 53–64. [CrossRef] [PubMed]

56. Toi, M.; Bando, H.; Ramachandran, C.; Melnick, S.J.; Imai, A.; Fife, R.S.; Carr, R.E.; Oikawa, T.; Lansky, E.P. Preliminary studies on the anti-angiogenic potential of pomegranate fractions in vitro and in vivo. *Angiogenesis* **2003**, *6*, 121–128. [CrossRef] [PubMed]

57. Han, B.; Fang, W.H.; Zhao, S.; Yang, Z.; Huang, B.X. Zinc sulfide nanoparticles improve skin regeneration. *Nanomed. Nanotech. Biol. Med.* **2020**, 102263. [CrossRef]

58. Aslam, M.N.; Lansky, E.P.; Varani, J. Pomegranate as a cosmeceutical source: Pomegranate fractions promote proliferation and procollagen synthesis and inhibit matrix metalloproteinase-1 production in human skin cells. *J. Ethnopharmacol.* **2006**, *103*, 311–318. [CrossRef]

59. Gren, M.S.; Mirastschijski, U. The release of zinc ions from and cytocompatibility of two zinc oxide dressings. *J. Wound Care.* **2004**, *13*, 367–369. [CrossRef]

60. Tandon, N.; Cimetta, E.; Villasante, A.; Kupferstein, N.; Southall, M.D.; Fassih, A.; Xie, J.; Sun, Y.; Vunjak-Novakovic, G. Galvanic microparticles increase migration of human dermal fibroblasts in a wound-healing model via reactive oxygen species pathway. *Exp. Cell Res.* **2014**, *320*, 79–91. [CrossRef]

61. Liang, C.-C.; Park, A.Y.; Guan, J.-L. In vitro scratch assay: A convenient and inexpensive method for analysis of cell migration in vitro. *Nat. Protoc.* **2007**, *2*, 329–333. [CrossRef]

62. Diegelmann, R.F.; Evans, M.C. Wound healing: An overview of acute, fibrotic and delayed healing. *Front. Biosci.* **2004**, *9*, 283–289. [CrossRef]

63. Kinane, D.F.; Stathopoulou, P.G.; Papapanou, P.N. Periodontal diseases. *Nat. Rev. Dis. Primers.* **2017**, *3*, 17038. [CrossRef]

Communication

Branched-Chain Fatty Acids as Mediators of the Activation of Hepatic Peroxisome Proliferator-Activated Receptor Alpha by a Fungal Lipid Extract

Garima Maheshwari [1,2], Robert Ringseis [1,*], Gaiping Wen [1], Denise K. Gessner [1], Johanna Rost [2], Marco A. Fraatz [2], Holger Zorn [2,3] and Klaus Eder [1]

[1]   Institute of Animal Nutrition and Nutrition Physiology, Justus Liebig University Giessen, Heinrich-Buff-Ring 26-32, 35392 Giessen, Germany; Garima.Maheshwari@lcb.chemie.uni-giessen.de (G.M.); gaiping.wen@ernaehrung.uni-giessen.de (G.W.); denise.gessner@ernaehrung.uni-giessen.de (D.K.G.); Klaus.Eder@ernaehrung.uni-giessen.de (K.E.)

[2]   Institute of Food Chemistry and Food Biotechnology, Justus Liebig University Giessen, Heinrich-Buff-Ring 17, 35392 Giessen, Germany; johanna.rost@gmx.de (J.R.); marco.fraatz@lcb.chemie.uni-giessen.de (M.A.F.); holger.zorn@lcb.chemie.uni-giessen.de (H.Z.)

[3]   Fraunhofer Institute for Molecular Biology and Applied Ecology, Winchester Str. 2, 35394 Giessen, Germany

*   Correspondence: robert.ringseis@ernaehrung.uni-giessen.de; Tel.: +49-641-993-9231

Received: 8 June 2020; Accepted: 28 August 2020; Published: 31 August 2020

Abstract: The study aimed to test the hypothesis that monomethyl branched-chain fatty acids (BCFAs) and a lipid extract of *Conidiobolus heterosporus* (CHLE), rich in monomethyl BCFAs, are able to activate the nuclear transcription factor peroxisome proliferator-activated receptor alpha (PPARalpha). Rat Fao cells were incubated with the monomethyl BCFAs 12-methyltridecanoic acid (MTriA), 12-methyltetradecanoic acid (MTA), isopalmitic acid (IPA) and 14-methylhexadecanoic acid (MHD), and the direct activation of PPARalpha was evaluated by reporter gene assay using a PPARalpha responsive reporter gene. Furthermore, Fao cells were incubated with different concentrations of the CHLE and PPARalpha activation was also evaluated by using the reporter gene assay, and by determining the mRNA concentrations of selected PPARalpha target genes by real-time RT-PCR. The reporter gene assay revealed that IPA and the CHLE, but not MTriA, MHD and MTA, activate the PPARalpha responsive reporter gene. CHLE dose-dependently increased mRNA concentrations of the PPARalpha target genes acyl-CoA oxidase (ACOX1), cytochrome P450 4A1 (CYP4A1), carnitine palmitoyltransferase 1A (CPT1A) and solute carrier family 22 (organic cation/carnitine transporter), member 5 (SLC22A5). In conclusion, the monomethyl BCFA IPA is a potent PPARalpha activator. CHLE activates PPARalpha-dependent gene expression in Fao cells, an effect that is possibly mediated by IPA.

Keywords: branched-chain fatty acids; *Conidiobolus heterosporus*; peroxisome proliferator-activated receptor α; lipid metabolism; fatty acid oxidation; hepatocyte

---

1. Introduction

The peroxisome proliferator-activated receptor alpha (PPARalpha) is a transcription factor that belongs to the superfamily of nuclear hormone receptors and is predominantly expressed in tissues with high rates of fatty acid oxidation. PPARalpha can be activated by micromolar concentrations of a variety of peroxisome proliferators including synthetic agonists, such as fibrates, a class of hypolipidemic drugs, and natural agonists, such as fatty acids (e.g., dietary ω-3 polyunsaturated fatty acids (PUFAs), like arachidonic acid (AA), linoleic acid (LA) or α-linolenic acid (ALA)), eicosanoids and their derivatives [1–3]. Upon activation, PPARalpha upregulates the expression of a large set of target genes involved in fatty acid uptake, transport and oxidation through binding to a specific sequence, called peroxisome proliferator response element (PPRE), in the regulatory region of these genes [4]. In addition, PPARalpha inhibits several proinflammatory genes through a negative crosstalk with the key regulator of inflammation, nuclear factor-kappa B, especially in the vasculature [5]. Owing to this, PPARalpha activators exhibit lipid-lowering and atheroprotective effects, thereby reducing the risk of developing cardiovascular diseases (CVDs) [6].

Apart from the abovementioned PPARalpha activators, the tetramethyl branched-chain, isoprenoid-derived fatty acids phytanic acid (3,7,11,15-tetramethylhexadecanoic acid) and pristanic acid (2,6,10,14-tetramethylpentadecanoic acid) as well as their CoA-thioesters were shown to be high-affinity natural ligands and potent activators of PPARalpha [7–10]. In addition, the feeding of phytol, which is metabolized to phytanic acid following absorption from the intestine, was found to activate PPARalpha in tissues of mice [11]. Branched-chain fatty acids (BCFAs) are primarily saturated fatty acids, both even-carbon- and odd-carbon-numbered, with one or more methyl branches, accordingly called mono- or multimethyl BCFAs, respectively. The branching is mainly at the penultimate (*iso*) or next to the penultimate carbon atom (*anteiso*). BCFAs belong to the minor fatty acids in food and are found in ruminant products (dairy and beef) due to synthesis by ruminal microorganisms [12,13]; they are also present at significant levels in fermented foods of animal (shrimp paste, fish sauce) and nonanimal origin (sauerkraut) [13,14]. In addition, BCFAs occur naturally in bacterial lipids and in several fungi [15,16]. As early as 1967, Tyrrell described the composition of fatty acids in various species of the fungal genus *Conidiobolus* and reported proportions of BCFAs of up to 73% of total fatty acids [17]. Among these *Conidiobolus* species, the *Conidiobolus heterosporus* Drechsler 1953, first described by Tyrrell in 1971, belonging to the phylum *Zygomycota*, contains 53% of the total fatty acids as BCFAs, with the main BCFA being 12-methyltridecanoic acid (MTriA, *iso*-C14:0) (33% of total fatty acids), 12-methyltetradecanoic acid (MTA, *anteiso*-C15:0) (13%), 14-methylpentadecanoic acid (=isopalmitic acid, IPA, *iso*-C16:0) (6%) and 14-methylhexadecanoic acid (MHD, *anteiso*-C17:0) (1%) [18]. In contrast to the multimethyl BCFAs phytanic acid or pristanic acid, the BCFA species contained in *Conidiobolus heterosporus*, like MTriA, MTA, IPA and MHD, are monomethyl BCFAs. Whether these monomethyl BCFAs are also able to activate PPARalpha, however, is currently unknown.

Therefore, the present study aimed to test the following two hypotheses: First, based on the structural similarity between multimethyl BCFAs and monomethyl BCFAs, we tested the hypothesis that monomethyl BCFAs are potent PPARalpha activators. Second, due to the high amount of monomethyl BCFAs in *Conidiobolus heterosporus* lipids, we tested the hypothesis that a lipid extract of *Conidiobolus heterosporus* (CHLE) also activates PPARalpha and thereby induces the expression of PPARalpha target genes. To test these two hypotheses, we treated rat Fao hepatoma cells, a frequently used cell line to investigate the ability of different substances to activate the PPARalpha pathway, with the monomethyl BCFAs MTriA, MTA, IPA and MHD, in isolated form and with various concentrations of the CHLE and tested their ability to activate PPARalpha by determining PPARalpha transactivation and induction of PPARalpha target genes. As a positive control, we used WY-14643, a synthetic agonist of PPARalpha, and two PUFAs, LA and ALA, both of which are naturally occurring PPARalpha agonists [2].

2. Materials and Methods

2.1. Submerged Cultivation of Conidiobolus heterosporus, Fungal Lipid Extraction and Preparation of Fatty Acid Methyl Esters

Conidiobolus heterosporus Drechsler 1953, which belongs to the phylum Zygomycota, order Entomophtorales, family Ancylistaceae and the genus *Conidiobolus*, was obtained from Centraalbureau voor Schimmelcultures (CBS 138.57; Utrecht, Netherlands), and grown in submerged cultures following a previously described method [19]. For the precultures, 100 mL of yeast extract (3 g/L) and malt extract (30 g/L) (YM) medium (Merck KGaA, Darmstadt, Germany) in 250 mL Erlenmeyer flasks was used. After 3 days, the precultures were homogenized for 30 s at 9800 rpm using an Ultra Turrax homogenizer (IKA, Staufen, Germany). For the main cultures, 20 mL of the homogenized suspension was used to inoculate 200 mL of YM medium in 500 mL Erlenmeyer flasks. The cultivations were carried out in an Ecotron incubation shaker (25 mm shaking diameter, 150 rpm, 24 °C) (Infors GmbH, Einsbach, Germany) for 5 days. For harvesting, the fungal mycelium was separated from the culture broth by vacuum filtration with a Buchner funnel (110 mm) and a DP 595 cellulose filter paper (Albet LabScience, Dassel, Germany).

Total lipids were extracted by the Soxhlet method. After saponification of 150 mg CHLE with 4 mL 0.5 M NaOH in methanol (80 °C, 10 min), free fatty acids were converted into their corresponding fatty acid methyl esters (FAMEs) by addition of 3.5 mL boron trifluoride-methanol solution (20%) and heating to 80 °C for 5 min as described previously [19]. After addition of 1 mL n-hexane and incubation at 80 °C for 1 min, 3 mL saturated NaCl solution was added. The organic phase was dried over anhydrous sodium sulfate overnight.

2.2. Analysis of Fatty Acid Composition

The fatty acid composition of the fungal lipid extract was determined by using an Agilent 7890A gas chromatograph (Agilent Technologies, Waldbronn, Germany), equipped with a model 5975C mass spectrometry detector (Agilent Technologies), under the following conditions: carrier gas, helium (5.0); constant flow rate, 1.2 mL/min; inlet temperature, 250 °C; split ratio, 10:1; septum purge flow rate, 3 mL/min; 30 m × 0.25 mm i.d., 0.25 μm BP21 (FFAP) column (SGE Europe Ltd, Milton Keynes, UK); temperature program, 40 °C (3 min) and 5 °C/min to 240°C (12 min); scan mode, TIC; scan range, m/z 33–400; electron ionization energy, 70 eV; source temperature, 230 °C; quadrupole temperature, 150 °C; transfer line temperature, 250 °C. The identification of the fatty acids was carried out by the comparison of retention indices with the reference standard Supelco 37 component FAME mix (Sigma-Aldrich, Taufkirchen, Germany). To determine the fatty acid composition, all FAME peak areas were summed up, set to 100%, and the results were expressed as a relative percentage for each fatty acid. The fatty acid composition of the CHLE is listed in Table 1.

2.3. Chemical Reagents

Ham's F12 medium, fetal calf serum (FCS), gentamycin and Trizol were purchased from Invitrogen (Karlsruhe, Germany). WY-14,643; the isolated fatty acids LA (≥99% pure), ALA (≥99% pure), IPA (≥98% pure), MTriA (≥98% pure), MHD (≥98% pure) and MTA (≥98% pure); and the MTT (3-(4,5-dimethylthiazole-2-yl)-2,5-diphenyltetrazolium bromide; Thiazole Blue) stock solution were obtained from Sigma-Aldrich (Steinheim, Germany).

2.4. Cell Culture

The rat hepatoma Fao cell line [20] was obtained from the European Collection of Cell Cultures (ECACC Cat. No. 89042701; Salisbury, UK) and grown in Ham's F12 medium supplemented with 10% FCS and 0.5% gentamycin at 37 °C and humidified atmosphere of 95% air and 5% CO_2. Fao cells were seeded either in 24-well culture plates (Greiner Bio-One, Frickenhausen, Germany) at a cell density of 2.1×10^5 per well for cell viability assay and qPCR analysis or in 96-well culture plates at a cell density of 17×10^4 per well.

2.5. Cell Treatments

After reaching 70–80% confluency, cells were treated with either CHLE, WY-14,643 or various isolated fatty acids (LA, ALA, MTriA, MTA, IPA or MHD) for 24 h at the concentrations indicated. Incubation media containing isolated fatty acids were prepared by diluting the fatty acid stock solutions (100 mM in ethanol) with low-serum Ham's F12 medium (0.5% FCS) as described previously [21]. Prior to adding CHLE into the incubation media, an aliquot of the stock solution (100 mM in ethanol, based on the molecular weight of an average triglyceride of 850) was evaporated under a nitrogen stream and dissolved in 0.03 N NaOH at room temperature to saponify triglycerides and bidistilled water.

Afterwards, the saponified CHLE was diluted in low-serum Ham's F12 medium (0.5% FCS) (pH 7-8). WY-14,643 was added to the low-serum medium from a 100 mM stock solution dissolved in DMSO. Cells treated with the vehicle alone (DMSO for WY-14,643, ethanol for isolated fatty acids, low-serum medium for CHLE) were used as controls and contained the same vehicle (ethanol) concentration (ethanol: 0.5 % (*v/v*); DMSO: 0.05% (*v/v*)). After addition of either CHLE, isolated fatty acids or WY-14,643 to the medium, the medium was gently mixed at RT to ensure complete solubility of the added substances. No signs of precipitation could be observed.

Table 1. Fatty acid composition of the *Conidiobolus heterosporus* lipid extract.

Fatty Acids [1]	Area (%)
Branched-chain fatty acids (BCFAs)	52.7
iso-C14:0 (MTriA)	33.0
anteiso-C15:0 (MTA)	13.1
iso-C16:0 (IPA)	5.9
anteiso-C17:0 (MHD)	0.7
Straight-chain fatty acids	39.5
C12:0	0.3
C13:0	1.4
C14:0	5.4
C14:1	0.1
C15:0	4.8
C16:0	7.8
C16:1	0.6
C17:0	0.4
C18:0	1.3
C18:1	1.7
C18:2	0.7
C18:3 (γ-linolenic acid)	0.8
C20:0	0.2
C20:1	0.2
C20:2	0.3
C20:3	1.1
C20:4	9.4
C20:5	0.4
C22:0	0.2
C22:1	0.3
C22:2	0.3
C23:0	0.1
C24:0	1.4
C24:1	0.3
Sum of saturated straight-chain fatty acids	23.3
Sum of monounsaturated straight-chain fatty acids	3.2
Sum of polyunsaturated straight-chain fatty acids	13.0
Total others	7.8

[1] Only fatty acid methyl esters in quantities greater 0.1% were considered.

2.6. Cell Viability Assay

Cell viability after treatment with either CHLE or isolated fatty acids was assessed by the MTT assay [22]. Briefly, after removing the incubation media, 5 mg/mL MTT stock solution dissolved in PBS (140 mM NaCl, 3 mM KCl, 8 mM Na_2HPO_4, 1.8 mM KH_2PO_4; pH 7.4) was added to each well, and cells were incubated at 37 °C for 4 h. Subsequently, the MTT/PBS solution was removed, and the formazan crystals generated during the incubation period were dissolved by adding isopropanol in 0.04 N HCl. After the crystals were completely dissolved, the solution was transferred into a 24-well plate, and its absorbance was measured at the wavelength of 540 nm using an Infinite 200 M microplate reader (Tecan, Männedorf, Switzerland). Cell viabilities are expressed as percentage of control cells, which was set to 100%.

2.7. Transient Transfection and Dual Luciferase Reporter Assay

After reaching 70–80% confluency, cells were transiently transfected with 50 ng of a 3X ACO-PPRE vector (containing three copies of consensus PPRE from the ACO promoter in front of a luciferase reporter gene) and pGL4.10 vector as negative control using FuGENE 6 transfection reagent (Roche diagnostics, Mannheim, Germany) for 12 h according to the manufacturer's protocol. Cells were also co-transfected with 5 ng of pGL4.74 Renilla luciferase (encoding the Renilla luciferase reporter gene; Promega, Mannheim, Germany), which was used as internal control reporter vector to normalize for differences in transfection efficiency (Promega). After transfection, cells were treated with either WY-14643; the isolated fatty acids LA, ALA, MTriA, MTA, IPA or MHD; or the CHLE at the concentrations indicated or vehicle alone for 24 h. Afterwards, cells were washed with PBS and lysed with lysis buffer (Promega). Luciferase activities were determined with Beetle-Juice and Renilla-Juice Kits from PJK (Kleinblittersdorf, Germany) in a Mithras LB940 luminometer (Berthold Technologies, Bad Wildbad, Germany). For control of background luminescence, Firefly and Renilla luciferase activities were also determined in the lysates of nontransfected control cells and subtracted from luminescence of transfected cells. Data were normalized for transfection efficiency by dividing Firefly luciferase activity of the 3X ACO-PPRE plasmid by that of Renilla luciferase activity of the co-transfected pGL4.74 Renilla luciferase plasmid. Results represent normalized luciferase activities and are shown relative to cells transfected with pGL4.10 vector and treated with the vehicle only, which was set to 1.

2.8. Quantitative Real-Time Reverse-Transcription Polymerase Chain Reaction (qPCR)

After treatment, the media were discarded, the cells were washed once with PBS and total RNA was isolated using Trizol reagent according to the manufacturer's protocol. Total RNA (1.2 µg) was reverse-transcribed using 60 U M-MuLV reverse transcriptase in a Biometra Thermal Cycler (Whatman Biometra, Göttingen, Germany). The cDNA was stored in aliquots at −20 °C. Subsequent real-time RT-PCR analysis of selected PPARalpha target genes and reference genes and calculation of gene expression data was performed as described recently in detail [23]. The normalization factor was calculated as the geometric mean of expression data of the three most stable out of six tested potential reference genes (CANX, MDH1, ACTB, RPL13, TOP1 and ATP5B). Means and SD were calculated from normalized expression data for cells of the same treatment group. The mean of the vehicle control cells was set to 1, and the mean and SD of the treated cells were scaled proportionally. Features of gene-specific primer pairs are listed in Supplementary Table S1.

2.9. Statistics

Statistical analysis was performed using the Minitab statistical software (Release 13, Minitab Inc., State College, PA, USA). Normal distribution of variables was tested with the Anderson–Darling test. Since all data showed normal distribution, the effects of different concentrations of CHLE, isolated fatty acids and WY-14643 in comparison to control treatment were analyzed by Student's *t*-test. Post-hoc analysis was performed using Fisher's multiple comparison test. Means were considered significantly different from control at $p < 0.05$. Data are means ± SD calculated from three independent experiments. In each independent experiment, all treatments were represented in 4 (qPCR) and 8 (reporter assay, MTT assay) wells (representing the numbers of technical replicates per treatment).

3. Results

3.1. Effects of Isolated Fatty Acids and CHLE on Cell Viability of Rat Fao Cells

Incubation with either isolated BCFAs or straight-chain fatty acids for 24 h did not impair cell viability up to a concentration of about 200 µM, with the exception of MTriA, as demonstrated by the

MTT assay (Figure 1A). Cell viability of Fao cells was, however, not reduced by 24 h incubation with the monomethyl BCFA MHD up a concentration 500 µM, the highest concentration tested (Figure 1A). Cell viability of Fao cells was not reduced by 24 h incubation with CHLE up to a concentration of 500 µM (Figure 1B)

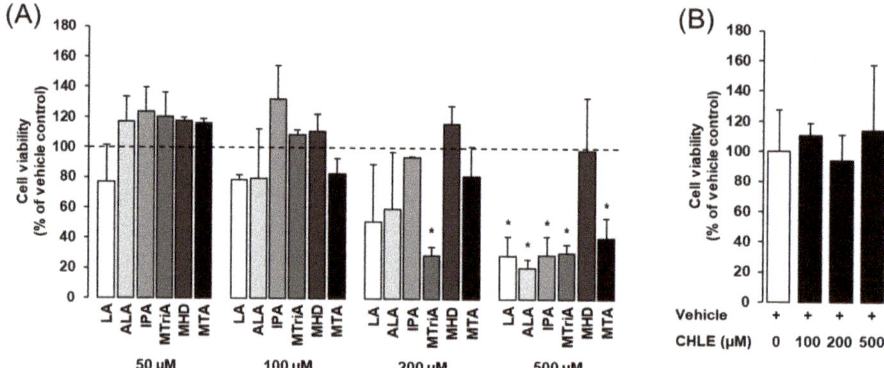

Figure 1. Effects of isolated fatty acids (**A**) and *Conidiobolus heterosporus* lipid extract (CHLE) (**B**) on cell viability of Fao cells. Fao cells were treated either without (vehicle) or with different concentrations of linoleic acid (LA), α-linolenic acid (ALA), isopalmitic acid (IPA), 12-methyltridecanoic acid (MTriA), 14-methylhexadecanoic acid (MHD), 12-methyltetradecanoic acid (MTA) or CHLE for 24 h, and cell viability was measured by the MTT assay. Bars represent means ± SD of three independent experiments. (**A**) The dashed line indicates the vehicle control (=100%). $*\ p < 0.05$ compared with vehicle control.

3.2. Effects of Isolated Fatty Acids and CHLE on PPARalpha Transactivation

Using a PPARalpha-responsive reporter gene, we studied the effect of CHLE and isolated fatty acids contained in the CHLE on PPARalpha transactivation. As shown in Figure 2, the synthetic PPARalpha agonist WY-14,643 markedly increased the luciferase activity of the PPARalpha-responsive reporter gene about 12-fold compared to vehicle control ($p < 0.05$). IPA, one of the main monomethyl BCFAs of the CHLE, increased the luciferase activity of the reporter gene about 3.5-fold ($p < 0.05$), while the other monomethyl BCFAs, namely MTriA, MHD and MTA, did not (Figure 2). Treatment of cells with CHLE increased the luciferase activity of the PPARalpha-responsive reporter gene about 4-fold compared to vehicle control ($p < 0.05$). The straight-chain fatty acids LA and ALA increased the luciferase activity of the reporter gene about 4- and 3-fold, respectively, compared to vehicle control ($p < 0.05$, Figure 2).

3.3. Effect of CHLE on the mRNA Concentration of PPARalpha Target Genes

Treatment with CHLE increased the mRNA concentrations of the PPARalpha target genes solute carrier family 22 (organic cation/carnitine transporter), member 5 (SLC22A5), acyl-CoA oxidase (ACOX1), carnitine palmitoyltransferase 1A (CPT1A) and cytochrome P450 4A1 (CYP4A1) in a dose-dependent manner ($p < 0.05$, Figure 3). The strongest increase was observed for CYP4A1, which was increased 6-fold by 500 µM CHLE compared to control ($p < 0.05$, Figure 3). As expected, the synthetic PPARalpha agonist WY-14,643 caused a markedly stronger induction of all PPARalpha target genes investigated when compared to CHLE ($p < 0.05$, Figure 3).

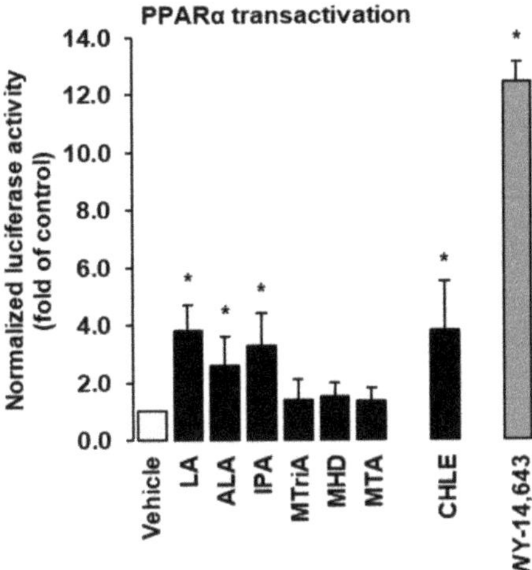

Figure 2. Effects of isolated fatty acids, *Conidiobolus heterosporus* lipid extract (CHLE) and synthetic PPARalpha agonist WY-14,643 on PPARalpha transactivation in Fao cells. Fao cells were transiently transfected with a 3X ACO-PPRE vector and a Renilla luciferase expression vector for normalization using FuGENE6. After transfection, cells were treated either without (vehicle) or with different isolated fatty acids (50 µM each), CHLE (500 µM) or WY-14,643 (50 µM) for 24 h. Afterwards, cells were lysed, and luciferase activities were determined by dual luciferase assay. Bars represent means ± SD of three independent experiments. * $p < 0.05$ compared with vehicle control.

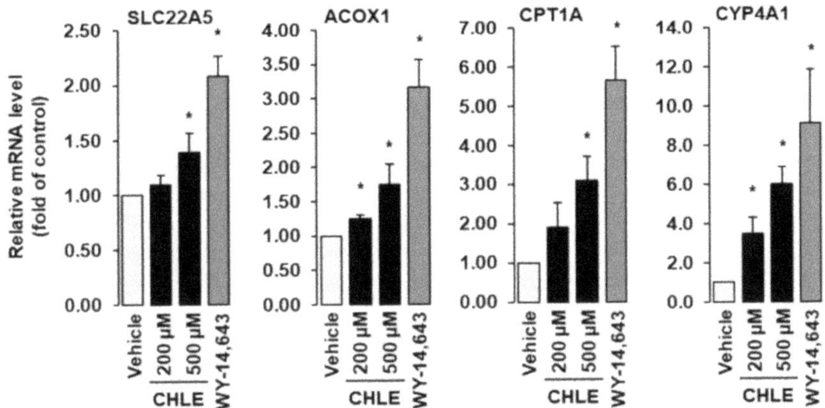

Figure 3. Effects of *Conidiobolus heterosporus* lipid extract (CHLE) on relative mRNA levels of the PPARalpha target genes SLC22A5, ACOX1, CPT1A and Cyp4A1 in Fao cells. Fao cells were treated either without (vehicle) or with CHLE at two different concentrations (200 and 500 µM) or WY-14,643 (50 µM) for 24 h, and mRNA levels were determined by qPCR. Bars represent means ± SD of three independent experiments. * $p < 0.05$ compared with vehicle control.

4. Discussion

In the present study, we tested the hypothesis that both isolated monomethyl BCFAs and CHLE, which is rich in monomethyl BCFAs, activate the nuclear transcription factor PPARalpha and CHLE induces the expression of PPARalpha target genes in rat Fao cells. Our data from the reporter gene assay clearly show that the monomethyl BCFA IPA and the CHLE are potent activators of PPARalpha. Moreover, we found that CHLE induces the expression of known PPARalpha target genes involved in fatty acid metabolism, like ACOX1, CPT1A, CYP4A1 and SLC22A5, in a dose-dependent manner. Compared with WY-14,643 the extent of activation of PPARalpha by IPA and CHLE was lower, but this is not surprising given that WY-14,643 is a high-affinity synthetic ligand of PPARalpha and such ligands typically cause a stronger stimulation of PPARalpha transactivation than naturally occurring ligands [2].

Several studies are available reporting that tetramethyl BCFAs, like phytanic acid, pristanic acid and their CoA-thioesters, as well as the synthetically produced short-chain BCFA valproic acid, cause PPARalpha activation [9,24,25]. However, to our knowledge, there is no study available exploring the effect of the monomethyl BCFAs MTriA, MTA, IPA and MHD on PPARalpha activation in isolated form. Thus, our study shows for the first time that the monomethyl BCFA IPA, which makes up about 6% of total fatty acids of CHLE, significantly stimulates PPARalpha transactivation, indicating that IPA is responsible for the effect of the CHLE. A few studies are available investigating the effect of a mixture of different BCFAs in laboratory animals. For instance, Shirouchi et al. [26] showed that feeding 5% porpoise oil, which contains 15.5% BCFAs (including the monomethyl BCFA IPA), to obese rats alleviates hepatic triglyceride accumulation and increases serum adiponectin levels, effects that are known to be mediated by PPARalpha activation [27]. The decrease of hepatic triglyceride accumulation in response to PPARalpha activation is mechanistically explained by the physiological function of the proteins encoded by many PPARalpha target genes, including those considered in the present study as indirect markers of PPARalpha activation. The enzymes encoded by ACOX1, CPT1A and CYP4A1 are directly involved in peroxisomal, mitochondrial and microsomal fatty oxidation pathways, respectively, and induction of these genes causes a decrease of hepatic triglyceride levels due to an increased oxidation of fatty acids [4]. SLC22A5 encodes the novel organic cation transporter OCTN2, which mediates carnitine uptake into tissues. Because carnitine is essential for the transport of long-chain fatty acids into the mitochondrial matrix, where mitochondrial fatty acid oxidation takes place, genetic defects of this carnitine transporter result in decreased intracellular carnitine levels and an impaired fatty acid oxidation [28]. A limitation of the abovementioned study [26] is that porpoise oil also contains ω-3 PUFAs, which are also known to be activators of PPARalpha. Thus, the contribution of BCFAs to the effect of porpoise oil in this study remains unclear. In another study, it was reported that rat pups receiving milk substituted with BCFAs (including IPA) showed an increased intestinal expression of the anti-inflammatory interleukin-10 (IL-10) compared with pups receiving no BCFAs and a reduced incidence of necrotizing enterocolitis [29], which is indicative of an anti-inflammatory action of BCFAs in the intestine. This effect might be also mediated by activation of PPARalpha by the BCFAs, because upregulation of IL-10 has been reported to occur in response to administration of WY-14,643 [30]. In addition, the anti-inflammatory effect of BCFAs in the neonatal rat intestine might be also explained by the well-known negative crosstalk of PPARalpha with critical inflammatory signaling pathways such as nuclear factor-kappa B [5].

In summary, the observation of the present in vitro study that the monomethyl BCFA IPA is an activator of PPARalpha may provide an explanation for the PPARalpha-mediated effects of BCFA-rich diets in in vivo studies [26,29]. Nevertheless, other fatty acids, particularly PUFAs like LA, AA and ALA (all of which were present in the CHLE), are naturally occurring PPARalpha activators [1,2]. Thus, not only LA and ALA, used as further positive controls in the reporter gene assay, but also AA may have also contributed to the PPARalpha-dependent effects of CHLE or BCFA-rich diets. In any case, the results of the present study suggest that dietary supplementation of CHLE might be a useful approach to induce the beneficial effects associated with PPARalpha activation like lipid lowering and reduction of CVD risk.

Considerations with regard to the use of BCFA-rich sources as a dietary supplement have to include the induction of possible adverse effects. In this context, it should be noted that various feeding studies in rats, lambs and steers did not observe any toxic or adverse effects after treatment with different BCFAs [31–33]. In contrast, adverse effects including teratogenic, hepatotoxic, neurotoxic and proapoptotic effects have been reported for valproic acid, which is also used as an antiepileptic drug, but only at extremely high concentrations [34,35]. Such effects might be also, at least partially, explained by the activation of PPARalpha, because administration of synthetic PPARalpha activators to rodents was found to cause hepatic peroxisome proliferation, hypertrophy, hyperplasia and even hepatocarcinogenesis [36], with this last condition being attributed to induction of oxidative stress and an imbalance between apoptosis and cell proliferation [37]. Apart from this, clinical observations in patients with peroxisomal disorders like Zellweger syndrome or Refsum's disease and studies employing corresponding mouse models have shown that elevated concentrations of BCFAs are associated with a high level of toxicity [38,39]. However, it has to be considered that in such cases the BCFA concentrations in blood markedly exceed those achieved by dietary supplementation. Proapoptotic effects were also found for naturally occurring BCFAs, like MTA [40,41]. Such proapoptotic effects however were found to occur in tumor cells, an effect that is probably beneficial. Collectively, it can be stated that more studies are required for the evaluation of safety aspects of BCFA-rich sources as food supplements.

5. Conclusions

This study shows for the first time that the monomethyl BCFA IPA and the CHLE are able to induce PPARalpha activation in cultured hepatocytes. These findings indicate that lipid extracts from fungi, such as CHLE, are a source of PPARalpha-activating biomolecules which might be considered as dietary supplements to induce beneficial effects associated with PPARalpha activation, like lipid lowering and reduction of CVD risk, provided that no safety concerns exist. In light of recent evidence showing that administration of BCFAs in neonatal rats increases the expression of anti-inflammatory cytokines and reduces the incidence of necrotizing enterocolitis, an anti-inflammatory action of BCFAs in the intestine might be postulated, which is likely mediated by the well-established negative crosstalk of PPARalpha activation with critical inflammatory signaling pathways. Regarding adverse effects reported from the use of synthetic PPARalpha activators in rodents, potential safety issues associated with BCFA-rich sources as food supplements have to be thoroughly evaluated.

Supplementary Materials: The following are available online at http://www.mdpi.com/2218-273X/10/9/1259/s1, Table S1: Characteristics of gene-specific primers used for qPCR analysis.

Author Contributions: Conceptualization, R.R., K.E. and H.Z.; formal analysis, G.M., G.W., D.K.G., J.R. and M.A.F.; writing—original draft preparation, R.R.; writing—review and editing, R.R. and K.E.; supervision, R.R.; project administration, K.E.; All authors have read and agreed to the published version of the manuscript.

Funding: The study was supported by the Hessen State Ministry of Higher Education, Research and the Arts (HMWK) via the LOEWE research center "Insect Biotechnology and Bioresources". G.M. was funded by the German Academic Exchange Service (DAAD; funding ID: 57243780).

Acknowledgments: The authors are grateful for the provision of the 3X ACO-PPRE vector by Prof. Dr. Sander Kersten, Department of Agrotechnology and Food Sciences, Chair of Nutrition, Metabolism and Genomics, University of Wageningen, The Netherlands.

Conflicts of Interest: The authors declare no conflict of interest.

References

1. Forman, B.M.; Chen, J.; Evans, R.M. Hypolipidemic drugs, polyunsaturated fatty acids, and eicosanoids are ligands for peroxisome proliferator-activated receptors alpha and delta. *Proc. Natl. Acad. Sci. USA* **1997**, *94*, 4312–4317. [CrossRef] [PubMed]
2. Krey, O.; Braissant, F.; Lhorset, E.; Kalkhoven, M.; Perroud, M.; Parker, M.G.; Wahli, W. Fatty acids, eicosanoids, and hypolipidemic agents identified as ligands of peroxisome proliferator-activated receptors by coactivator-dependent receptor ligand assay. *Mol. Endocrinol.* **1997**, *11*, 779–791. [CrossRef] [PubMed]
3. Kersten, S.; Wahli, W. Peroxisome proliferator activated receptor agonists. *EXS* **2000**, *89*, 141–151. [PubMed]
4. Mandard, S.; Müller, M.; Kersten, S. Peroxisome proliferator-activated receptor alpha target genes. *Cell. Mol. Life Sci.* **2004**, *61*, 393–416. [CrossRef] [PubMed]

5. Marx, N.; Duez, H.; Fruchart, J.C.; Staels, B. Peroxisome proliferator-activated receptors and atherogenesis: Regulators of gene expression in vascular cells. *Circ. Res.* **2004**, *94*, 1168–1178. [CrossRef]

6. Loomba, R.S.; Arora, R. Prevention of cardiovascular disease utilizing fibrates—A pooled meta-analysis. *Am. J. Ther.* **2010**, *17*, 182–188. [CrossRef]

7. Ellinghaus, P.; Wolfrum, C.; Assmann, G.; Spener, F.; Seedorf, U. Phytanic acid activates the peroxisome proliferator-activated receptor alpha (PPARalpha) in sterol carrier protein 2-/sterol carrier protein x-deficient mice. *J. Biol. Chem.* **1999**, *274*, 2766–2772. [CrossRef]

8. Wolfrum, C.; Ellinghaus, P.; Fobker, M.; Seedorf, U.; Assmann, G.; Börchers, T.; Spener, F. Phytanic acid is ligand and transcriptional activator of murine liver fatty acid binding protein. *J. Lipid. Res.* **1999**, *40*, 708–714.

9. Zomer, A.W.; van Der Burg, B.; Jansen, G.A.; Wanders, R.J.; Poll-The, B.T.; van Der Saag, P.T. Pristanic acid and phytanic acid: Naturally occurring ligands for the nuclear receptor peroxisome proliferator-activated receptor alpha. *J. Lipid. Res.* **2000**, *41*, 1801–1807.

10. Hostetler, H.A.; Kier, A.B.; Schroeder, F. Very-long-chain and branched-chain fatty acyl-CoAs are high affinity ligands for the peroxisome proliferator-activated receptor alpha (PPARalpha). *Biochemistry* **2006**, *45*, 7669–7681. [CrossRef]

11. Gloerich, J.N.; van Vlies, G.A.; Jansen, S.; Denis, J.P.; Ruiter, J.P.N.; van Werkhoven, M.A.; Duran, M.; Vaz, F.M.; Wanders, R.J.A.; Ferdinandusse, S. A phytol-enriched diet induces changes in fatty acid metabolism in mice both via PPARalpha-dependent and—Independent pathways. *J. Lipid. Res.* **2005**, *46*, 716–726. [CrossRef] [PubMed]

12. Bravo-Lamas, L.; Barron, L.J.R.; Farmer, L.; Aldai, N. Fatty acid composition of intramuscular fat and odour-active compounds of lamb commercialized in northern Spain. *Meat Sci.* **2018**, *139*, 231–238. [CrossRef] [PubMed]

13. Ran-Ressler, R.R.; Bae, S.; Lawrence, P.; Wang, D.H.; Brenna, J.T. Branched-chain fatty acid content of foods and estimated intake in the USA. *Br. J. Nutr.* **2014**, *112*, 565–572. [CrossRef] [PubMed]

14. Wang, D.H.; Yang, Y.; Wang, Z.; Lawrence, P.; Worobo, R.W.; Brenna, J.T. High levels of branched chain fatty acids in nātto and other Asian fermented foods. *Food Chem.* **2019**, *286*, 428–433. [CrossRef] [PubMed]

15. Ratnayakem, W.M.; Olsson, B.; Ackman, R.G. Novel branched-chain fatty acids in certain fish oils. *Lipids* **1989**, *24*, 630–637. [CrossRef]

16. Kates, M. Bacterial lipids. *Adv. Lipid. Res.* **1964**, *2*, 17–90.

17. Tyrrell, D. The fatty acid compositions of 17 Entomophthora isolates. *Can. J. Microbiol.* **1967**, *13*, 755–760. [CrossRef]

18. Tyrrell, D. The fatty acid composition of some Entomophthoraceae. *Can. J. Microbiol.* **1971**, *17*, 1115–1118. [CrossRef]

19. Fraatz, M.A.; Goldmann, M.; Geissler, T.; Gross, E.; Backes, M.; Hilmer, J.M.; Ley, J.; Rost, J.; Francke, A.; Zorn, H. Biotechnological production of methyl-branched aldehydes. *J. Agric. Food Chem.* **2016**, *66*, 2387–2392. [CrossRef]

20. Deschatrette, J.; Weiss, M.C. Characterization of differentiated clones from a rat hepatoma. *Biochimie* **1974**, *56*, 1603–1611. [CrossRef]

21. König, B.; Eder, K. Differential action of 13-HPODE on PPARalpha downstream genes in rat Fao and human HepG2 hepatoma cell lines. *J. Nutr. Biochem.* **2006**, *17*, 410–418. [CrossRef] [PubMed]

22. Mosmann, T. Rapid colorimetric assay for cellular growth and survival: Application to proliferation and cytotoxicity assays. *J. Immunol. Methods* **1983**, *65*, 55–63. [CrossRef]

23. Ringseis, R.; Rosenbaum, S.; Gessner, D.K.; Herges, L.; Kubens, J.F.; Mooren, F.C.; Krüger, K.; Eder, K. Supplementing obese Zucker rats with niacin induces the transition of glycolytic to oxidative skeletal muscle fibers. *J. Nutr.* **2013**, *143*, 125–131. [CrossRef] [PubMed]

24. Hanhoff, T.; Wolfrum, C.; Ellinghaus, P.; Seedorf, U.; Spener, F. Pristanic acid is activator of peroxisome proliferator activated receptor α. *Eur. J. Lipid. Sci. Technol.* **2001**, *103*, 75–80. [CrossRef]

25. Lampen, A.; Carlberg, C.; Nau, H. Peroxisome proliferator-activated receptor delta is a specific sensor for teratogenic valproic acid derivatives. *Eur. J. Pharmacol.* **2001**, *431*, 25–33. [CrossRef]

26. Shirouchi, B.; Nagao, K.; Furuya, K.; Nagai, T.; Ichioka, K.; Tokairin, S.; Iida, Y.; Yanagita, T. Physiological functions of iso-type short-chain fatty acid and omega 3 polyunsaturated fatty acids containing oil in obese OLETF rats. *J. Oleo Sci.* **2010**, *59*, 299–305. [CrossRef]

27. Castillero, E.; Martín, A.I.; Nieto-Bona, M.P.; Fernández-Galaz, C.; López-Menduiña, M.; Villanúa, M.Á.; López-Calderón, A. Fenofibrate administration to arthritic rats increases adiponectin and leptin and prevents oxidative muscle wasting. *Endocr. Connect.* **2012**, *1*, 1–12. [CrossRef]

28. Wang, Y.; Ye, J.; Ganapathy, V.; Longo, N. Mutations in the organic cation/carnitine transporter OCTN2 in primary carnitine deficiency. *Proc. Natl. Acad. Sci. USA* **1999**, *96*, 2356–2360. [CrossRef]

29. Ran-Ressler, R.R.; Khailova, L.; Arganbright, K.M.; Adkins-Rieck, C.K.; Jouni, Z.E.; Koren, O.; Ley, R.E.; Brenna, J.T.; Dvorak, B. Branched chain fatty acids reduce the incidence of necrotizing enterocolitis and alter gastrointestinal microbial ecology in a neonatal rat model. *PLoS ONE* **2011**, *6*, e29032. [CrossRef]

30. Yanagisawa, J.; Shiraishi, T.; Iwasaki, A.; Maekawa, S.; Higuchi, T.; Hiratuka, M.; Tanaka, T.; Shibaguchi, H.; Kuroki, M.; Shirakusa, T. PPARalpha ligand WY14643 reduced acute rejection after rat lung transplantation with the upregulation of IL-4, IL-10 and TGFbeta mRNA expression. *J. Heart Lung Transplant.* **2009**, *28*, 1172–1179. [CrossRef]

31. Deetz, L.E.; Richardson, C.R.; Pritchard, R.H.; Preston, R.L. Feedlot performance and carcass characteristics of steers fed diets containing ammonium salts of the branched-chain fatty acids and valeric acid. *J. Anim. Sci.* **1985**, *61*, 1539–1549. [CrossRef] [PubMed]

32. Cline, T.R.; Garrigus, U.S.; Hatfield, E.E. Addition of branched- and straight-chain volatile fatty acids to purified lamb diets and effects on utilization of certain dietary components. *J. Anim. Sci.* **1966**, *25*, 734–739. [CrossRef] [PubMed]

33. Smith, A.; Lough, A.K.; Earl, C.R. Fatty acid composition of tissue lipids of rats given dietary branched-chain fatty acids. *Proc. Nutr. Soc.* **1978**, *37*, 76.

34. Chateauvieux, S.; Morceau, F.; Dicato, M.; Diederich, M. Molecular and therapeutic potential and toxicity of valproic acid. *J. Biomed. Biotechnol.* **2010**, *2010*. [CrossRef] [PubMed]

35. Driever, P.H.; Knüpfer, M.M.; Cinatl, J.; Wolff, J.E. Valproic acid for the treatment of pediatric malignant glioma. *Klin. Padiatr.* **1999**, *211*, 323–328. [CrossRef]

36. Reddy, J.K.; Azarnoff, D.L.; Hignite, C.E. Hypolipidaemic hepatic peroxisome proliferators form a novel class of chemical carcinogens. *Nature* **1980**, *283*, 397–398. [CrossRef]

37. Peters, J.M.; Cheung, C.; Gonzalez, F.J. Peroxisome proliferator-activated receptor-α and liver cancer: Where do we stand? *J. Mol. Med.* **2005**, *83*, 774–785. [CrossRef]

38. Fan, C.Y.; Pan, J.; Chu, R.; Lee, D.; Kluckman, K.D.; Usuda, N.; Singh, I.; Yeldandi, A.V.; Rao, M.S.; Maeda, N.; et al. Hepatocellular and hepatic peroxisomal alterations in mice with a disrupted peroxisomal fatty acyl-coenzyme A oxidase gene. *J. Biol. Chem.* **1996**, *271*, 24698–24710. [CrossRef]

39. Idel, S.; Ellinghaus, P.; Wolfrum, C.; Nofer, J.R.; Gloerich, J.; Assmann, G.; Spener, F.; Seedorf, U. Branched chain fatty acids induce nitric oxide-dependent apoptosis in vascular smooth muscle cells. *J. Biol. Chem.* **2002**, *277*, 49319–49325. [CrossRef]

40. Yang, P.; Collin, P.; Madden, T.; Chan, D.; Sweeney-Gotsch, B.; McConkey, D.; Newman, R.A. Inhibition of proliferation of PC3 cells by the branched-chain fatty acid, 12-methyltetradecanoic acid, is associated with inhibition of 5-lipoxygenase. *Prostate* **2003**, *55*, 281–291. [CrossRef]

41. Yang, Z.; Liu, S.; Chen, X.; Chen, H.; Huang, M.; Zheng, J. Induction of apoptotic cell death and in vivo growth inhibition of human cancer cells by a saturated branched-chain fatty acid, 13-methyltetradecanoic acid. *Cancer. Res.* **2000**, *60*, 505–509. [PubMed]

MDPI

St. Alban-Anlage 66

4052 Basel

Switzerland

Tel. +41 61 683 77 34

Fax +41 61 302 89 18

www.mdpi.com

Biomolecules Editorial Office

E-mail: biomolecules@mdpi.com

www.mdpi.com/journal/biomolecules